Endorsements

"Much has been written about the mechanics of modern AI – what it is, what it isn't, and how to build and train one – but this fascinating book is among the few that explain what it means for the enterprise – where it fits in to business needs, how it can be integrated into existing systems, and how it can create opportunities for new value."

—Grady Booch, IBM Fellow, Chief Scientist for Software Engineering

"I recommend *Beyond Algorithms: Delivering AI for Business* to anyone that is responsible for leading and transitioning AI to achieve and maintain a competitive advantage. The book addresses many of the important issues faced in the application of AI to businesses. It is written in an easy-to-read style, while providing enough technical content that many people working in the field can capture the main messages."

—David R. Martinez, MIT Instructor and Laboratory Fellow

"Books about AI too often tend towards the extremes – either praising or condemning unconditionally. *Beyond Algorithms* offers a more nuanced, nontechnical introduction from experienced practitioners. Filled with guiding principles and real-world examples, this book offers a balanced view of AI's strengths and limitations and provides practical advice for problem selection, project management, and expectation setting for AI initiatives."

—Emily Riederer, Senior Analytics Manager at Capital One

"*Beyond Algorithms* is one of those wonderful story-driven IT books that distils decades worth of real-world experience into intoxicating wisdom that is refreshingly easy to consume. The book plows through and around the hype of the 'all powerful' deep neural nets to provide an engineering approach for the field. It won't tell you how to establish the Singularity, but its 'A to I checklist', accuracy and monitoring advice, and 'doability method' could make you an AI delivery hero."

—Richard Hopkins, FREng FIET CEng, Former President of IBM Academy of Technology

"Important and timely. A practical guide for realizing business value through AI capabilities! Rather than focusing on hard to access details of the mathematics and algorithms behind modern AI, this book provides a roadmap for a diversity of stakeholders to realize AI business solutions that are reliable, responsible, and sustainable."

—Matthew Gaston, AI Engineering Evangelist

Beyond Algorithms

Beyond Algorithms

Delivering AI for Business

James Luke
David Porter
Padmanabhan Santhanam

CRC Press is an imprint of the
Taylor & Francis Group, an **informa** business
A CHAPMAN & HALL BOOK

First Edition published 2022
by CRC Press
6000 Broken Sound Parkway NW, Suite 300, Boca Raton, FL 33487-2742

and by CRC Press
4 Park Square, Milton Park, Abingdon, Oxon, OX14 4RN

CRC Press is an imprint of Taylor & Francis Group, LLC

Library of Congress Cataloging-in-Publication Data
Names: Luke, James, author. | Porter, David (Data architect), author. |
Santhanam, Padmanabhan, author.
Title: Beyond algorithms : delivering AI for business / James Luke, David
Porter, Padmanabhan Santhanam.
Description: First edition. | Boca Raton : CRC Press, 2022. |
Includes bibliographical references and index.
Identifiers: LCCN 2021056146 | ISBN 9780367622411 (hardback) |
ISBN 9780367613266 (paperback) | ISBN 9781003108498 (ebook)
Subjects: LCSH: Business—Data processing. |
Artificial intelligence—Industrial applications. |
Electronic data processing—Management.
Classification: LCC HF5548.2 .L843 2022 |
DDC 658/.05—dc23/eng/20220110
LC record available at https://lccn.loc.gov/2021056146

ISBN: 978-0-367-62241-1 (hbk)
ISBN: 978-0-367-61326-6 (pbk)
ISBN: 978-1-003-10849-8 (ebk)

DOI: 10.1201/9781003108498

Typeset in Minion Pro
by codeMantra

Contents

Authors

Dr. James Luke For as long as I can remember, I have been dreaming of Thinking Machines. As a child, I was fascinated by the prospect of creating intelligent machines and desperate to understand how such machines could be created. As an adult, I have focused my entire career, first in IBM and now at Roke, on building real artificial intelligence (AI) systems that solve real problems.

It may surprise some to hear that for most of my career there was little interest, from any quarter, in AI. In fact, I remember being told by at least one senior executive, in the late 1990s, that if I wanted to be successful in my career, I had to stop going on about AI. I still meet him every so often … but he's even more senior now … and I'm ashamed to admit that I haven't had the courage to remind him of his terrible career advice.

The recent explosion of interest in AI is both exciting and terrifying for me. Exciting as, for the first time in my career, there is genuine interest in the mainstream adoption of AI. Terrifying as, with so much expectation, there is a real risk that society will become disillusioned if projects fail to deliver.

In this book, I want to share my experience with you. Quite simply, I want to spare you the terrible sense of pain that comes with a failed project by sharing the experience that myself and others have gained over many years. To use a simple analogy, I often feel that many newcomers to the domain of AI are learning to fly … by crashing. I'd like to teach you to fly safely and without the need to experience a crash.

David Porter I started my degree at the University of Greenwich with a vague idea I wanted to do something in IT, somewhere the magic happened, and I graduated in 1995 with a burning passion for all things data and analytics. I have been lucky enough to work in Data Science, ever since. I have held senior consultancy roles at SAS Software, Detica/BAE Systems and now IBM. Early on in my consultancy career, I chose to focus on counter-fraud and law enforcement systems, I have never regretted that choice. My specialisation has allowed me to work with governments and organisations all over the world, it turns out financial crime is both universal and relentless. I didn't know then, but it's turned into one of those rare things, a job for life. Career highlights include co-inventing the NetReveal graph analytics software which in turn enabled the UKs first Insurance Fraud Bureau to launch and designing the conceptual architecture for what was to become the UK's ground breaking tax compliance system Connect. Job security hasn't been the only upside, pitting my analytic wits against, sometimes brilliant (but mostly lucky) criminals has sustained my insatiable curiosity and fascination for every new AI development, in the hope I can use it to sock it to the bad guys. As I often say, to anyone that will listen, it's like being paid to play chess. I joined IBM in 2016, enticed by the Watson story, could AI be used to catch crooks? It took a while for me to convert my maths brain to understand Natural Language Processing (NLP), but it was totally worth the effort, who knew words could be used to predict things? I've been putting NLP to good use ever since. NLP unlocks the value buried in the written data that every organisation collects, but no one has time to read. We truly are at the start of a new data gold rush. My ambition in this book is to help you get at that gold.

Dr. Padmanabhan Santhanam After graduate school and a brief career in low-temperature physics, I joined a team in IBM to work in software engineering research some 28 years ago. The goal was to improve the quality of commercial software produced and the productivity of software engineers who created it. I realised that the software development was a combination of computer science, some engineering, sociology and culture. I spent two decades working on leading-edge tools that covered a wide range of topics such as software defect analysis, project metrics, model-based testing and human–computer interaction. One of my proud accomplishments was a tool that automated ways to assess the maturity of a software project by analyzing various project artefacts. It was deployed across hundreds of software development teams in IBM with much success. Almost 15 years ago, we leveraged the NLP technology (that was created for the IBM Watson Jeopardy Project) to analyse textual use cases (in English and Japanese) in software projects. We were able to extract expected software behaviour and create functional test cases and other key artefacts automatically. Since 2014, I have worked on various aspects of AI technical strategy for IBM Research and our clients.

My personal interest is both in the use of AI for engineering traditional software systems and the emerging field of AI Engineering i.e. how to engineer trustworthy AI systems. In this book, I want to bring some practical perspectives on AI Engineering.

Acknowledgements

THIS BOOK IS TRULY a labour of love over the past 2 years! James Luke had sensed the need for this book based on his experiences with clients and had even started writing drafts of some chapters. With the usual demands at work, he needed partners to make it a reality. The collaboration on the book started on 14 September 2019 when James sent an email to the other two authors asking for interest in the project. From the very first meeting, we all agreed on the need for the book, what the book should be about and who the target audience must be. Thus started the long journey. To keep the consistency of the narrative, we decided that all of us will contribute to all the chapters, instead of divvying up the chapters among us. Given the broad audience we had in mind, the concept of using vignettes as side narratives for different purposes was also very appealing.

To give the book the authenticity of different voices and experiences, we recruited a few of our IBM colleagues to write vignettes on various related topics.

Author	Topic	Chapter
Murray Campbell	Chess Programs: Then and Now	4
Laura Chiticariu	Choosing between Different Algorithms for NLP Task	6
Daphne Coates	Bias in Development of AI	5
Richard Hairsine	Virtual Agent Assistant	5
Roy Hepper	C5-Rule-Based Model	3
Kai Mumford	Neural Network Model	3
Michael Nicholson	Exam Scandal in the UK	5

We are indebted to them for their enthusiasm and contributions to enrich the contents of this book.

We discovered our publisher, Randi Cohen, at Taylor & Francis in early 2020. She has been patient with us over the months, and has provided feedback and steady encouragement at every opportunity.

We thank Annemarie McLaren for bringing her fresh eyes to review the contents of this book and giving us detailed suggestions for improvement. Francesca Rossi helped to clarify the articulation of trustworthy AI and the role of ethics in developing AI systems.

In addition, PS would like to thank Zach Lemnios and Glenn Holmes for their enthusiastic management support for his contribution to this book and Hema Santhanam for many useful discussions on various sections of the book while being home bound during the COVID pandemic.

In the end, we hope that we have provided some clarity on what it takes to select and run a successful AI project in business with the current state of the technology and are looking forward to hearing from you on how we did in helping you overcome the common pitfalls in adopting AI for business.

James Luke
David Porter
Padmanabhan Santhanam
November 2021

Prologue

The CEO has declared that our new business will be the lead in Artificial Intelligence within 3 months. He read it in the newspaper, everyone is doing it, it has got to be good; we must do it!

Looks like we may need some data to make it work. No problem we have loads of data? Between the data in our legacy systems and the internet, which is free right? There should be plenty. Let us start with what we can find to build a prototype in the next month. We can work out the GDPR issues later.

I know you are going to say the data format we have isn't quite right! And some documents are in foreign languages, that the team doesn't speak, but that shouldn't be a problem because they're easy to translate online, right? Sure, some of our corporate data is not really digital ... our documents are mainly typed with handwritten notes; and so what if a few of the documents were burnt (only at the edges) in a fire, that's ok, we shouldn't worry, AI can deal with all that. Did you see there was an article in the paper last week about how AI is already working on damaged documents in the National Museum? the CEO says, if it's good enough for them...

With all that data, we just need some good Machine Learning algorithms to make sense of it. There has to be gold in there somewhere, the AI will find it for us. We had some interns put together a Machine Learning app in just a few days, so our programmers should have no trouble putting a production system together.

Our CTO thinks our current infrastructure and DevOps processes should be more than enough to create the first version of the AI application in three months. We will deal with any other issues later. ok?

Oh, just a reminder, do not bother our business subject matter experts yet. They're busy with our clients. We can always get them to talk to the programmers later when there is something to see. Besides, AI should be able to figure this stuff out on its own. I hear accuracy of these algorithms can be 80% without too much trouble and that should be good enough for our business.

See you in 3 months for the big launch!

If any part of the above resonates with your current Artificial Intelligence project, then you need to read this book!

DOI: 10.1201/9781003108498-1

If all the above resonates, then you need to find a new job and STILL read this book.

If none of the above resonates, and you are successfully rolling out new Artificial Intelligence applications without a problem, then please get in touch so that you can help us co-author the next edition of this book!

CHAPTER 1

Why This Book?

In the years before the Second World War, officials in the British Air Ministry became increasingly concerned about the risks to aircraft of a death ray. In addition to offering a £1,000 prize to anyone who could demonstrate an effective death ray, they also consulted the scientist Robert Watson-Watt. Watt, after collaborating with his colleague Skip Wilkins, determined that a death ray was not feasible. However, Watt and Wilkins did propose bouncing radio waves off aircraft as a means of detecting and tracking them. The search for a death ray inadvertently led to the invention of radar.

AI IS EVERYWHERE

The moral of this story [1] is simple. When it comes to any form of new technology, the destination you arrive at may not be the one you were planning for. As a society, we are embarking on arguably the most exciting technological journey; the quest to improve the way we live and, especially, the way we work through the practical use of Artificial Intelligence (AI). Reading the daily media reports and the recent AI Index report [2], many would argue that we are already using AI extensively in our personal and professional lives.

One reason that AI is seen as ubiquitous is that there are many high-profile and successful applications that do have a major impact on our daily lives. Many of these applications are driven by large web companies such as Google, Amazon and Facebook and enable impressive applications such as search, question answering, and image classification. In our online shopping, we experience AI-enabled profiling and targeted advertising continually and can be forgiven for thinking that the AI knows us better than we know ourselves. We no longer need to learn a foreign language as online translation tools allow us to read foreign language text or order a drink anywhere in the world. When we call our bank or book a holiday, we can count on dealing with virtual agents that seem to understand our conversation as if they were human. So, what's the problem?

DOI: 10.1201/9781003108498-2

ENTERPRISE APPLICATIONS

Enterprise applications are different from these popular consumer web applications, and they really do run the world in both the commercial and public sectors. In the commercial sector, enterprise applications are used in retail sales, insurance, finance, telecommunications, transportation, manufacturing, and many other industries. In the public sector, government agencies deploy applications to support areas including Law Enforcement, Social Security Administration, Internal Revenue Service, Health Services and National Defense. With so much AI already available and working for our benefit, you may be surprised to know that not only are actual enterprise applications of AI still in their infancy, but that *the majority of enterprise AI projects fail*. As evidence, recent analyst reports [3–6] estimate that at most 20%–40% success rate for the adoption of AI to create business value. This supports the assertion that moving AI from a proof-of-concept to a real business solution is not a trivial exercise.

While the consumer web applications mentioned above are extensive in their reach and impact, they represent just a tiny fraction of real Information Technology (IT) applications around the world. Hidden under the covers of the organisations that we rely on for our day-to-day lives are tens of thousands of applications. Individually, these applications may appear much smaller in reach and impact than web search or online shopping, but, in reality, they are really critical to our lives. These applications perform all the critical functions of the modern world from managing our prescriptions to evaluating life insurance risks, controlling city traffic, managing bank accounts and scheduling the maintenance of trains and buses. There is a vast number of such enterprise applications, and many could benefit from the application of AI. However, many well-intended AI projects still underestimate the extra complexity of their delivery in an enterprise setting, and often even after stellar early success, still fail to deliver actual business benefit. In creating these enterprise applications, we need to recognise that they are very different from consumer web applications and that delivering AI in the enterprise is different from delivering AI in a web company.

Consider, for example, a consumer web application such as one of the personal assistants that we all now have in our homes and use in our everyday activities. These assistants are designed to answer the most frequently asked questions such as, "is it going to rain today?" To answer these questions, the developers provide specific services and then capture data from millions of users to discover all the possible ways the question could be asked.

But what about answering general knowledge questions? While we all love to be impressed by the power of our online assistants, answering general knowledge questions is considerably easier than answering enterprise questions. "Who was the British Prime Minister during the Suez Crisis?" requires a factual response. To find the answer to such a transactional question, the technology can exploit the massive levels of information redundancy on the internet. There are tens, if not hundreds of thousands of documents online about the Suez Crisis. This information redundancy means that it is possible to use simple correlation algorithms to identify the correlation between the terms "Suez crisis", "Prime Minister" and "Anthony Eden". All the web AI has to do is take the statistically strongest answer. The internet also includes many trusted sources of data, including news agencies, educational organisations and encyclopaedic sites that aim to provide validated and trustworthy information. In tuning their algorithms, the web companies can use the feedback from millions of users to spot and correct errors.

In other words, consumer web applications tend to focus on quite simple, common tasks that can be performed with huge volumes of information, from many (often trusted) sources, while using feedback from tens of millions of users to tune the algorithms. In the enterprise, the questions asked are rarely that simple and the volumes of information are much, much lower. Often information is contradictory and may take a skilled user to assess and really understand. Alternatively, the required information may not exist at all. As for the number of people involved, in an enterprise, far fewer people ask far more complex questions and the differences between questions may be subtle, but important. Our ability to capture high volume feedback is limited.

Finally, there is one further and very significant difference between applying AI in an enterprise setting and applying AI for consumer web applications. We are much more forgiving of errors in web AI; it is mostly of low consequence if a web AI brings back a wrong answer. If we ask, "Alexa play my favorite tune" and the response is, "THE MONTH OF JUNE HAS 30 DAYS", it's another thing to laugh about at dinner parties. However, we will be far less forgiving if a police application leads to the arrest of the wrong person or a medical AI leads to a misdiagnosis.

The excitement generated by AI web applications does still add value to enterprise applications. First, it changes the way we might think about enterprise applications by placing greater emphasis on ease of access and simplicity. Graduates joining modern enterprises expect web technology and web style user experiences in their life at work. Second, it drives innovation and helps push forward the business cases for enterprise applications of AI. Of course, such an endeavour will succeed only if the AI can deliver on the expectations assumed in the business case. In this respect, we must recognise that the domain of AI has a track record of failed delivery.

AI WINTERS

There have been two periods in the history of AI where dashed expectations have led the industry to lose all confidence in AI. The perception of AI was so poor that these periods were called 'AI Winters'. During these periods, funding for new AI endeavours all but disappeared and there was widespread disillusionment, some would say cynicism, about AI.

ORIGINS OF ARTIFICIAL INTELLIGENCE RESEARCH

The Beginning

The modern formulation of AI started with Alan Turing's famous paper, "Computing Machinery and Intelligence" [1], where he discussed the concept of computing machines emulating human intelligence in some context, giving rise to the term "Turing Test" for machines. The next big event was the 1956 Dartmouth Summer Workshop [2] organised by John McCarthy, Marvin Minsky, Nathaniel Rochester and Claude Shannon when the term 'Artificial Intelligence' was introduced as an interdisciplinary area of research. Two key people who could not attend the event were, Alan Turing who had died in 1954, and the computing visionary, John von Neumann who was already seriously ill. Even though the workshop did not produce anything specific, it gave the participants the motivation to approach AI from their different perspectives over the next many decades.

What Is AI?

John McCarthy defined [3] AI as, "the science and engineering of making intelligent machines" and defined "intelligence" as "the computational part of the ability to achieve goals in the world." Marvin Minsky offered [4] a similar definition, "the science of making machines do things that would require intelligence if done by men". The Encyclopedia Britannica currently defines AI as, "the ability of a digital computer or computer-controlled robot to perform tasks commonly associated with intelligent beings".

These definitions all share a common theme in that they refer to performing tasks that would normally require or be associated with human intelligence. There is clearly a paradox in this definition. Pamela McCorduck [5] describes this: "Practical AI successes, computational programs that actually achieved intelligent behavior, were soon assimilated into whatever application domain they were found to be useful, and became silent partners alongside other problem-solving approaches, which left AI researchers to deal only with the "failures," the tough nuts that couldn't yet be cracked." There is actually a name for this! It is called the "AI Effect" and summarised in Larry Tesler's theorem [6], "Intelligence is whatever machines haven't done yet". This is because the society wants to associate intelligence only with humans and does not want to admit that human tasks can indeed be performed by machines!

REFERENCES

1. A.M. Turing, "Computing machinery and intelligence," *Mind, New Series*, 59(236), pp. 433–460 (October 1950).
2. "Dartmouth summer research project on Artificial Intelligence," http://jmc.stanford.edu/articles/dartmouth/dartmouth.pdf.
3. J. McCarthy, "What is AI", http://jmc.stanford.edu/articles/whatisai/whatisai.pdf.
4. M. Minsky, *Semantic Information Processing*, (Cambridge: MIT Press, 2003).
5. P. McCorduck, *Machines Who Think*, (London, UK: Routledge, 2004).
6. D. Hofstadter, quoted Larry Tesler differently as "AI is whatever hasn't been done yet" in "Gödel, Escher, Bach: An eternal golden braid," (1980).

The first AI Winter occurred between 1974 and 1980 following the publication of a report by Sir James Lighthill [7] criticising the failure of AI research to meet its objectives and challenging the ability of many algorithms to work on real problems. Funding for AI research has been cut across UK universities and across the world.

In fact, AI's fortunes seemed to be on the up again in the mid-1980s when investment banks found that neural networks and genetic algorithms seemed to be able to predict stock prices better than humans. A stampede of activity took place as banks competed to get the upper hand with better more sophisticated automated trading algorithms. What could possibly go wrong? In the rush to get rich, IT architects failed to acknowledge the critical weakness of neural nets; they need historical precedent in their training data for predictions. They can be unpredictable when applied to previously unseen situations. The stock price boom, made possible by AI, was (you guessed it) unprecedented. By this point everybody trusted the algorithms, when they said, "Sell!" the bank sold, when they said "Buy!", it bought, no questions asked. The global financial crash of 1987 aka "Black Monday" was enabled by a chain reaction of AI trading algorithms going off the rails [8].

Unsurprisingly, this was a major factor in triggering the Second AI Winter (1987–1993) with increasing disillusionment in expert systems and the collapse of funding in specialist AI hardware. Either side of these major AI Winters, there were multiple periods of less significant disillusionment when AI failed to meet over-hyped expectations.

WHAT IS DIFFERENT NOW?

While there are some who believe that the current excitement about AI is a precursor to the next "Winter", there are a few reasons to be optimistic. After all, we've learned our lessons the hard way. The main reason being that, after years of considerable investment in IT, our society has built an extensive digital infrastructure that is awash with data. Twenty years ago, we did not have access to this volume of customer, or passenger, or citizen data and all efforts were focused on establishing the basic IT infrastructure. The potential to exploit this data is clear, and the business opportunities are quantifiable. Since there are not enough skilled humans to even look at all the data, we do need more automation and intelligence in the way we do analysis. At the same time, continuing performance improvements in computer processors, increases in on-board memory and storage capacities, and the availability of 'elastic' computing resources on the Cloud are key enablers of AI. This last point is noteworthy since many of the algorithms attributed with the current excitement in AI have been understood, yet impractical to use, for decades. It is only the increase in volumes of data and processing power that are now making their application practical. Not surprisingly, the availability of data and computing power has injected new energy and innovation in the algorithm creators.

The simple fact is that businesses, and the society at large, are now ready to exploit the wealth of data and IT infrastructure that exists for advantage. Those who can process data in intelligent ways will perform well and their financial returns can be huge. In addition, over the past decade, machine learning algorithms have steadily improved in performance for doing narrow well-defined tasks (e.g. image recognition, simple question-answering, machine translation, etc.) and can perform at accuracies comparable to or better than humans. There is no reason that businesses cannot leverage these advancements to solve the data deluge problem they face.

PROCEED WITH CAUTION!

In short, with all the progress made in the past decade, another AI Winter is extremely unlikely. However, just the rush to exploit the new machine learning techniques in pretty much every domain conceivable, has already exposed some serious cracks in the current AI technology [9–11]. The fact remains that far too many AI projects in business still fail past the prototyping stage. Not only do they fail, but the reasons for their failure are often misdiagnosed. There is a huge tendency to attribute failure of a project to the failure of the AI. It is too easy to declare that the AI isn't intelligent enough. The reality is that often projects are not chosen well to meet the current capabilities of AI technology and the practical aspects of engineering the application are not adequately addressed to meet the business challenge. The current excitement about AI provides us with a massive opportunity. For the first time in decades, there is a genuine desire to embrace AI and to make the systems we interact with every day more intelligent and more responsive to our needs. We simply cannot afford to miss this opportunity. We can deliver on the current promise of AI, only if we understand how to engineer and deliver AI solutions. As the title of this book suggests, there is more to building a successful AI application beyond the shiny algorithms of the day.

DELIVERING AI SOLUTIONS

As it stands today, there is way too much focus on algorithms, and not enough attention paid to engineering real solutions. Millions of researchers in Academia and Industry around the world with deep expertise in AI are working on algorithms and the science behind them. There are far fewer engineers who have specialised in how to define and build complete reliable end-to-end solutions. To adopt an automotive analogy, we have a huge investment in the science of fuels and brilliant researchers coming up with more and more efficient fuel compounds. However, we lack the overall systems engineering skills to build useful and enjoyable automobiles. While algorithms are critical to advancing AI, we also need to embrace other disciplines to build complete solutions. We need Data Scientists, Business Analysts, Ethicists, Performance Engineers, Systems Integrators, Test Specialists and User Experience Designers, all working towards a common goal under an overall engineering method. Such a framework does not exist yet.

BETTER UNDERSTANDING OF AI IS CRITICAL FOR SOCIETY

One significant feature of AI that differentiates the discipline from conventional technologies is the fact that everyone in society will both impact and be impacted by the decisions made by AI systems. These systems will have a symbiotic relationship with users since they provide data explicitly or implicitly through their interactions. AI applications will learn from interactions with users in a way that doesn't exist in conventional IT systems. This societal impact means that a far larger proportion of society will need to understand how

AI really works and how to work with it, not just a small group of brilliant scientists. This is different from the past, when users did not really need to understand how the technology behind the application worked. As a society, a huge cross section of people will need to understand AI in a way that they haven't had to previously.

TARGET AUDIENCE FOR THE BOOK

Numerous technical books have been written about the science of AI targeting researchers. There are also many books for data scientists to learn tools and techniques for developing algorithms. We are starting to see some books that provide high-level introductions to AI and guidance aimed at managers and consultants. However, in real AI projects, technology capabilities and the business impact are very much intertwined, due to the extreme reliance of data, critical choice of the domain applicability and the expected quality & trust objectives. That's the purpose of this book. We want to broaden the discussion on what it takes to define, engineer, and deliver end-to-end AI solutions. In addition, AI applications in business have a long list of stakeholders: investors, business sponsors, domain specialists, engineers, customers, regulatory agencies, media, etc. The success of the application depends on the perception of these stakeholders, and their needs should be considered as an integral part of the development process so that late surprises/shocks are avoided. In short, this book should be of interest to anyone interested in applying AI in real business applications in a commercial or public enterprise.

AN OUTLINE OF THE BOOK

This book aims to help anyone who may touch AI in some way to understand the domain and how to build AI systems. To achieve this:

- We will explain the core concepts of AI using real practical examples, with a focus on the systems engineering and the business change aspects of building solutions.
- For the technically inclined, there are many technical deep dives scattered throughout the book. These can be skipped over by the less technical reader.
- For the philosophers, we've included the odd thought experiments designed to challenge your brain and destroy your weekends.
- Finally, there are case studies of real AI systems and projects that will hopefully help you understand what works … and what doesn't.

Chapter 2 gives a broad overview of examples of AI applications and discusses key aspects of building and sustaining them. Chapter 3 introduces various approaches to algorithms and their applicability in practice. Chapter 4 helps you to select the right AI project and avoid the bad ones. Chapters 5 and 6 aim to define and measure business value of AI. Chapters 7 and 8 address the doability aspect of projects with detailed discussion of the data usage and specific challenges in real projects. Chapter 9 helps to evaluate your business idea in terms of specific questions related to value and doability and take actions accordingly. Chapter 10 deals with the not-so-boring engineering aspects of creating and maintaining AI systems. Chapter 11 is a view of the future of AI in business to get you prepared for what is coming.

So, let's get started on this exciting journey and make AI real.

REFERENCES

1. "Robert Watson-Watt and the triumph of radar," https://blog.sciencemuseum.org.uk/robert-watson-watt-and-the-triumph-of-radar/.
2. 2021 AI index report: https://aiindex.stanford.edu/wp-content/uploads/2021/11/2021-AI-Index-Report_Master.pdf
3. KPMG 2019 Report: "AI transforming the enterprise".
4. O'Reilly 2019 Report: "AI adoption in the enterprise".
5. Databricks 2018 Report: "Enterprise AI adoption".
6. MIT Sloan-BCG Research Report: "Winning with AI".
7. J. Lighthill (1973): "Artificial intelligence: a general survey," *Artificial Intelligence: a Paper Symposium*, Science Research Council, London.
8. F. Norris, "A computer lesson still unlearned," *The New York Times*, (October 18, 2012): https://www.nytimes.com/2012/10/19/business/a-computer-lesson-from-1987-still-unlearned-by-wall-street.html.
9. G. Marcus and E. Davis, *Rebooting AI*, Pantheon Books, New York (2019).
10. N.C. Thompson, et al., "Deep learning's diminishing returns," *IEEE Spectrum*, 58(10), pp. 50–55 (October 2021).
11. C.Q. Choi, "7 revealing ways AIs fail," *IEEE Spectrum*, 58(10), pp. 42–47 (October 2021).

CHAPTER 2

Building Applications

No way! Why would we pay for a translator when we can use a free online service?

The Requirements Document was complex and, to make matters worse, it was written in Arabic which no one on the team spoke. We had just two weeks to respond and it was a must win deal. Our manager point blank refused to pay for a human translator, so we resorted to the online translation tool of a well-known web search company.

I watched as one of the team copied the first requirement from the "Administration and Management" section and pasted it into the translation tool. The translated question brought an instant smile to my face! It said simply, "what is the point of management?"

James Luke

In many respects, Artificial Intelligence (AI) applications are no different from any other type of complex systems. But in some respects, they're very different. Later in this book, we'll highlight the similarities and differences in the system development activities of AI and typical information technology (IT) systems. In this chapter, we'll explain what makes AI applications very different from traditional business applications. Then we'll review some of the most famous applications in the history of AI and discuss what we can learn from them. We will discuss some existing examples of business applications that incorporate AI and the underlying architectural considerations that you may want to think about in building your AI application.

WHAT'S DIFFERENT ABOUT AI WHEN BUILDING AN APPLICATION?

Data Decides the Application Behaviour

First and foremost, it's all about the data! In conventional applications, software code is developed to a specification and data is just used to test that the resulting functionality is correct. In AI applications, especially those developed using Machine Learning, the data plays a critical role in defining the functionality. In conventional systems, data is an input

DOI: 10.1201/9781003108498-3

to the program. In AI applications, data is encapsulated together with the code in a way that actually determines the functional behaviour.

One aspect of data that needs to be considered upfront is the relationship between the training data and the production data. Almost always, we see the need for extensive data cleansing (see Chapter 7) in order to create effective training and test data. Remember that any data cleansing needed in producing training and test data will also be required, consistently each and every day, in the production environment. If a crucial stage of the data cleansing for training data involves a human manually and laboriously correcting some fields, then that person is going to be needed in the production environment and they will need to move a lot faster if you are planning on building a sub-second process! Seriously, if you have to make lots of manual interventions at the training data stage, you are going to need to find a way to automate that data cleansing routine before you can go live with an operational application.

Beyond the data, there are many other critical considerations that must be understood. While these next considerations may not apply to all applications, many will apply in most cases.

Making Mistakes

As AI applications are designed to perform tasks normally requiring humans then, by definition, there are distinctly human aspects to their functionality, such as making judgement calls in decisions. You will not only need to accept that they will not always make the right decision but will also need to accept that it may not be possible to determine what the right decision is. This challenge will be compounded if you choose to use AI technologies that are not able to explain their reasoning.

Accepting that AI applications will make mistakes is not easy for human beings to do! Consider, for example, the use of AI in enabling driverless vehicles. At the time of writing, just a small number of fatalities were reported in accidents involving driverless vehicles, but each of them evoked a huge media reaction. Yet in 2019, nearly 40,000 people were killed by cars in the US alone without any significant media reaction. It seems that the public is far more accepting of human error than machine error.

Ability to Generalise

In Machine Learning systems, we deliberately develop systems that achieve a percentage accuracy against test data. There is a continuous contention within the Machine Learning community regarding accuracy versus generalisation. Our aim in Machine Learning is to create a system that can generalise in the same way humans do. Continuing with our driving theme, in the UK, people are taught to drive right-handed vehicles on the left-hand side of the road. They learn a particular set of road signs and road layouts. Once qualified, they are able to take their driving license to Europe or the US and are able to drive left-handed vehicles reasonably safely on the opposite side of the road with different types of signs, intersections and road layouts.

Our ability to generalise is extremely powerful and is perhaps a true test of human intelligence. It is important because it allows the system, either human or machine, to operate in an unfamiliar environment. Generalisation enables a pilot to adapt to manage a serious technical fault or a fire chief to develop a plan for a disaster that has never been previously encountered. Generalisation enables a judge to make a decision on a case that represents a serious legal and ethical dilemma. Generalisation is critical in delivering the ability to deal with a previously unseen scenario.

So, in developing Machine Learning systems, we are constantly trying to understand whether the system is actually generalising or whether it is simply memorising all the training cases; the latter is often referred to as either "over-fitting" or "learning the deck". If the system has "learnt the deck", then it will not work well in previously unseen situations. One of the great indicators that the system has learnt the deck is a near-perfect performance against training and test data. For this reason, Machine Learning algorithms are designed and tuned not to achieve perfect performance against training and test data.

In other words, we must deliberately design systems that fail test cases and are therefore deploying into operational use systems that are expected to make mistakes.

Need for Experimentation

A further point to note is that this is not a static situation. AI technology will continue to evolve with the delivery of new algorithms and access to training data may increase; the training data may increase in volume and the actual data itself may evolve and change with time. As a consequence, the performance, in terms of accuracy, of an AI component may change almost continually. In cases where multiple technology suppliers are competing, there is a reasonable chance that an accuracy arms race will emerge with each supplier taking turns to out-perform the other. This continual evolution means that evaluating and selecting technologies as part of a strategic architecture may be a mistake; instead, architects needs to design in such a way that it is easy to swap in and swap out technologies with a continual process of evaluation.

Knowing the Right Decision

In many applications, it may be possible to determine if a mistake has been made. However, increasingly, AI technologies will be applied in cases where it is more difficult to determine whether or not a mistake has been made. For example, in a rare medical case, an AI application may recommend one course of treatment whilst an expert doctor recommends an alternative. Whichever course of action is followed, it may not be possible to determine whether the alternative would have been better. If the patient recovers, there may be the possibility that the alternative treatment may have led to a more rapid and better recovery. If the patient dies, there may be no way of knowing whether they would have survived with the alternative treatment.

AI Ethics

When dealing with such judgement calls, especially those involving human life, ethical and trust issues become paramount. We are accustomed to seeing ethical debates in the media about AI applications. Often these debates focus on whether it is ethical to develop a particular application. From an architectural and development perspective, the ethical considerations run deeper. Given the ambiguity inherent in AI systems, resulting from the accuracy considerations and the fact that many decisions will be judgement calls, how do we ensure that an AI application is competent to perform its task? Note the word competent is being used in place of the more scientific "fit for purpose" engineering term. "Fit for purpose" implies a formal test process as applied in conventional engineering. With AI applications, we have already explained why such testing is not effective. Instead, we need to think about "competence" in the same way that we would refer to the competence of a professional person.

AI Accountability

Let's consider a common question asked about AI applications ... who is responsible when it goes wrong? If a driverless vehicle makes a mistake that results in a human death, who is responsible? Can the AI be deemed responsible ... and perhaps in a future of sentient AI droids would it be deactivated? Is the owner of the AI responsible or, more likely, is it the developer? To a certain extent, this is not a new problem, and society has pretty much figured this stuff out. If an aircraft crashes, there is an investigation that looks at all aspects of the process, from the training of the pilots to the design of the aircraft to the standards mandated in the industry and beyond. We suspect it will be the same with AI! Given that it may not be possible to exhaustively test all aspects of an AI application, there will be an ethical responsibility on the developer to ensure best practice was adopted in the development process. For example, was the right amount of training data used and was the level of testing reasonable for the complexity and risk associated with the application? The developer may even need to preserve the actual training data so that retrospective investigations can be run. For example, "did the driverless vehicle training data include sufficient dogs running across the road, snow covered lane markers?"

Determinism in AI

In considering best practice for AI, as opposed to conventional software engineering, the subject of determinism is particularly interesting. Determinism is considered an essential element of conventional software engineering and ensures that a system always functions in exactly the same way for a given set of input data. Quite simply, if you put the same input data into the application, you should get the same output! In AI systems, it may not be quite that simple. For example, the AI model generated by some machine learning algorithms depends on the order in which the training data is presented to the algorithm. If you change the order, then you get a different model and data that was previously classified correctly may now be classified incorrectly and vice versa. As these algorithms require considerable computational power and take a long time to run, we use parallel processing to speed up the training process. Depending on the memory and number of processing cores available, the order in which the data is presented may change and, therefore, subtly different models may be generated for the same data.

Impacting the Operational Environment

Finally, AI applications often impact their environment to a greater extent than their conventional predecessors. If we look at the stock exchange, AI is being used in automated trading systems that directly impact the environment in which they operate. AI applications buy or sell shares that cause changes in the share price that cause the AI to buy or sell shares. It should not be a surprise therefore that we experience unexpected feedback loops and maelstroms when these scenarios cascade out of control. Whilst these risks exist in any automated system, with or without AI, the use of AI means they are harder to test and prevent.

From the discussion above, can we all agree that AI Applications are sufficiently different from conventional applications? Good ... then let's continue our journey by looking at the spirits of AI applications: past, present and future.

PROMINENT AI APPLICATIONS OF THE LAST SEVEN DECADES

Before we delve into the current trends in AI applications, let us look at the list of AI applications that caught the public fascination in the last seven decades! We divided the list into four broad groups: (i) Board Games, (ii) Natural Language Understanding (NLU), (iii) Expert Systems and (iv) Autonomy. *The vignettes shown on the side give more details on them.* What can we learn from looking at these famous applications?

Board Games

This shouldn't really be a surprise! Game playing naturally provides a rich environment in which to develop and test AI capabilities. The reasons for this are simple. First, the scope of the problem is naturally limited to the domain of the game, with clear rules on what is allowed and what constitutes a win. Second, assessing performance is easy, since the games fit the human paradigm of winners and losers. Third, it is easy to explain what is actually accomplished by the machine to the population at large. Finally, game playing is extremely low risk in comparison with the real world. All these factors make playing board games a very attractive proposition for AI researchers.

We just want to highlight the key technology advancement that resulted from these gaming applications. Arthur Samuel [1] introduced the phrase "Machine Learning" in the 1950s to explain how his program playing Checkers learnt the game. It took a little while, till the early 1990s when Gerald Tesauro created TD-Gammon [2], which learnt by playing against itself (using reinforcement learning) to reach championship level in Backgammon. Then came Deep Blue [3] from IBM in 1997 that captured the imagination of both AI researchers and Chess fans by beating world chess champion Garry Kasparov in a regulation match. This system used both custom hardware and AI learning from historical games and player styles. In 2016, the AlphaGo system [4] from Google Deep Mind that uses DNNs convincingly beat the world champion Lee Sedol in the Chinese board game Go. The current version of this system called AlphaGo Zero is even better, with no need for human input, beyond the game rules. The low risk in game playing gives those developing AI for games a massive advantage over those developing real-world AI at a time when the virtual world and real worlds are converging. As simulation games become more and more realistic, we should not be surprised to see AI developed initially in games being applied in the real world.

BOARD GAMES

AI playing board games with humans has been an exciting aspect of demonstrating the sophistication of the AI to the community at large. Here is a short summary of some famous AI projects playing board games in the last seven decades.

- Checkers Program (1951–1959)
  Arthur Samuel [1] at IBM used IBM 701 and 704 machines to create a program that played checkers. He introduced the phrase 'Machine Learning' in the literature for the first time. He used two different learning algorithms to train the program and came to the conclusion, "a computer can be programmed so that it will learn to play a better game of checkers than can be played by the person who wrote the program". He demonstrated the program on television on 24 February 1956.
- TD-Gammon (1993)
  TD-Gammon was a neural network-based program created by Gerald Tesauro [2] at IBM Research that trained itself to play backgammon, by playing against itself. In this project, Tesauro advanced the field of reinforcement learning, a technique that provides reward to the AI, indicating how good or bad its output was. The goal is to encourage optimal actions by AI, leading to maximal reward. In cases where the reward is given at the end of a long sequence of inputs and outputs (e.g. in a game with many moves before the result is known), AI has to figure out how to apportion credit or blame to each of the various inputs and outputs leading to the ultimate reward. This is known as the "temporal credit assignment" problem. TD-Gammon used 'Temporal Difference' (TD) method to address this problem, hence the name. TD-Gammon was able to play backgammon at championship levels.
- Deep Blue Chess (1997)
  IBM's Deep Blue [3] was the first computer to defeat a reigning human world chess champion, Garry Kasparov, in a regulation six-game match by two wins, one loss and three draws. The machine strategy relied on search algorithms over possible moves, leveraging knowledge from a database of 700,000 Grandmaster games in terms of game opening strategies and an 'Extended Book' that included factors such as the number of times a move has been played by Grandmasters and skill of the player making the move. On the hardware side, the system had a 30-way machine with 30 RISC/6000 processors and 480 custom chess chips and operated at the sustained speed of 200 million chess positions per second or about 8 Tera (10^{12}) operations per sec.
- AlphaGo (2016)
  In 2016, Google Deep Mind's AlphaGo program [4] beat the world champion Lee Sedol, four games to one in the Chinese board game Go. The game is played on a 19 × 19 grid which is much bigger than 8×8 for chess. Even though the rules are simple, the game is very complex. For each player's turn, there is an average of 200 possible moves through most of the game, and hence, it takes a very large number of calculations to anticipate even a few moves ahead. AlphaGo evaluated

positions and selected moves using deep neural networks (DNNs). These neural networks were trained by supervised learning from human expert moves and by reinforcement learning from playing against itself. After the success of AlphaGo, there is a more recent version of the program called *AlphaZero*, which was developed with NO human input beyond the game rules. AlphaZero won 100–0, playing against the AlphaGo program, the previous winner.

REFERENCES

1. A. L. Samuel, "Some studies in machine learning using the game of checkers," *IBM Journal of Research and Development*, pp. 211–229 (1959).
2. G. Tesauro, "Temporal difference learning and TD-gammon," *Communications of the ACM*, 38, pp. 58–68 (1995).
3. M. Campbell, "Knowledge discovery in deep blue," *Communications of the ACM*, 42, pp. 65–67 (1999); https://www.youtube.com/watch?v=KF6sLCeBj0s.
4. D. Silver et al., "Mastering the game of go without human knowledge," *Nature*, 550, pp. 354–359 (2017).

Natural Language Understanding (NLU) Challenges

Our natural desire to create an AI that we can interact with as if it were a human is as old as the field of AI itself. In fact, the Turing Test is based on this very concept. ELIZA [5] was the first attempt to create a virtual agent to interact with humans using language, although only using the context of just the user input, with no awareness about the real world. SHRDLU [6] was a more complex system to carry out operations in its own restricted environment based on human commands expressed in textual input with sophisticated syntax. These early virtual agents were the fore runners of the virtual agents that are increasingly part of modern customer contact centres, with many of us cheerfully interacting, often unknowingly with AI text/chat bots every day. IBM's Watson [7] playing Jeopardy! was significant in that it represented the first significant application of AI to unstructured data and open-domain Deep Question-Answering. Unlike the game playing projects we discussed above, the domain of relevant knowledge was largely unconstrained. The questions were also in much more nuanced format that are hard for even humans to resolve. IBM's more recent entry into advancing NLU came with the Project Debater [8], which went way beyond answering questions! It was able to aggregate pro and con arguments for open-domain topics by reading billons of sentences and put together a cohesive presentation for humans to evaluate. This painted the vision of a future where the machine will do all the grunt work of extracting useful information from vast number (millions!) of documents and present it in a human digestible form. Augmenting human intelligence, indeed! Recent advances [9] with the use of deep learning (DL) techniques for multi-level automatic feature representation of text are making a major impact in NLU by machines.

NATURAL LANGUAGE UNDERSTANDING

Interacting with humans naturally requires understanding the human language. This was a major topic in the history of AI. Here are four projects that show the evolution of the technology over the decades.

- ELIZA: The First ChatBot (1966)
  Joseph Weizenbaum [1] at MIT created a system that made certain kinds of natural language conversation using written text between humans and computer possible. At a high level, the machine looked for the presence of a keyword in the human input, and if found, the sentence was transformed according to a rule associated with the keyword and displayed back to the user. The conversation with the machine was easily comparable to a dialog with a psychiatrist, since the machine responded just to the user input, *without knowing anything about the real world.*
- SHRDLU (1968–1970)
  The name comes from the string 'Etaoin shrdlu', which represents the 12 most commonly used letters in the English language, by the approximate order of frequency. Terry Winograd [2] created SHRDLU program at MIT to carry on a dialog (via teletype) between a person and a simulated "robot" arm in a tabletop world of toy objects on a screen. It could answer questions, carry out commands to move the objects around and incorporate new facts about its world. The program was popular among the AI community because it "combined a sophisticated syntax analysis with a fairly general deductive system, operating in a 'world' with visible analogs of perception and action".
- Watson-Jeopardy! (2007–2011)
  IBM Watson was an open-domain Deep Question-Answering system [3] that beat the two highest ranked human players, Ken Jennings and Brad Rutter, in a nationally televised two-game Jeopardy! match in 2011. The contestants are expected to understand clues in (often highly nuanced) English and buzz with answers with enough confidence before the other contestants do, in about 3 seconds, with penalties for wrong answers. Like the humans, the machine had no access to external data sources during the game. The relevant knowledge was contained in 6.8 million documents (derived from historical Jeopardy! games and archival data sources such as Wikipedia & Encyclopedia Britannica) with added annotations, taking 50-GigaBytes of storage. The system had a highly parallelised architecture integrating over 100 different analytic components into a single application deployed across 71 machines, each having 32 CPU cores. In addition to knowing the likely answer, game playing strategy (e.g. selecting topics, placing bets, etc.) was also critical for its success.
- Project Debater (2019)
  In 2019, IBM Project Debater [4] faced Harish Natarajan, who has a world record for most debate competition victories, in a live debate on the resolution "We should subsidise preschool". Even though the machine lost the debate that

day, it won the points for enriching the knowledge of the audience by a large margin. Project Debater's knowledge base consisted of around 10 billion sentences, taken from newspapers, journals and archival sources. The system had to identify the contents relevant for the topic of the debate, both pro and con, and pull together a coherent argument supporting its position using precise language in a clear and persuasive way. A good rebuttal is always the hardest part of a debate. Project Debater applies many techniques to anticipate and identify the opponent's arguments and then responds with claims and evidence that counter these arguments. The major accomplishment of Project Debater was the compilation of different ideas from its corpus on both sides of the resolution and creating a coherent presentation in real time. This clearly demonstrated the possibilities of human and machine working together to perform better than either of them.

REFERENCES

1. J. Weizenbaum, "ELIZA: a computer program for the study of natural language communication between man and machine," *Communications of the ACM*, 9, pp. 36–45 (1966).
2. T. Winograd, *Understanding Natural Language* Academic Press, (1972).
3. D. A. Ferrucci, "This is Watson," *IBM Journal of Research and Development*, 56 (2012) and the rest of the Issue.
4. N. Slonim, et al. "An autonomous debating system," *Nature*, 591, pp. 379–384 (2021).

Expert Systems

From the 1970s to the early 1990s, building systems to emulate human knowledge was a big endeavour in various domains. This required definition of domain specific rules by humans who are experts in the domain, hence the name, Expert Systems. MYCIN [10] was meant to be an assistant to a physician to diagnose and treat patients with bacterial infections. DENDRAL project [11] dealt with the problem of determining the molecular structure from the results of mass spectrometry experiments. AARON: The Robotic Artist [12] was created to emulate a painter who produced physical, representational artwork. These and many other applications of that time suffered from the same basic problem: The creation of the systems needed experts and vast amount of domain knowledge and they could not scale to other domains. The manual effort needed to create them and the practical difficulty of maintaining them with evolving knowledge in any domain were the other reasons that the expert systems approach as the primary method to build AI systems went out of favour. Fundamentally, these systems tried to do things decades ago that would be significant value to the society. Clearly, medicine is an area of huge potential value; however, it is also a domain where the challenges of successful safe applications are massive. Creation of "Deep Fake" images of today may be the current incarnation of what AARON tried to do all those years ago.

EXPERT SYSTEMS

In our zest to make AI systems emulate humans, considerable effort over at least three decades was devoted to transferring human knowledge to machines. Here are three famous examples of these 'Expert Systems'.

- MYCIN (1972–1979)
  MYCIN was a rule-based expert system [1] developed at Stanford University, designed to assist physicians in the diagnosis and treatment for patients with bacterial infections. In addition to the consultation aspects, MYCIN also contained an explanation system which answered simple questions to justify its advice or educate the user. The system's knowledge was encoded in the form of about 350 rules that captured the clinical decision criteria of infectious disease experts. The architecture of MYCIN allowed the system to dissect its own reasoning and facilitated easy modification of the knowledge base. Even though the system was competitive when compared to human experts, it was never used in real clinical settings.
- The DENDRAL Project (1965–1993)
  DENDRAL was a collection of programs [2] developed at Stanford University that addressed the important problem in organic chemistry of determining the organisation of the chemical atoms in specific molecules. In order to do this, they collected rules that chemists use to describe the results of mass spectrometry. They later added specialised postprocessing knowledge and procedures to deal with data beyond mass spectra. The program was assessed by experts by testing the predictions on structures not in the training set. It succeeded in rediscovering known rules of spectrometry that had already been published, as well as discovering new rules.
- AARON: The Robotic Artist (1975–1995)
  Harold Cohen at the University of California, San Diego, created AARON, a robotic action painter [3] that could produce physical, representational artwork. It used a model for synthesising a set of objects (e.g. people and plants) and chose its subjects, composition and palette entirely autonomously. While the paintings have some elements of randomness, the style and aesthetics of the paintings were designed and programmed by humans.

REFERENCES

1. W. Van Melle, "MYCIN: a knowledge-based consultation program for infectious disease diagnosis," *International Journal of Man-Machine Studies*, 10, pp. 313–322 (1978).
2. R. K. Lindsay, B. G. Buchanan, E. A. Feigenbaum and J. Lederberg, *Applications of Artificial Intelligence for Organic Chemistry: The DENDRAL Project* McGraw-Hill, New York, (1980).
3. C. Garcia, "Harold Cohen and AARON—A 40-year collaboration," (2016). https://computerhistory.org/blog/harold-cohen-and-aaron-a-40-year-collaboration/.

Autonomy

It is hard to believe that the first instance of a physical robot that could react to the environment not explicitly guided by humans was demonstrated in 1949 by William Grey Walter [13]. Over many decades since then, the field of robotics has evolved significantly. You can visit the vignette on "Advances in Robotics" in this chapter for a summary of the uses of robots in practice today. The quest for autonomous vehicles also had gone on for decades before the DARPA Grand Challenge of 2004 [14] that posed a challenging military scenario. Even though no one really won the competition, the event brought various stakeholders to a common goal that has since proven instrumental in the evolution of the driverless vehicle technology of today.

AUTONOMY

Human desire for the AI to have a physical form and be independent from the human control has persisted for many decades. Here are two projects that capture the essence of the autonomy across many decades. Please visit our vignette on 'Robotics' in this chapter for various uses of robots today.

- ELMER & ELSIE Robots (1948–1949)

  Neuroscientist William Grey Walter, who played a pivotal role in the invention of electroencephalograms (EEG) in the1930s, created two mechanical robots, ELMER (Electro-Mechanical Robot) and ELSIE (Electro Light Sensitive with Internal and External stability) [1], during his tenure at the Burden Neurological Institute in Bristol, England. The robots were able to respond to stimulation by light and by contact with obstacles. They were able to *learn* about their physical environments and could recharge themselves automatically when their batteries were running low (much like iRobot's ROOMBA of today). They demonstrated random variability and independence in behaviour, not explicitly guided by humans.
- DARPA Grand Challenge in Autonomous Vehicles (2004)

  Defense Advanced Research Projects Agency (DARPA) organised the Grand Challenge [2] to stimulate the development of autonomous vehicle technologies in the US to support the military requirements, with a prize money of $1 million to the winner. The competition course was from Barstow, CA, to Primm, NV, of about 150 miles that included some rugged terrain. Of the 15 qualified participants, no one actually completed the challenge. However, the event showed the possibilities of potential approaches to achieve the goal and created a community to sustain the necessary research and the vision.

REFERENCES

1. W. G. Walter, "An imitation of life," *Scientific American*, 182, pp. 42–45 (1950).
2. R. Behringer, "The DARPA grand challenge – autonomous ground vehicles in the desert," *IFAC Proceedings*, 37, pp. 904–909 (2004).

ADVANCES IN ROBOTICS

In 1917, a Czech author Joseph Capek introduced the term 'automat' to refer to an artificial worker in one of his short stories. In 1921, his brother Karel Capek wrote the play "RUR" (Rossum's Universal Robots) [1], which was derived from the Czech word 'robota' for 'forced labor'. Thus, the Capek brothers introduced 'robots' to popular language. Isaac Asimov captured the imagination of so many of us through short stories in science fiction [2] and with his three laws for robots:

1. A robot may not injure a human being or through inaction allow a human to come to harm.
2. A robot must obey orders given to it by humans except when doing so conflicts with the first law.
3. A robot must protect its own existence as long as this does not conflict with the first or second law.

In 1958, General Motors brought robots to reality when it introduced the Unimate [3] to assist in automobile production. With its use on the assembly line in 1961, the application of robotics in industry began. Over decades, there has been widespread use of robots for various purposes. In all cases, robotics has one of the following goals:

- Remove repetitive work from humans.
- Improve upon human capabilities.
- Operate in hostile/hazardous environments not suitable for humans.

Category/Mobility	Domain	Example
Industrial/fixed	Manufacturing	**YASKAWA**/Welding, Painting robots, **KAWASAKI**/Assembly, Material handling robots.
Service/fixed	Medical	**Intuitive Surgical Inc./Da Vinci**: Assists surgeons to manipulate tiny surgical instruments through small incisions in the patient.
Service/mobile	Domestic	**iRobot Corp/Roomba**: An inexpensive disk-shaped machine to vacuum the floor that learns the map of the room.
Service/mobile	Hospitality	**SoftBank Robotics/Pepper**: Social humanoid robot that can recognise faces and basic human emotions.
Service/mobile	Space	**NASA/Mars Pathfinder Sojourner Rover:** Equipped with advanced sensors and ability to make independent decisions. **NASA/Robonaut:** Humanoid robot with skills to assist human astronauts.
Service/mobile	Commercial	**Boston Dynamics/Legged Robots**: Designed to move in difficult terrains and unstructured environments.

To perform any complex function, robots need appropriate sensors (visual, infrared, etc.) depending on the intended purpose. Mobility is another consideration. Fixed robots have a clear reference coordinate system for their operation. Mobile robots, however, are expected to move around and perform tasks in a more open environment that may not be precisely known and hence need to depend on their sensors to compute their location and orientation. Due to the significantly different mechanisms for navigation, robot designs for the three operating environments, i.e. aquatic, terrestrial and aerial, are very different. The table below shows some examples of popular uses of robots, grouped into two broad categories [4]: (i) Industrial Robots that work in well-defined environments (ii) Service Robots that support humans in their tasks in various domains.

REFERENCES

1. K. Capek, *R.U.R. (Rossum's Universal Robots)*, Penguin Group, New York, (2004).
2. I. Asimov, "Runaround," In: *Astounding Science Fiction*, Street & Smith, New York (March 1942).
3. http://www.robothalloffame.org/inductees/03inductees/unimate.html.
4. M. Ben-Ari and F. Mondada, *Elements of Robotics*, Springer, Cham. (2018) https://doi.org/10.1007/978-3-319-62533-1_1.

AI OR NO AI?

Having looked at some of the most famous historical applications of AI, let's now consider where AI is being applied today and how it's impacting our lives. The first point to note is that the use of the term AI in a product name doesn't really mean the product uses AI. If we are to believe the media and advertising, every single product from a vacuum cleaner to a clothes peg is using AI and on the verge of demonstrating sentience.

Whilst we should be cautious about believing that AI is ubiquitous, there is an element of emergent AI behaviour that is creeping into society as a result of the Internet of Things. As we fill our lives and daily activities with more and more devices, we are generating more and more data that can both enable and be exploited by AI. Within this environment, even the most simple of algorithms can start to generate an appearance of intelligence and even sentience. For example, after commuting to work, you may get off a train and return to your car parked in the station car park. As you climb into your car, your smart phone flashes up a message saying, "7 minutes drive to home with light traffic on the High street".

The algorithms that analyse your movements to understand the location of your home, where you parked your car and the fact that you are likely to drive home (because you do so every day at this time) are really quite simple. These algorithms are enabled by the mobile device you carry that tracks your every movement. Traffic sensors allow us to determine in real time the density of traffic and route planners allow the rapid calculation of optimum routes. Bringing all of these capabilities together creates a really simple but effective and useful application with the appearance of some form of intelligence. Purists would of course argue that this is not AI because the capability is being delivered by a machine and does not, therefore, require human intelligence. This is a classic example of what is known as the AI effect. A problem is only considered an AI problem until it has been solved at which point it is no longer considered to be AI.

Let's take this a step further. In addition to this simple tracking application, a completely different developer may put together a completely separate application. This second application analyzes your shopping patterns and fuses this data with a sensor in your bin to determine that you're running low on milk. You only bought two pints at the weekend

THOUGHT EXPERIMENT – SENTIENT WITHOUT AI

The year is 2047 and our lives haven't really changed that much.

People still go to work, make friends, argue, socialise, have children and do all the other things that make life normal … except we get a lot of help! Technology is ubiquitous and billions of simple algorithms leverage data from trillions of devices and sensors to make life better.

You no longer worry about a shopping list because, before you run out of beer, a drone automatically delivers a few more bottles. As the weather warms up for the summer, the delivery is mainly made up of lager that tastes great from the fridge on a hot day. In the depths of winter, a good old English real ale will be more suitable.

The delay on the underground that would make you late for work is no longer a problem. Your personal assistant wakes you a few minutes earlier to make sure you're on time.

The news articles read to you whilst commuting are customised to your personal interests and adjusted to reflect your state of mind on any particular day.

In everything you do, a set of very simple algorithms ensure your life is fun, relaxing and easy. It's almost as if there is a global sentience hidden in the depths of the vast network of computers and algorithms that make everything so good.

Collectively, do all these algorithms … working together … represent a form of sentience?

and one of the empty cartons was thrown away 2 days ago. As you have to drive home past a garage and you need petrol, the application reminds you to buy milk at the same time. Again, it's a really simple algorithm, enabled by location services together with some new sensors and the Internet of Things; nothing intelligent about that application so no AI.

The point is, however, that if we keep adding sensors and devices and very simple algorithms, then more and more of these simple but cool functions are going to emerge and the Internet of Things is going to appear to be more and more intelligent … even though the underlying algorithms are really quite simple and not in any way considered, at least by purists, as real AI.

THE PRESENT – THE DOMINANCE OF THE WEB

Given the impact of the internet on society, it should be no surprise that many of the most well-known AI applications today are based on, and developed by, the large web companies. For the average person on the street, their perception of AI is driven largely by the publicity of the big web companies. These companies are doing incredible things and have achieved so much in transforming the way we live and work. Internet technology companies (Google, Facebook, Amazon, etc.) are in the business of creating mobile and web applications to support e-commerce and social media for their global customers on the internet, typically in hundreds of millions or even over a billion users. Due to their business models and dominating internet technology platforms, they are able to collect vast amounts of user data through their applications with no direct cost to their users. Due to the practical utility of the applications, users are willing to give access to transactional and sometimes personal data, without much thought. Their business models (typically based on advertising or referral fees) rely on their ability to monetise the vast amounts of collected data and AI is the only viable option for them. So, it is no coincidence that these companies are investing heavily in AI skills and infrastructure. In addition, due to the nature of these applications and their licensing terms, these companies have complete control of their contents, underlying data and deployment infrastructure. Intrinsically, the data collected across the global population is considered by them to be public domain knowledge. The web companies are then able to license the use of this acquired public domain knowledge through various services to enterprise customers. Governmental policies and regulations have been slow to catch up on these practices due to their relative novelty, this may well change as the public become more conscious of their digital privacy.

However, when it comes to AI, we need to remember that many of the most successful applications are actually quite basic. Consider, for example, simple question answering. One very successful technique in simple question answering is to correlate the co-existence of terms in a large text corpus. They don't come much larger than the World Wide Web. Let's go back to the use case we touched on briefly in Chapter 1; a question answering application being asked the question, "when was Neil Armstrong born?"

If you take the billions (trillions?) of pages of content on the web, there are going to be a very large number of documents about the first man on the moon. A significant number of these documents will include phrases such as:

- "Neil Armstrong (born 5th August 1930) was the first man to walk on the moon",
- "The first man to walk on the moon was Neil Armstrong who was born in the US on the 5th August 1930",
- "Born on the 5th August 1930, Neil Armstrong trained as a pilot ...".

The sheer volume of content on the web means that there will be hundreds, if not thousands, of pages that include statements similar to those above. By writing a relatively simple algorithm that correlates dates with names and the word "born", it is relatively simple to identify dates of birth for famous people. By generalising the algorithm to look for any verb, the correlation algorithm could identify when famous people died, graduated, married, divorced and a host of other life events.

This simple correlation algorithm works for two reasons. Firstly, the question we're asking is simple! We're asking for a single fact which is a lot more simple than asking a computer to explain the causes of the Civil War or the reasons behind the economic crash of 2009. Secondly, there is a massive amount of redundancy in web data (especially about common knowledge and famous people). Neil Armstrong's date of birth exists in thousands of places expressed in many different ways. That's why the correlation algorithm works!

Correlation algorithms are far less successful when applied to more sparse data using more complex language. For example, in a police intelligence database, there could be a single sentence stating, "John Edgar cannot be the brother of Steve Edgar because he was born on 12th July 1978 which was 2 years before the latters' parents first met". If such a sentence is the only sentence in the entire database that mentions the Edgar brothers (or non-brothers), there is no way a correlation algorithm can work!

In addition, the web companies are able to leverage vast human resources to provide feedback on their AI decisions. A web company wanting to develop an image classifier can leverage their vast user base to score the performance of the systems to rapidly develop a huge set of training data.

In developing these amazing AI web services, the web companies benefit to a large extent in the same way as those early game playing researchers. The risk and cost of an error is relatively low ... perhaps even non-existent! That's possibly a bit unfair as there is clearly a massive commercial return on getting things right ... there must therefore be a loss if you get things wrong. However, the failure to generate the optimum amount of advertising revenue does not equate to the cost of making a mistake in other domains such as medicine or defence or arresting the wrong Edgar.

In their core business, the web companies are using a very specific set of techniques to answer simple problems using massive information redundancy with feedback from huge numbers of users. That is very different, almost opposite in fact, to the challenge faced in many enterprises. As we grow beyond these commodity web applications and move into the enterprise, we need to understand that the risks, consequences and technical challenges are far greater, and the tools and approach taken by a web company may not always work out of the box for your business problem!

Let us summarise some of the prominent consumer facing AI applications heavily influenced by the web companies. This helps to compare their functions and usage with typical enterprise applications to run a business.

Technology	What Does it Do?	Examples
Search engines	User enters a few words and the system finds sources that are relevant to the user input and ranks them in decreasing order of relevance.	Google/Search, Microsoft/Bing, Yahoo, etc.
Question-answer system	User types a question and the system finds the best answer to the question based on the corpus of the data available to the system.	Amazon/Alexa, Apple/Siri, Google/Home, Microsoft/Cortana, etc.
Speech-to-text	User speaks to the device that transcribes the speech to text.	Apple/Siri, Samsung/Bixby, "Dictate" in Microsoft Word
Text-to-speech	User selects textual content and the device produces a speech output.	Accessibility feature in Apple & Android phones
Translation	User gives text in one language and the system translates that into a target language.	Google Translate, Microsoft Bing Translator
Embodied cognition	Physical objects get endowed with interaction capabilities with humans and the interaction involves physical (e.g. gestures) and/or sensory aspects (vision, speech, etc.).	Amazon/Alexa, Apple/Siri, Google/Home, Microsoft/Kinect, etc.

(*Continued*)

Technology	What Does it Do?	Examples
Text classification	System analyzes textual content to classify them into different categories.	Spam filters, Smart e-mail categorisation
Image classification	System analyzes images to classify them into different categories.	Pinterest, Snapchat, Google Photos
Recommender	Based on prior choices of the user or of others 'similar' to the user, the system recommends a 'new' option user may like (e.g. movies, restaurants, 'friend', etc.).	Amazon/Products, Netflix/ Movies, Facebook/Friends, etc.
Navigation/ planning	Given the destination and the current location, system recommends a set of steps to reach the destination under various options and constraints.	Google Maps, GPS Systems, Ride Share apps (e.g. Uber, Lift).

They all share some common attributes:

- Many thousands to many millions of users.
- Perform a simple task.
- Cost of mistakes is not high.
- No explicit cost to using the applications.
- Users are willing to share their data with the application owners.
- Application owners and their web platforms dictate the user behaviour.
- Revenue for application owners does not come directly from the applications but from other related endeavours (e.g. advertising).

As you will see below, these are very different attributes from those of an enterprise AI application that is supporting a business function in a specific domain (e.g. retail, banking, insurance, manufacturing, etc.) with a direct connection to the business impact.

THE FUTURE – THE ENTERPRISE STRIKES BACK

Challenges of an Enterprise

In contrast, to a web company, let us consider a traditional enterprise. This may be a commercial company (e.g. banking, retail, insurance, manufacturing, etc.) whose business relies on selling goods and/or services to their customers. Or, the enterprise can be a non-profit (e.g. healthcare, emergency relief, food distribution, etc.) or a governmental organisation (e.g. law and order, transportation, tax collection, etc.) providing essential services to the society. These enterprises may also own 'brick and mortar' store fronts in addition to their presence on the internet. *Their investment in the information technology and AI is only to help them succeed in their primary business objectives.* Since the technology investment is to support the enterprise's business goals, there are many other stakeholders (e.g. various lines of businesses, customers, sales teams, distribution chains, business partners, etc.). Any AI application has to get vetted with these stakeholders and accepted before they can be successful. The business processes and the supporting resources already exist, and therefore, any investment in an AI has to be thoroughly evaluated for business value in terms of quality and efficiency. Since the customer satisfaction is paramount, there is a clear need for correctness and consistency in their applications. The consequence of following the recommendation from an AI system to the enterprise can be severe (e.g. processing bank loans, identifying potential criminals, recommending medical treatments, etc.) compared to getting movie or purchase recommendations on the internet or social media. In addition, there may be government regulations and auditability requirements on their business practices (e.g. Sarbanes–Oxley Act in banking, HIPAA in healthcare, GDPR in EU, etc.). The application requirements can be unique and industry and/or organisation specific, requiring a good understanding of the various data sources, not controlled by common data models or governance. Due to concerns about competitive advantage or privacy regulations, these enterprises may not want to share their data with third party vendors to build AI models for their purpose. All these considerations point to much higher complexity in building enterprise applications compared to the popular applications from the web technology companies.

In most businesses, we are trying to answer very complex (and mission critical) questions, requiring the fusion of complex data from multiple sources, using multiple AI components with much less feedback from users. It's not just the questions that are more complex! It's also the fact that it's harder to know when the answer is right or wrong. If an online search engine answers a question about the Beatles incorrectly, it will be easily recognisable as incorrect and, if we're honest, it doesn't really matter. In the enterprise, it does matter and it's harder to know when the system is wrong.

An Insurance Example

Consider, for example, an insurance company that wants to automate the analysis of medical records in order to assess life insurance risk. One such solution, developed in 2012, required three different AI components integrated into an intelligent business process. The first stage used Optical Character Recognition (OCR) to transform hard copy documents into machine-readable text. The text was then processed by an entity extraction tool that identified key medical events and terms such as "myocardial infarction" and "patient smokes more than 20 cigarettes a day". Finally, a risk model used the output of the entity extractor to make the final risk assessment for review by the medial underwriter.

This solution required the integration of these three different AI components. The data was incredibly complex with records going back more than 50 years in some cases. The historical nature of the data caused new challenges as medical knowledge continually changes. Fibromyalgia, a condition which is said to affect 3%–6% of the population was only first defined in 1976. Prior to 1976, it was described under many headings including fibrositis. Finally, feedback on accuracy came from a small number of medical underwriters.

This is just one single example that demonstrates the challenge of applying AI in the enterprise. However, in terms of complexity, this underwriting solution is still relatively simple.

A Mismatch of Expectations

Over the next few years, we will be delivering AI solutions of increasing sophistication and we're going to need to develop new tools and methods to do so. Take a moment to think about conventional engineering and how we build incredibly complicated machines. Airliners, nuclear submarines, satellites and many other amazing feats of engineering are designed and built by teams of engineers using sophisticated tools and processes.

Just consider for a moment the Computer-Aided Design (CAD) tools available to civil, or nuclear, or aeronautical engineers. Then consider the sophistication of the processes for selecting and managing components in large engineering projects.

Now, think about the size of a typical AI project and the limitations of the tooling available. Most AI projects are typically very small; in our experience, project teams can comprise only a small number of AI specialists supported by programmers to help with data access, integration and such like. Whilst tooling is improving with the growth in data science as a discipline, we still rely heavily on spreadsheets and simple scripting. Delivering thinking machines has to be one of the greatest engineering challenges ever attempted, yet we are many years behind conventional engineering in terms of our ambition for tackling complexity and the tooling and processes we need to do so.

Before you drop this book and run away in fear ... it's not all bad news. We may have a long way to go in developing our tooling and processes; however, we have already started. There are things you can do to bring a professional engineering approach to your project ... so please read on!

Understanding the Enterprise Solution

To really understand the complexity of an enterprise solution, it is worth considering a number of features:

- **Business Relevance**: Will the investment in the AI support your business and make more than it costs? "If you build it, they will come" applications or novelty AI built just for the sake of it will still have value for a web company it often showcases the art of the possible or is simply fun and will attract consumers to browse more – both valid business outcomes. It is not so easy for a typical business, corporate enterprises with a duty to shareholders to tie any investment to specific measurable business outcomes before they spend out on an IT project.
- **Stakeholder Agreement**: Will the key stakeholders see the AI as a good thing? A web company and a traditional business or government department have very different stakeholders with different perspectives of what is good for the business. Would you as a customer stakeholder of Google stop using it as a search engine if it occasionally

returned poor search results, would you feel the same if your bank AI occasionally gave you someone else's financial data?

- **Application Complexity**: Can you make a profit from a simple AI or will you need a moonshot project to find success? There is a lot of hype around AI, especially around how easy it is. Luckily it is, for the most part easy to use, but that doesn't mean it will be a good fit to every business problem you'd dearly like to solve. AI fits some problems well and others really badly, AI applications are always data dependent. This book will help you spot the viable and valuable opportunities. Businesses, contemplating their first AI, often underestimate the complexity that will be required to meet their "simple" requirement.
- **Correctness & Consistency**: Do you need to guarantee that the AI is giving you the best answer every time or is some ambiguity/error acceptable? You can either pick a business problem that the users can easily distinguish and skip erroneous results, such as a web search for a new golf club, bringing back adverts for Volkswagen cars, or prepare to invest, to chase down those edge cases that could ruin your plans.
- **Decision Consequence**: What will you do when (not if) the AI makes the wrong decision? Put simply, the stakes are often higher for an AI in a business setting, versus a web demonstrator system, that is not trying to solve a specific business problem but rather indicate how it might be possible.

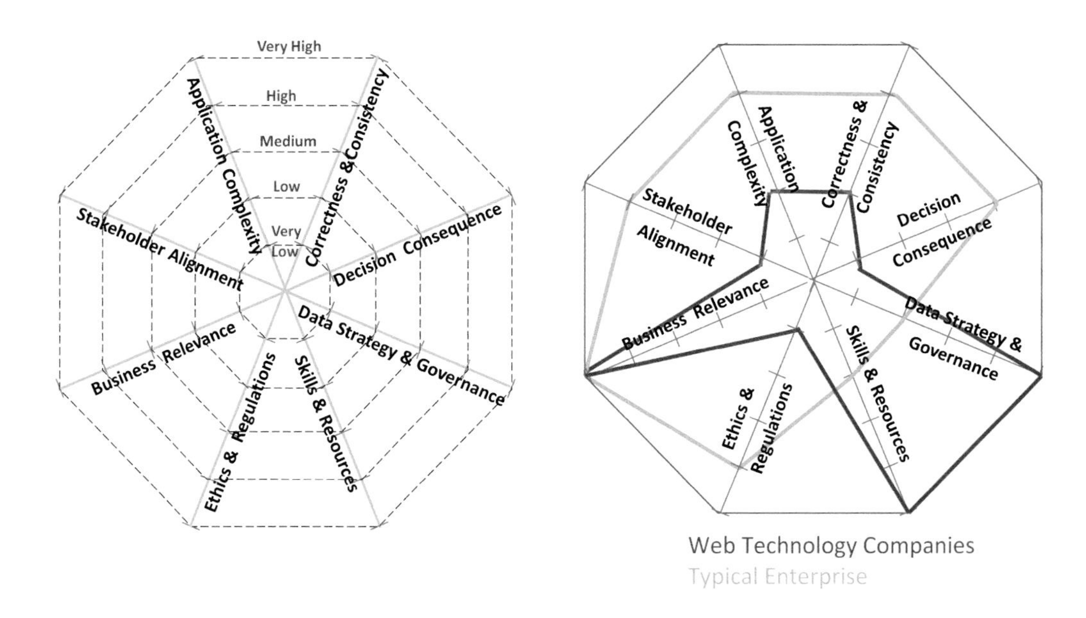

FIGURE 2.1 Figure on the left describes a simple rubric of eight metrics to assess an AI project. Each metric is rated from very low to very high. Figure on the right compares the metrics for an application in web technology companies (in red) to a business solution for a typical enterprise (in blue).

- **Data Strategy & Governance**: Can you guarantee a steady supply of source data, will you need to archive it in case something goes wrong and you need to rebuild your system? Do you have a person/team ready to take on the care of the AI. Hint: AI projects need skilled maintenance throughout their lifecycle.
- **Skills & Resources**: Do you have enough AI skilled staff, both for the development and the ongoing maintenance?
- **Ethics & Regulations**: Are you breaking any laws or will your AI cause a Public Relations issue?

Figure 2.1 provides a radar chart showing these features and comparing a typical enterprise application with those of internet technology companies.

As you can see, enterprise and web AI applications are very different. The hard part is understanding and accepting that point. Once it is accepted, then good management and engineering practice can be applied to ensure the challenges of the enterprise are properly addressed. The challenges are all manageable once the decision has been taken to manage them.

EXAMPLES OF REAL ENTERPRISE APPLICATIONS

The goal of this section is to give some examples of real enterprise AI applications. We want to make sure that it is possible for you to find out more details about them in the public domain, if you are interested. We have chosen papers at the Innovative Applications of AI (IAAI) conference held annually by the American Association of Artificial Intelligence (AAAI). These applications are also deployed at the enterprise in a real setting so that they are validated to be useful in a business context, beyond just exploratory AI technology. The other advantage of this forum is that it represents the wide range of AI technologies in practice, not just artificial neural networks which are the *shiny objects* of the day. A very insightful review by Smith & Eckroth [15] describes the evolution of AI applications represented at the IAAI conference over three decades and the lessons learnt for building future AI applications.

The other topic we want to address in this discussion is the complexity of the AI application. Clearly, if there are more AI components, the complexity of the solution will be higher; it also matters how the multiple AI components are connected in the solution. We will also discuss other factors that contribute to the complexity at the end of this section. The other thing to remember is that the AI component is only a part of the AI application, since in general, you may need other non-AI components to assemble the AI application.

The simplest and the most common *AI Application* comprises just one *AI Component* (Figure 2.2). The *AI Component* itself requires no information other than that provided in the input.

HOW HARD CAN A SINGLE COMPONENT BE?

Whilst an Application based on a single Component is easy to build and deploy from an architecture perspective, don't underestimate the challenge of delivering even a simple AI Component.

A few years ago, I was asked if I could help a radio station with a "simple" problem. The radio station used to receive huge numbers of text messages during their shows. The interaction between the audience and the show hosts via the text messages was an increasingly important aspect of the shows. It was fun for the audience and the hosts were brilliant and creating a sense of 'banter'. The number of text messages increased and increased … so much so that it was impossible for the host or the production staff to read every message. Instead, they scrolled past the host on a display and the host just called out any that caught his or her attention at the time.

During one show, a producer glanced at the message display and saw a rather disturbing message from an individual claiming that he was going to commit a very serious, and unpleasant, crime. On that occasion, the police were informed and appropriate action was taken. However, the radio station was concerned that it could happen again and not be spotted. They asked if we could build a system to scan messages to identify potential criminal behaviour. To be helpful, they even provided a list of specific terms that they were concerned about.

It only took a few minutes to demonstrate the challenge they faced. When scanning text messages to a radio station for criminal terminology, you would be surprised to see messages such as …

"Hey – just bumped my girlfriend's car … she's gonna kill me tonight!"

Since then, we have seen the large web companies battle with the challenge of identifying fake news and illegal content. It's a difficult problem because we're a long way from developing AI that can match our human ability to understand context, irony, humour and a whole raft of other human characteristics that exist in the content we generate.

James Luke

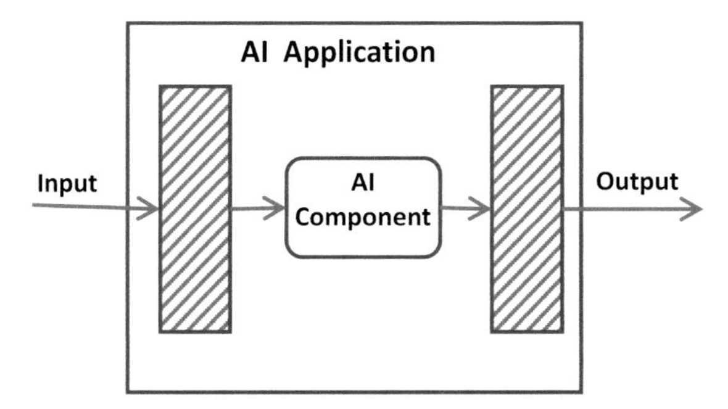

FIGURE 2.2 AI application with one AI component. Shaded boxes are non-AI components for dealing with inputs and outputs in general.

Personalized Categorization of Financial Transactions (Intuit Inc.)

Business Problem

Financial accounting requires that every business transaction must be categorised using a customisable filing system, called the Chart of Accounts (CoA). For millions of accounting software (QuickBooks) users, even at 3 seconds per task, this would represent over a thousand person-years in a year!

What Does the AI Do?

A recommender system [16] that answers the question, "Which CoA account of this user is most likely to apply to this financial transaction given the set of accounts to which this transaction has been assigned by other users?" In contrast to systems that use a single shared set of categories across all users, this system allows customisable categories.

Business Results

Since November 2016, the AI service is deployed to several million small businesses in US, UK, Australia, Canada, India and France, involving billions of financial transactions each year. The AI service results in 56% fewer uncategorised transactions and 28% fewer errors, resulting in a substantial reduction in the number of errors and the manual work needed.

Defect Finder for Weld Nuts in a Manufacturing Company (Frontec Co, Ltd.)

Business Problem

Factory produces critical weld nuts for use by companies in the manufacturing industry [17]. Quality inspection of the weld nuts was manual, subjected to individual decisions and employee state, such as fatigue. Since even an apparent defect resulted in expensive investigations, it became necessary to automate the inspection. The speed of quality inspection had to be less than 0.2 seconds per weld nut with at least 95% defect classification accuracy.

What Does the AI Do?

After some preprocessing of the images, the AI component (a Convolutional Neural Network) classifies each weld nut as either non-defective or into one of the seven defective categories, with an accuracy over 99% and response time of about 0.14 seconds.

Business Results

The resulting embedded classification system was integrated with an existing vision inspector environment. Since the deployment on November 2018, the expected monetary benefit is US$20,000 per month, a combination of labour costs and failure costs from prior manually performed quality inspections.

Procurement Demand Aggregation (A*STAR)

Business Problem

A large governmental research organisation in Singapore with annual procurement expenditure in hundreds of millions of dollars wanted to aggregate the procurement to reduce expenses by (i) lowering bulk prices, (ii) larger vendor tendering, (iii) lowering shipping and handling fees and (iv) reducing legal and administration overheads [18].

What Does the AI Do?

The relevant data is represented by a graphical model relating procured items and target vendors. A novel clustering approach mines the graph efficiently to reveal potential demand aggregation patterns. This is a good example of unsupervised learning technique (see Chapter 3). Results on past benchmark cases were as follows: (i) All aggregation opportunities identified by the engine were correct; (ii) engine correctly identified 71% of the past aggregation exercises that were transformed into bulk tenders and (iii) 81% of the newly identified cases were deemed useful cases for potential bulk tender contracts in the future.

Business Results

A proof-of-concept prototype was developed in 2017, and a refined version was rolled out in 2019. The annual cost savings from the true positive contracts spotted so far are estimated to be 7 million Singapore dollars.

The next set of application examples involves two AI components that can be connected in different ways depending on the application architecture. The components can also represent different types of underlying algorithms, as will be evident below.

Enterprise Risk Management (IBM)

Business Problem

Complex business environments require companies to anticipate potential risks and be prepared with suitable mitigation plans [19]. This helps organisations to move from reactive to proactive postures when foresight of possible future events can be used to shape a better outcome. Typical scenarios address geopolitical situations (e.g. Brexit, Covid-19, etc.). This is typically a manual, highly labour-intensive process limited by human imagination and knowledge, involving dozens of experts and hundreds to thousands of person-hours.

What Does the AI Do?

IBM Scenario Planning Advisor helps to address this problem by using two AI components. (See Figure 2.3.) (i) **A Causal Extraction component** that takes a set of risk forces and a set of authoritative documents (e.g. books, reports, articles, etc.) in the domain of interest as input and automatically creates causal models using Natural Language Processing techniques. This is followed by manual step to clean up and improve the quality of the causal relations and validate implications. (ii) **AI Planning component** that takes the causal model and forces of interest and creates many scenarios consisting of hundreds or thousands of plans.

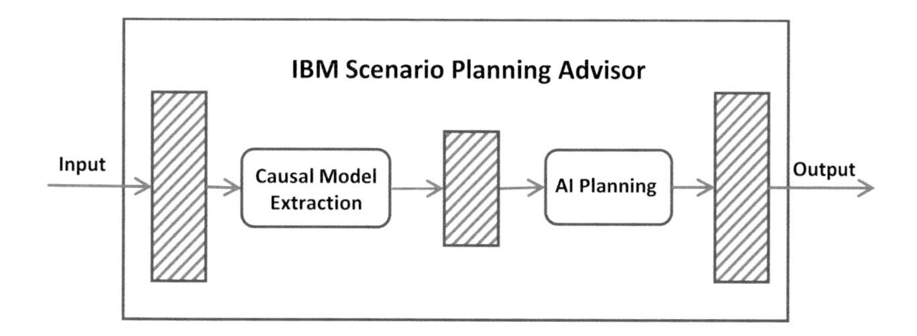

FIGURE 2.3 Two AI components in series. Shaded boxes are non-AI components.

Business Results

IBM Scenario Planning Advisor is currently deployed to support 70 IBM finance teams and an external client. A careful user study shows that the tool can generate a model and ten scenarios in about 11 hours compared to about 3800 person-hours for people to achieve the same.

Life Insurance Underwriting Application (Mass Mutual)

Business Problem

Life insurance is a critical need for the financial securities of individuals and their families [20]. In order to price their products competitively, while maintaining business financial strength, life insurance companies have to assess mortality risk of individuals accurately. The traditional underwriting process to assess this risk relies on manually examining an applicant's health, behavioural, and financial profile. Can AI improve the accuracy of the current underwriting process?

What Does the AI Do?

The underwriting practice had five risk classes and cohorts defined by sex and 5-year age bands. The AI application had two AI components (see Figure 2.4). (i) A simulator that modelled the historical distribution of risk classes by cohorts. (ii) A Random Survival Forest model that takes the data representing applicant's health, behavioural and financial

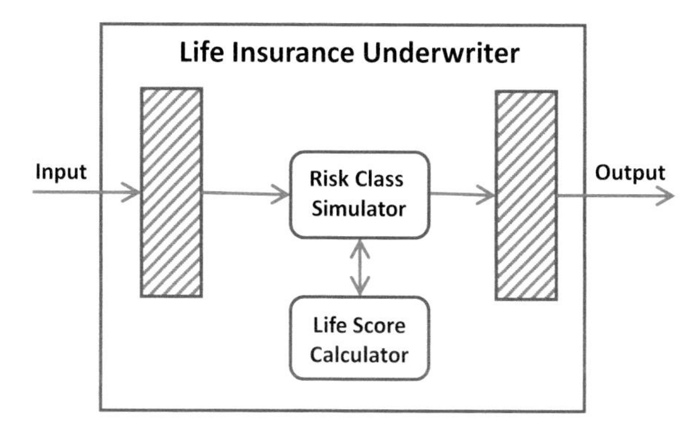

FIGURE 2.4 Two components in the Insurance underwriting application.

profile as input and produces a single, standardised life score for each individual. In the end, AI underwriter had a life score and risk class for each applicant.

Business Results

The new approach resulted in a 6% reduction in claims in the healthiest pool of applicants. Over a 2-year period, this system reduced the time to issue policies by >25% and increased customer acceptance by >30% for offers made with only light manual review. Overall savings were millions of dollars in operational efficiency behind tens of billions of dollars of policy holder benefits.

Next let us consider applications that have three or more AI components. These components can be used in different ways in the architecture based on the application flow.

Chatbots

The best examples are popular Chatbots such as Apple/Siri, Amazon/Alexa, Google/Home or Microsoft/Cortana used in the simple question answering mode. Figure 2.5 represents the high-level view of a speech-based Chatbot. When the user speaks "Who is the president of the United States?", it gets transcribed by the "Speech-to-Text" service into the corresponding text. The text gets passed as input to the next AI task "Question-Answering", which is another service in the popular domain from the companies, which brings the answer as "Joe Biden". The answer gets passed to the "Text-to-Speech" service, which renders the answer to the user. For simplicity, we have represented 'Question-Answering' as one AI component in Figure 2.5, but in reality, it will consist of many subcomponents [21]. There is a user input analysis component that extracts user intent ("Name") and entity identification ("President of the United States"). It may also perform additional tasks such as user sentiment analysis. There is a dialogue management component that manages ambiguities, errors & information retrieval and a response generation component which prepares the textual content to be delivered to the user.

Multi-Skill Agents

This next example is an agent that can answer questions in multiple domains such as Travel, Weather and Food [22]. Bot designers call these domains 'skills'. The basic idea

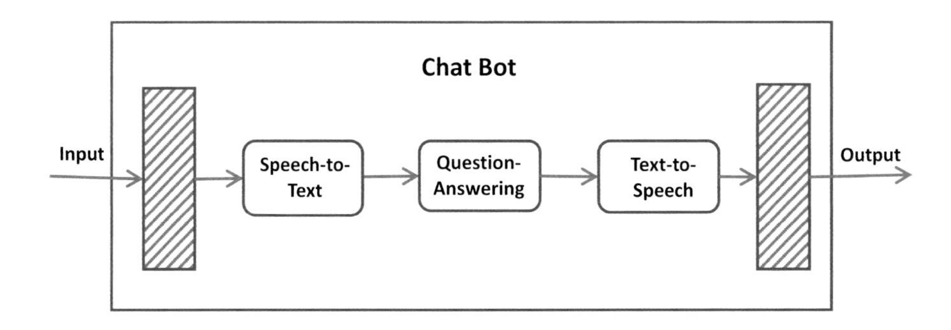

FIGURE 2.5 Three components in series making up the basic Chatbot.

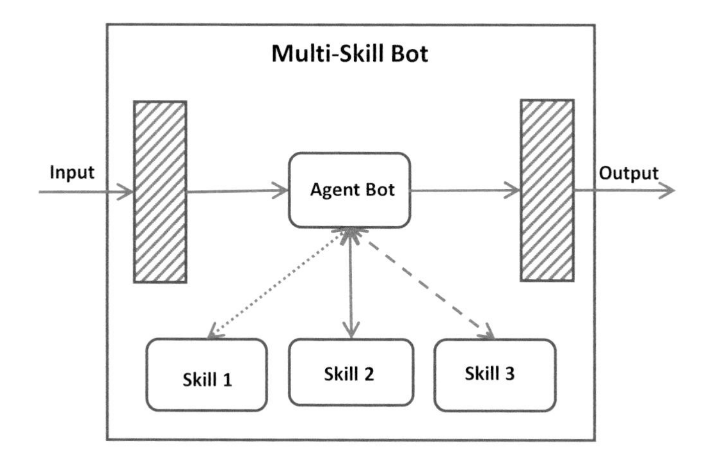

FIGURE 2.6 Multi-skill bot with parallel AI components for three different skills.

is that each bot specialises in one skill (i.e. answering questions in one domain). In this example (see Figure 2.6), the input comes as text and the agent bot determines the intent of the input message from user and directs the flow to the specific bot with sufficient details and collects the response of the bot as output to the user. Clearly, this pattern can be combined with the Chatbot design above to create a voice-driven application.

Applications Processing Different Data Streams in Parallel

Here we want to discuss applications that require many AI components that may use different types of data in parallel, and overall output of the applications requires the aggregation of knowledge across the different data sources. This is called 'fusion' in the AI community. To highlight the underlying technology components and the practical challenges, we pick the High-Five (Highlights from Intelligent Video Engine) system from IBM [23] that automates the curation of highlights in sports tournaments.

Business Problem

In a typical sports tournament (e.g. Masters Golf, Wimbledon Tennis, US Open Tennis, etc.) there are numerous matches played between many dozens of players over many days. The production of sports highlight summary capturing the most exciting moments of the tournament is an essential task for broadcast media. For even 1 day of the tournament, this process may require manual viewing of hundreds of cumulative hours of video and selecting a few minutes of highlights appealing to a large audience, in time for the evening summary report. This is obviously very labour intensive. Can AI do better?

What Does the AI Do?

High-Five fuses information from visual cues from players, via physical manifestations (such as high-fives and fist bumps) and facial expressions (e.g. tense, smiling), audio cues from the environment (e.g. cheering of spectators, tones and descriptions of commentators)

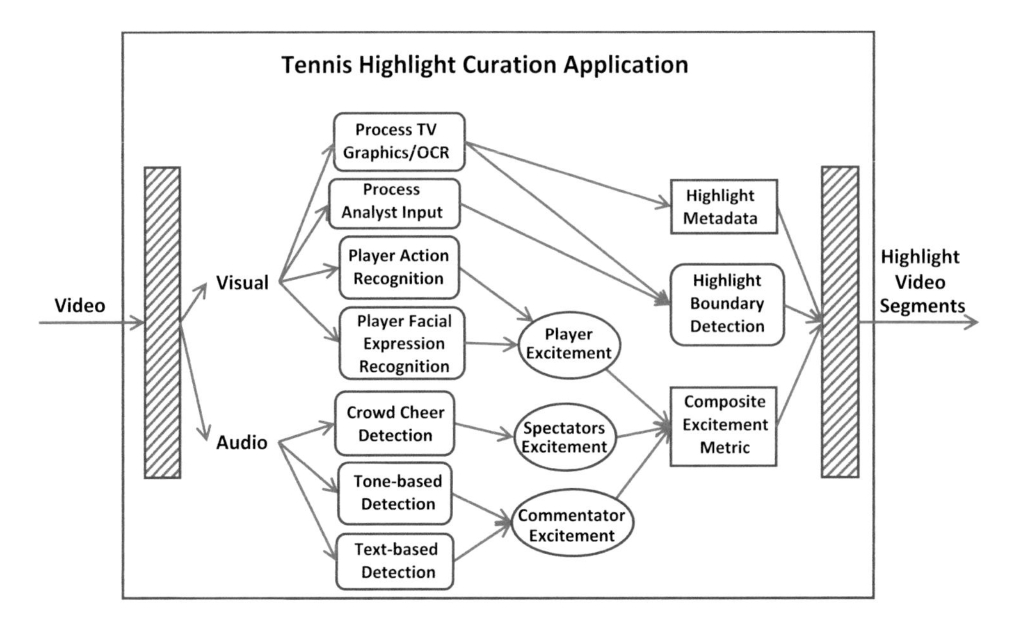

FIGURE 2.7 AI components in IBM's High-Five system to select video highlights from the video inputs of sports tournaments. The system processes visual, audio and text data in parallel to identify the video segments of interest. The rectangles with rounded corners are AI components, ovals are evaluation steps and rectangles are summary information for the output.

along with game analytics to determine the most interesting moments of a game. High-Five also accurately identifies the start and end of the video clips with additional metadata, such as the player's name and game context, or analysts input allowing personalised content summarisation and retrieval. Figure 2.7 gives a high-level architecture of the High-Five system.

Business Results

The system was demonstrated at the 2017 Masters Golf tournament and at the 2017 Wimbledon and US Open tennis tournaments through the course of the sports events. For the 2017 Masters, 54% of the clips selected by the system overlapped with the official highlights reels. User studies showed that 90% of the non-overlapping ones were of the same quality of the official clips for the 2017 Masters. The automatic selection of highlights of 2017 Wimbledon and 2017 US Open agreed with human preferences 80% and 84.2% of the time, respectively.

WHERE DO YOU INTRODUCE AI?

Now that we have discussed a few real examples of AI components in business applications, it is important to understand how to select the task for the AI to do. This should be done even before any investment is made to create one. You will almost certainly have business processes for critical areas such as sales, shipping, customer relations management, billing and procurement. Each of these areas has many tasks that have to be completed

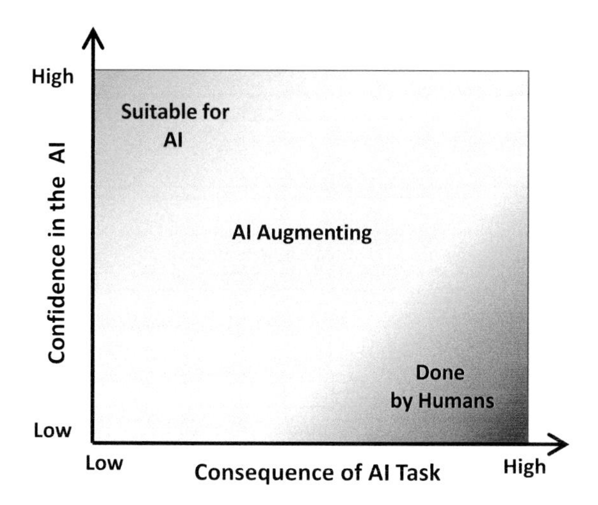

FIGURE 2.8 AI or humans? The spectrum of applicability. (Adapted from "Artificial Intelligence: Short History, Present Developments, and Future Outlook: Final Report", MIT Lincoln Lab (2019).)

in order to meet your business objectives. You need to think about where you can benefit the most by the introduction of AI and its potential to please or displease the user. As an example, if you have live customer support only during a certain time window during a day and only for the weekdays, you can consider introducing a customer service bot that can provide some support during the off hours. You do not even have to match the full functionality of your live agents, that could come over time, but customers will see it as a bonus and are likely to put up with its shortcomings as it provides an out of hours service. If you were to introduce this limited service as a replacement for your live agents the customers would rightly see it as a degrade in service, it might have a negative effect on your business. Once you start having some experience with the performance of the bot, you can consider introducing the bot to take care of simple informational tasks so that you can free your employees to attend to more demanding tasks in customer support.

Figure 2.8 gives a broader guidance to this concept by considering the confidence in the AI to do the task versus the consequence of the AI task. When the consequence of the AI task is low (e.g. restaurant recommendation) and you have high confidence in AI doing the task, this is the best situation to introduce AI. When the confidence in the AI is low and the consequence of the task is high (e.g. risk of human life, significant revenue loss, etc.), the choice is clearly a human. For tasks in the middle of this spectrum, augmenting humans with AI is the best approach.

ACTIVITIES IN CREATING AN AI APPLICATION

In this section, we want to discuss the various development activities typically involved in creating an enterprise AI application. Figure 2.9 gives an overview of the activities and the tasks within these activities. We discuss these briefly here, since they are discussed in detail in other chapters of the book.

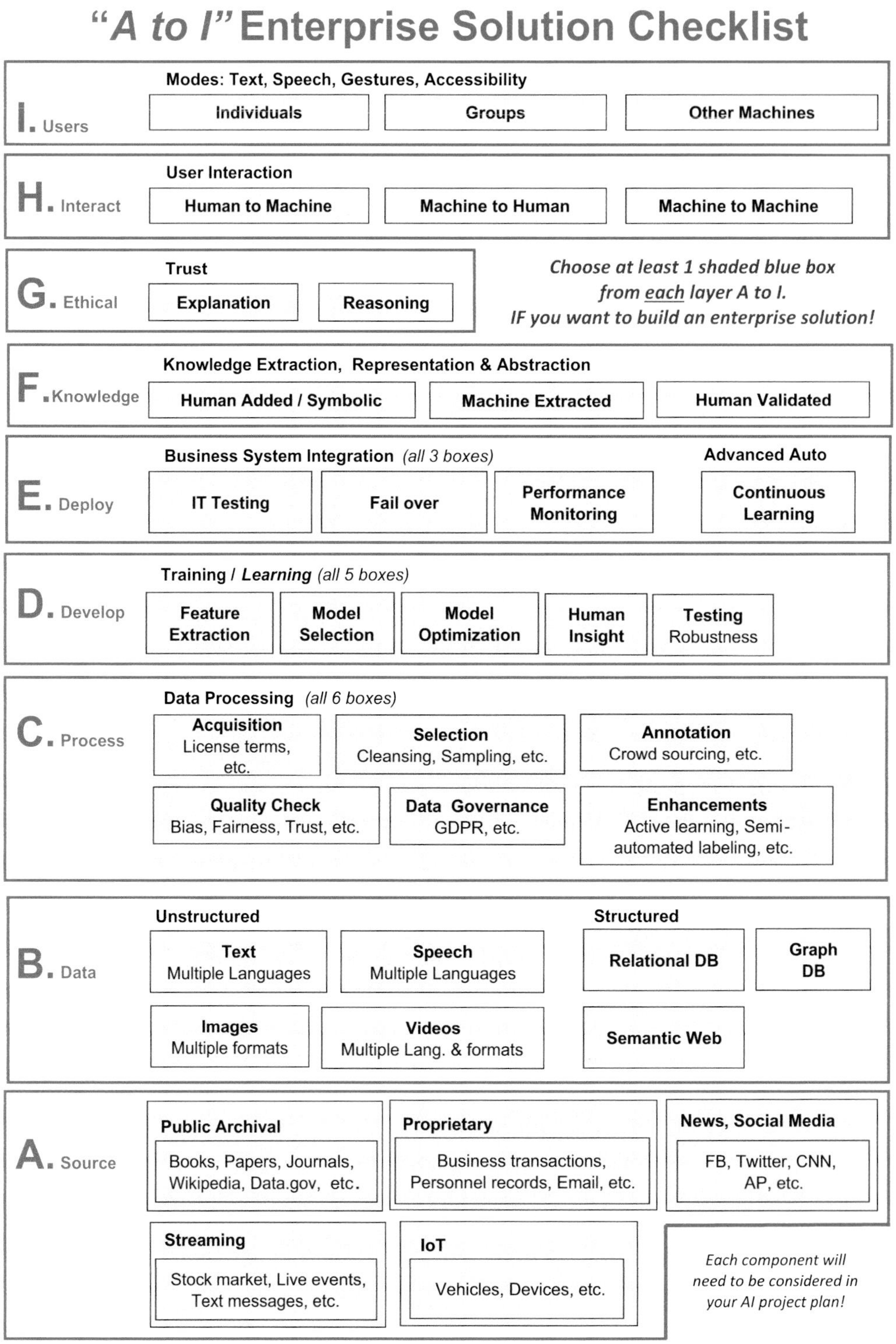

FIGURE 2.9 Activity diagram. An overview of the various tasks to build AI applications grouped by the Activity labels on the left. An AI application will need one or more tasks from each activity layer.

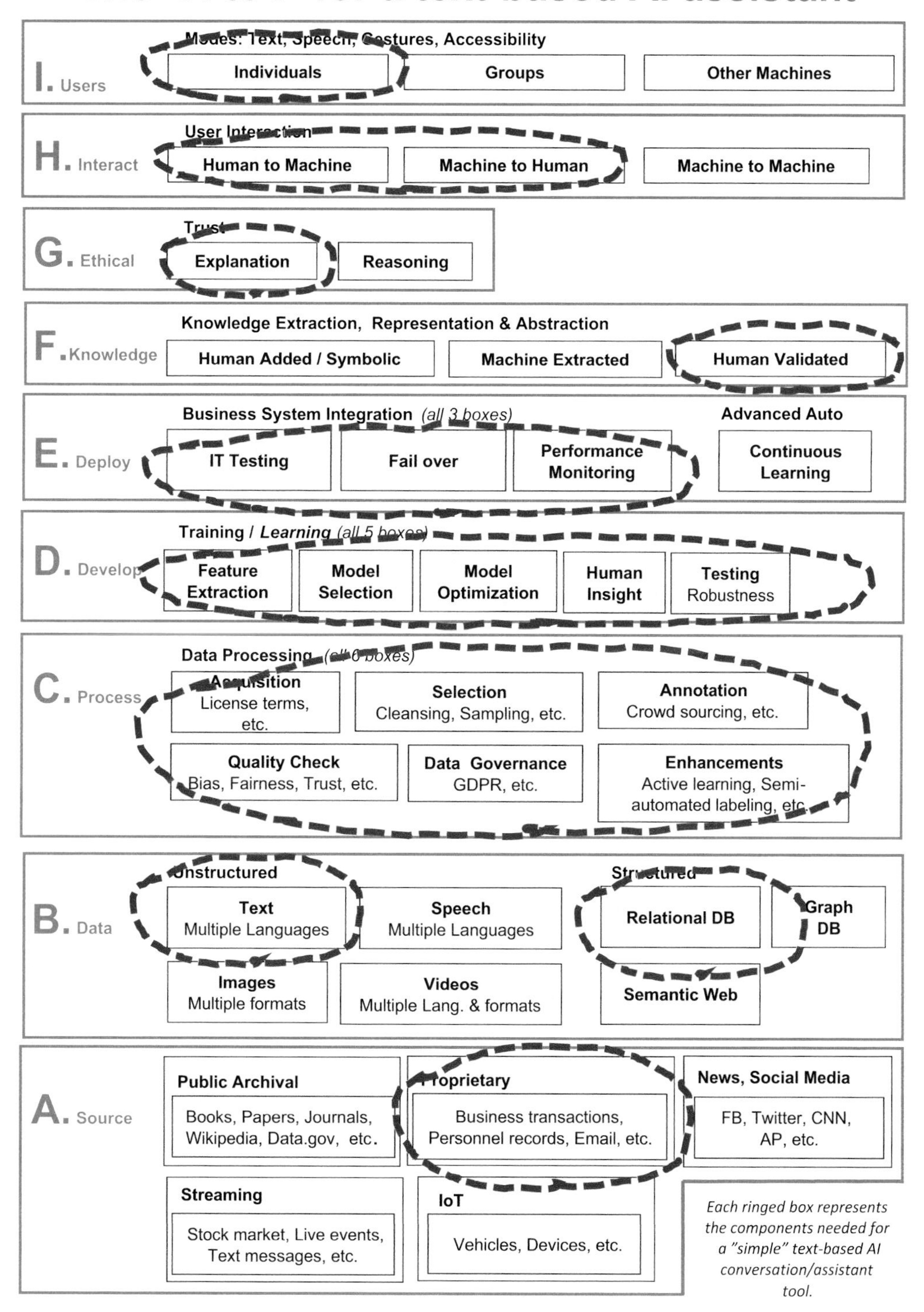

FIGURE 2.10 Activities needed for a text-based AI assistant tool.

The bottom three layers A, B and C have to do with the choice of the data sources, data types and the data workflow that needs to be supported during the AI application lifecycle. These are discussed in Chapter 7. AI model management, layers D & E, includes all aspects of model creation, validation and sustenance through the application deployment (Chapters 2 and 10). Human "Knowledge" tasks in layer F refer to the ability of the application creators to understand and validate the knowledge captured in the AI and its limitations to be able to add suitable insights (e.g. guardrails in terms of rules for allowed application behaviour). The ethical layer G is to make sure that the application meets the trust expectations of the users of the application in terms of explainability, reasoning and AI governance (Chapter 5). User interaction models, layer H, include the consideration of various aspects how users will interact with the application in terms of input and output modes (text, speech, gestures, etc.) and fault-tolerant designs to recover from potential failures in the user-AI communication. The top layer just recognises the fact that the users of the AI application can be one or more individuals or for that matter, another machine. To be considered complete, an enterprise application has to be designed with consideration of at least one component (represented by the blue shaded boxes) from every row.

An Example of Using the Activity Diagram

Figure 2.10 highlights the specific activities in Figure 2.9 related to building a simple text based conversational assistant tool for a typical business enterprise.

COMPLEXITY OF AI APPLICATIONS

Given the various options for implementing enterprise AI applications, it is worthwhile discussing what makes an application complex. Some of the attributes are unique to AI while others are generic attributes that contribute to any application complexity.

Number of AI Components

Generally speaking, more AI components in the system, more complicated it will be. Most *AI Applications* that include *AI Components* operate with some form of data flow where data enters at one end and an answer is produced at another.

Interdependence of AI Components

If your AI application does rely on multiple AI components, then you need to think about the interdependence between them. In a simple system, where data flows out of one AI component into another AI component, what happens to any errors? Generally, errors tend to be compounded when they flow through systems so it is really important you understand how the different sub-components will interact. It's only by doing so that you can ensure the correct level of accuracy at each stage.

Event-Driven Systems

However, in some cases, our *AI Applications* are required to interact in real time with their environment. In such cases, the *AI Components* may be integrated using more of an event-driven architecture. Clearly, such architectures are significantly more complex to develop and test than the simpler data processing integrations.

In conventional data processing systems, the challenges of deadlocks or race conditions are well understood. These problems are normally overcome with careful design and extensive testing. However, with AI Applications, it may not be possible to ensure completeness of testing. After all, AI Applications tend to generalise and use judgement; they are far less deterministic than conventional systems. As a result, there is potentially a greater risk of deadlock or race conditions. The best way to address this risk may be to deploy appropriate monitoring tools and ensure that the application is capable of being interrupted.

Context Aware Systems

AI Components become more difficult to develop when they need to retain some form of contextual knowledge or state. It often surprises people to learn that some of the most successful AI Applications delivered to date have no concept of state. For example, when Deep Blue defeated Gary Kasparov, each move was analyzed as a completely stand-alone problem. There was no concept of an end-to-end game in Deep Blue's logic, and the historical behaviour of the opponent was not considered. Deep Blue simply analyzed the game as it was there and then. Maintaining some form of state that describes the current situation, and how the system arrived at this situation, adds significant complexity to an *AI Component*.

Continuous Learning

One area that really excites *AI* Junkies, *Algorithm Addicts* and Nerds in general is the thought of continuous, interactive learning. The idea that the *AI* is continually learning and improving really excites those who dream of thinking machines. Sadly, it introduces a whole new level of complexity and risk, as a system owner how will you know it's learning the right things? The most notable example of this to hit the press was of course the Microsoft Virtual Agent Tay [24] that was taught highly inappropriate language by its user community.

Federation of AI Applications

The final factor that introduces complexity to an *AI Application* is when we decide to adopt a system of systems where decision-making is federated out to multiple *AI Applications*. One example of this would be to build a super search engine that took a user's search term and sent it to the top three search providers before fusing the results into a single result set. Such approaches are also used in safety critical systems such as auto-pilots where there are perhaps three separate collision detection systems and one over-arching *Application* that arbitrates their separate decisions.

Note that the complicating factors we talk about above are not a hierarchy. They are not mutually exclusive. So, it is possible to have just about any permutation of these features in your AI Application.

The day-to-day operations of every enterprise involve many different business processes and many of those processes could benefit from the application of AI. This raises the very serious challenge of how we scale our ability to apply AI when there are so many opportunities. Our ability to meet this challenge is limited by the complexity of these solutions and the skills required to deliver them, hence, the need to rapidly grow skills in the application of AI to real-world enterprise problems.

ARCHITECTURAL AND ENGINEERING CONSIDERATIONS

The use of AI within an application introduces a number of interesting new considerations. Three that are of particular interest are the need to enable the continuous evaluation of capabilities, the need to manage dependencies between AI components and the DevOps challenge of managing AI models.

Starting with the continuous evaluation of capabilities, we have previously talked about the dynamic nature of AI systems. With continuous evolution and development of both AI algorithms and the data available to systems, it is simply neither sensible nor logical to select a particular AI component and hard wire it into an application for 5, 10 years or longer. Given the prevalence of (Cloud) service-based architectures, a more logical approach would be to leverage multiple services. These services could be used in parallel with arbitration or, if parallel use is prohibited by cost, used individually. Either way, the services being used should be subject to continuous and automated evaluation (Figure 2.11).

Our second consideration relates to the use of multiple AI components within an application. Multiple AI components may be integrated according to a number of patterns. However, they are integrated, the fact is that they will impact and depend on one another. In these multi-component architectures, we need to consider the potential impact of these dependencies:

- Will errors cascade through the system?
- Will there be version dependencies between models? For example, if two AI components are integrated with Version 2.0 of Model A only work with Version 3.4 of Model B?

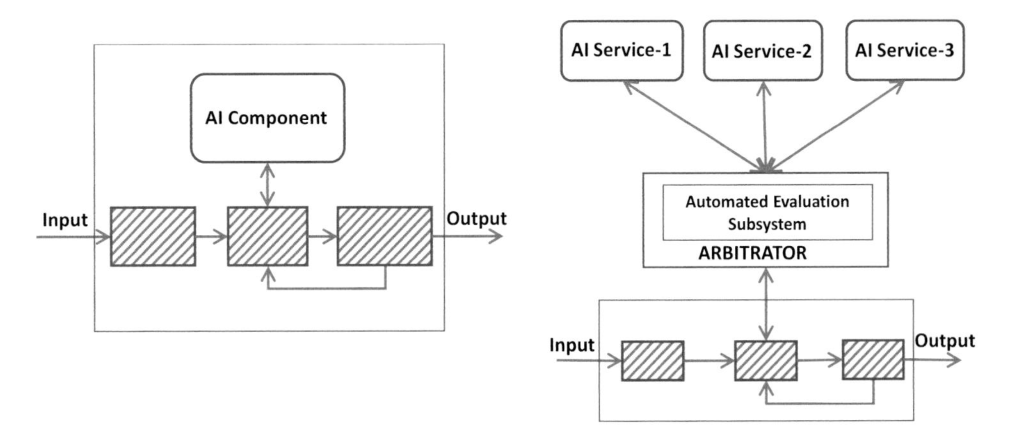

FIGURE 2.11 On the left is the conventional 'Static Integration' where one AI Component is retained for several years. On the right is the future proof 'Dynamic Integration', where multiple services are continually being evaluated with the optimum configuration in use at any given time.

- How can we understand the impact of changing an individual component on the overall performance of the system? If we have a limited budget to invest in optimisation, which component should we focus on?

Managing the configuration of applications becomes considerably more challenging with AI applications. The importance of versioning and version management in conventional software development is well understood. Best practices exist to ensure that we understand the hardware and software dependencies required for a particular version of an application. We also have well proven methods for testing and managing defects that are all linked to version management. In AI applications, we introduce two significant new elements. We introduce the concept of an *AI Model* and we have the data that was used to develop and test that model.

In addition to introducing these two additional elements, we are dramatically changing the development lifecycle and timelines because the data and the resulting model can be changed almost continuously. It would not be unreasonable to develop a new model ever day or even every hour. In fact, it is possible to configure an AI system to learn continually so that the model is constantly changing. Changing the model directly changes the functionality of the system and therefore raises the question of how we perform basic tasks such as defect tracking?

To give you an idea of the scale of the challenge, let's imagine a Machine Learning model was used by an insurance company to determine the insurance risk of an individual. This risk assessment directly impacts the decision to insure and the price at which insurance is offered. In our imaginary scenario, the company uses a set of real customer data to develop a model that is deployed into the operational production environment.

After a period of days, a Customer complains to the regulatory authorities and the company is asked to explain why an insurance application was rejected. Explainability can be a challenge with some Machine Learning approaches (this will be discussed further in Chapter 5). However, in this case, the insurance company is able to share their training data with the regulator and point at specific cases in the training data where the profile of the Customer is a close match. The regulator upholds the company's decision and rejects the Customer complaint.

So far so good … except! The very next day, a group of disgruntled Customers exercise their rights under General Data Protection Regulation (GDPR) to request that their data is deleted by the company. The company obliges and, in the process, deletes a number of training cases from the training set. As the ML model can be considered a summary of the training data, the company also feels that they should delete the ML model. From this point forwards, they cannot investigate any historical complaints from the time period when that model was operational. In fact, they have no way of even re-creating system behaviour from that period.

How do we deal with this challenge? Well, the answer is a combination of careful data management, the possibly controversial use of synthetic data (developed as part of an AI factory) and sophisticated DevOps that will be discussed further in Chapter 10.

TIME & COST TO VALUE

Over a 9-month period, I observed one Client start three separate AI proofs of concept all related to entity extraction.

Each evaluation was a significant undertaking requiring a four- or five-person team, several cloud-based servers, subscriptions to entity extraction services and considerable effort to define test data, configure services and evaluate the accuracy of each different option. Each of these projects focused on just one single AI application.

With the same Client, I personally ran three short workshops to discuss their AI needs. In these three workshops, we identified 25 separate potential AI entity extraction projects ... and we weren't even scratching the surface. Every time we led a workshop, more and more potential projects were identified.

The number of potential AI applications in this particular Client could probably be counted in the hundreds ... yet it took 9 months to conduct proofs of concept for just three!

We simply can't scale the adoption of AI if the time and effort required to undertake a proof of concept is huge. We need to be more efficient in running proofs of concept so that we can test more ideas more quickly, identify the ones that are feasible, park (note I said 'park' not 'scrap') the ones that don't work (yet) and make progress on those that do!

One way to do that is to create an organisation that is both skilled in evaluating and delivering AI and ensure that it is equipped with the right tools and the right infrastructure to do the job.

James Luke

THREE STAGES OF AN ENTERPRISE AI APPLICATION

In the previous sections in this chapter, we have given you some ideas on various practical aspects of building an AI application. However, developing the application is only the first step. We need to deploy the application and 'sustain' the application to realise the business value.

There are many moving parts. First, we need to explain our use of the verb 'sustain'! Why are we talking about 'sustain' and not 'maintain'?

All applications require maintenance so the idea of developing an application, deploying it to a production environment and then maintaining it is easy to understand. However, AI applications differ from conventional applications because they are so heavily data dependent. As explained earlier, in an AI application, the functionality of the application is directly coupled to the data used in development and the data observed in production. The AI Model needs to be continually monitored and updated to ensure it is performing correctly. We're calling that sustenance!

Whilst Figure 2.12 describes the basic and somewhat obvious process, we really need to drill down into the detail. Note this is a process that describes the basic steps that need to be considered ... it's not an architecture!

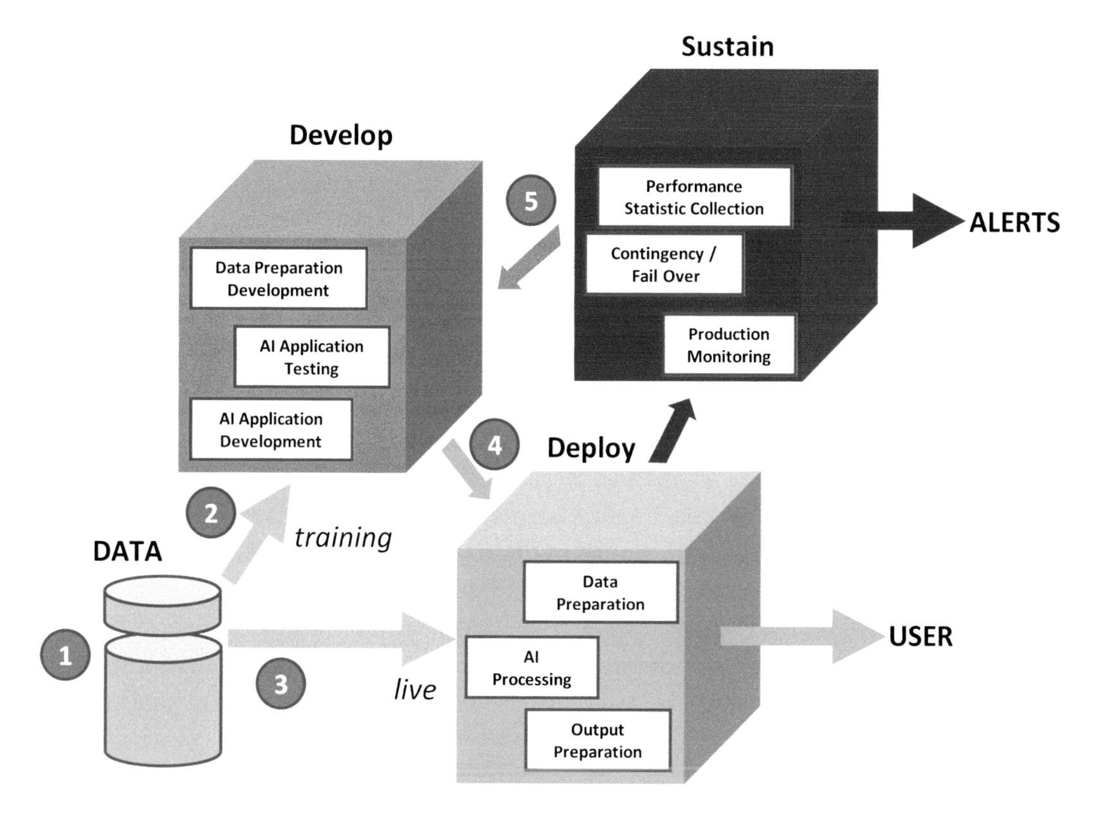

FIGURE 2.12 Three stages of enterprise AI applications.

What Are the Key Aspects to Understand?

1. It's all about the data … we talk about that in Chapter 7, and the key point is that you're going to need to take data from the real world for use in the development process. The more real-world data you can collect, the better; however, in reality you will almost certainly need to filter out a sub-set of the data. You need an effective strategy to ensure the data used for development is truly representative of the data the AI Application will encounter in the wild.

2. Unless you are extremely lucky, the data you use for development will not be suitable out of the box. It will require a lot of preparation including basic cleansing, preprocessing, fusion with other sources, de-duplication and a whole host of other activities to make it usable. In the development environment, this will be done manually under the supervision of Data Scientists. Whatever you do to clean up the data in the Development environment needs to be done in the Deployed environment. If a Data Scientist manually cleans up the data for development, you're going to need to automate whatever they do and deploy it into the Deployed environment.

3. Just as you filtered a subset of real-world data for use in Development, you're going to need to filter off a subset of the Development data for Testing. We talk about this further in Chapter 6 (Ensuring It Works & How Do You Know).

4. When the Development team have finished building an AI Application, it needs to be put into the Deployed environment where it will be used to process real-world data and often interoperate with existing corporate systems. This often involves different IT skills to the ones possessed by your data scientist developers – your team will need to grow and the code, etc. will need to pass your corporate standards for safety/ethics/archival etc.

5. Finally, you need to sustain and monitor your deployed capability. This monitoring goes far beyond that normally applied to a conventional application. In a conventional application, the monitoring is basically waiting for faults to appear. In an AI application, the monitoring needs to look for anomalies in the data processing that may indicate the AI is operating outside its intended scope. For example, if the AI Application is designed to analyze the English language, then the monitoring will look for indications that non-English language content has been fed into the system. This would be indicated by a change in the statistics relating to the number of entities and relationships extracted. You need to work on the business processes around what to do when the system starts to underperform. Note in the real world this can be years after the original code was written, there will be no guarantee your original developers are still around. If you have developed your solution with rigour and good documentation, and have a dedicated model monitoring/sustenance team, picking up a 'flagging' model and retraining it with the latest data, life will be easy. If you hacked something together just to get something working, and decided documenting was too expensive… you could be in for a costly surprise.

ENABLING ENTERPRISE SOLUTIONS AT SCALE

To deliver successful AI in the enterprise, it is important to develop a strategy that covers much more than just the technology. In fact, an overall AI strategy should consider these four elements: (i) making good decisions, (ii) understanding the value of AI, (iii) going beyond proofs of concept and (iv) delivering AI at scale. We discuss in this section what actions (see Figure 2.13) you need to take in order to execute on these four strategic elements.

Education

The first and perhaps the most important is Education. That's why you're reading this book so at this point we must, very sincerely and with straight faces, ask you to recommend this book to your colleagues, friends, family, fellow commuters and anyone you happen to walk near or past in the street. It's your solemn duty to do so for the greater good of humanity.

In all seriousness though, education is particularly important with reference to Domain Understanding. One hard-earned lesson from delivering real AI solutions is that domain knowledge is a major critical success factor. No IT consultant or engineer can understand your business as well as you do! It's therefore really important for you to acquire sufficient

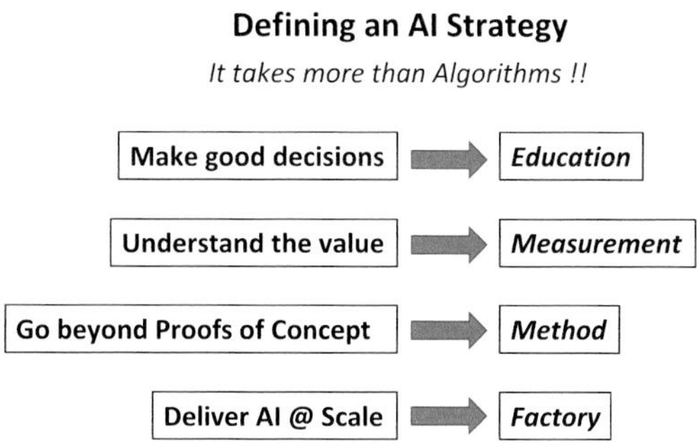

Defining an AI Strategy

It takes more than Algorithms !!

Make good decisions	→	*Education*
Understand the value	→	*Measurement*
Go beyond Proofs of Concept	→	*Method*
Deliver AI @ Scale	→	*Factory*

FIGURE 2.13 Mapping strategy to action.

AI knowledge amongst the domain experts in your enterprise to be able to form joint teams with AI suppliers in order to ensure success for your AI projects.

Measurement

Beyond education, you need to understand the value that AI brings to the business. This can be measured directly, in terms of overall business impact, or indirectly in terms of enabling business impact. To clarify the difference between the two, an AI that enables a lawyer to find relevant case information more quickly has a measurable impact in terms of the time saved and the improvement in efficiency. However, an AI that automatically translates foreign language documents into a native language as an input to the aforementioned search tool has an indirect benefit. It may not be possible to quantify this benefit in terms of overall business impact; however, it is a critical enabler for a solution where the impact can be directly measured. Measurability and value are massive subjects and will be discussed in detail in Chapters 5 and 6 (*Business Value & Impact* and *Ensuring It Works & How Do You Know*).

Method

Now that you're educating your team and measuring business value, you can start to understand where to apply AI for optimum benefit. Selecting the right project is another critical success factor and will be discussed in the following chapter (imaginatively called Know Where to Start – Select the Right Project). One anecdotal observation from business consultants about AI is that the domain seems to comprise a never-ending series of Proofs of Concept (POC). This has rather amusingly been referred to by one of our colleagues as "perpetual beta". Selecting the right project is essential in escaping "perpetual beta".

Factory

Finally, you need to put in place a development organisation, which we refer to as a Factory, aimed at reducing the cost of evaluating and delivering real AI Applications. The AI Factory brings together people (skills), assets and tools from multiple suppliers with a process to enable the rapid development and deployment of operational capability (including ongoing operational support). The AI Factory is intended to enable the delivery of capabilities from an ecosystem of suppliers such that the optimum capability is available to the authority.

The AI Factory comprises:

- Ready-to-go infrastructure so that an AI proof of concept can be started immediately without the need to provision new Cloud servers and infrastructure.
- Assets, developed on previous projects, that can be used out of the box or adapted as the base of new capabilities. This re-use of capability is a major driver of the efficiency of scale enabled through the AI Factory.
- Tools for automated evaluation, testing and continual monitoring of capabilities. For example, if multiple different suppliers offer a particular service then the AI Factory can use test tools to routinely evaluate which specific service performs best and use this service. This ability to continually monitor and switch services ensures the best possible quality of service in an area where technology is continually evolving.
- Tools and infrastructure for deployment to multiple production environments. The AI Factory can run services and pipelines for missions or deploy pipelines onto mission-owned environments.
- Content for development and evaluation of capabilities.
- Technologies and services from multiple suppliers already provisioned such that we can enable a capability rapidly.

The aim of the AI Factory is to reduce the time, cost and effort to evaluate and deploy new AI capabilities. When a requirement is identified, the AI Factory first looks to existing capabilities (e.g. developed for other missions with the same, or similar, requirements) and makes these services available. In the event that the existing capability is not suitable, the next question is whether an existing capability can be adapted. If not, then a new capability can be developed and evaluated. Even when this is required, the capability development should be more rapid due to the existence of existing infrastructure, services, test and automation tooling.

One of the major inhibitors to the successful delivery of AI projects is the upfront cost of evaluating possible projects.

Invariably, once a possible project has been identified, the first question that needs to be asked is whether the AI can do the job. At that point, a proof of concept is proposed and a whole series of logistics need to be considered. What data is needed for the POC? What technologies will be evaluated? Are there multiple suppliers that need to be compared? What hardware will the evaluation run on? Who is going to label any training data (remember not all AI projects are Machine Learning)? How will we actually compare the output of the AI with any test data?

In short, conducting a POC is not a small task and it's not uncommon for what appeared to be a simple evaluation to grow into a project lasting several months and costing tens, or even hundreds, of thousands of pounds. The costs of running such POCs can soon become an inhibitor ... especially when the first attempt is unsuccessful. Beyond the POC, remember that this is not a static situation. The initial POC may prove the feasibility of applying AI; however, maintaining and optimising the application is a continual process. Imagine developing an AI application and identifying three potential technology suppliers. Each supplier is continually developing and enhancing the service. The service that performs best in today's evaluation actually performs worst in 3 months' time. If we are to realise the full potential of AI, applications need to be continually maintained with an ongoing evaluation of technology options. That requires skills, infrastructure and tooling.

In deciding whether or not to adopt an AI Factory model, organisations need to give serious consideration to the scale and form of their AI ambitions. When brainstorming potential AI applications, it is not unusual to identify 20 or more ideas. The cost-to-POC issue causes organisations to focus on just one or two. That's a serious mismatch and demonstrates the braking effect of the cost-to-POC issue. There is a chance that your organisation only needs to develop one single AI application and, therefore, the cost of developing an AI factory cannot be justified. In that case, you are the exception. In most cases, there is a need to develop and maintain large numbers of applications and the initial investment in the Factory will result in both considerable savings and an acceleration in your ability to transform your business through AI.

IN SUMMARY – ARE YOU READY TO START BUILDING APPLICATIONS?

We have covered a fair bit about building AI applications in this chapter. Here is a summary:

- Enterprise AI applications are very different from conventional software systems.
- Public perception of AI applications is mostly defined by consumer applications of the web companies.
- Enterprise applications have very different challenges from the consumer applications on the web, in terms of business relevance, data availability and stakeholder expectations.
- We described many examples of real enterprise applications and the role of AI in them.
- Based on the confidence in the AI and the consequence of the task, it is important to select the appropriate task for the use of AI.
- Creation of an AI application requires a careful selection & execution of activities from data selection to the user interface.
- It is important to be aware of various elements that can increase the complexity of an AI application.
- Managing all three stages of an AI application (i.e. Develop, Deploy and Sustain) is critical for its success.
- Delivering AI applications at scale needs education, selecting the right projects, a measurement framework and an AI factory.

It's exciting ... but we're only just getting started. We will develop various concepts in the upcoming chapters to understand the unique challenges of real enterprise AI applications and how to address them in practice.

REFERENCES

1. A. L. Samuel, "Some studies in machine learning using the game of checkers," *IBM Journal of Research and Development*, pp. 211–229 (1959).
2. G. Tesauro, "Temporal difference learning and TD-gammon," *Communications of the ACM*, 38, pp. 58–68 (1995).
3. M. Campbell, "Knowledge discovery in deep blue," *Communications of the ACM*, 42, pp. 65–67 (1999); https://www.youtube.com/watch?v=KF6sLCeBj0s.
4. D. Silver et al., "Mastering the game of go without human knowledge," *Nature*, 550, pp. 354–359 (2017).
5. J. Weizenbaum, "ELIZA: a computer program for the study of natural language communication between man and machine," *Communications of the ACM*, 9, pp. 36–45 (1966).
6. T. Winograd, *Understanding Natural Language* Academic Press, (1972).
7. D. A. Ferrucci, "This is Watson," *IBM Journal of Research and Development*, 56 (2012) and the rest of the Issue.
8. N. Slonim, et al. "An autonomous debating system," *Nature*, 591, pp. 379–384 (2021).
9. T. Young, et al. "Recent trends in deep learning based natural language processing," *IEEE Computational Intelligence Magazine*, 13(3), pp. 55–75 (2018).
10. W. Van Melle, "MYCIN: a knowledge-based consultation program for infectious disease diagnosis," *International Journal of Man-Machine Studies*, 10, pp. 313–322 (1978).
11. R. K. Lindsay, B. G. Buchanan, E. A. Feigenbaum and J. Lederberg, *Applications of Artificial Intelligence for Organic Chemistry: The DENDRAL Project* McGraw-Hill, New York, (1980).
12. C. Garcia, "Harold Cohen and AARON—A 40-year collaboration," (2016). https://computerhistory.org/blog/harold-cohen-and-aaron-a-40-year-collaboration/.
13. W. G. Walter, "An imitation of life," *Scientific American*, 182, pp. 42–45 (1950).
14. R. Behringer, "The DARPA grand challenge - autonomous ground vehicles in the desert," *IFAC Proceedings*, 37, pp. 904–909 (2004).
15. R. G. Smith and J. Eckroth, "Building AI applications: yesterday, today, and tomorrow," *AI Magazine*, 38(1), pp. 6–22 (2017).
16. C. Lesner, et al., "Large scale personalized categorization of financial transactions," *The Thirty-First AAAI Conference on Innovative Applications of Artificial Intelligence (IAAI–19)*, pp. 9365–9372.
17. K. J. Lee, et al., "Embedding convolution neural network-based defect finder for deployed vision inspector in manufacturing company Frontec," *The Thirty-Second Innovative Applications of Artificial Intelligence Conference (IAAI–20)*, pp. 13164–13171.
18. E. Shaham, et al., "Using unsupervised learning for data-driven procurement demand aggregation," *The Thirty-Third Innovative Applications of Artificial Intelligence Conference (IAAI–21)*, Vol. 2., 2021.
19. M. Feblowitz, et al., "IBM scenario planning advisor: a neuro-symbolic ERM solution," *Proceedings of the Demonstration Track at the 35th Conference on Artificial Intelligence (AAAI-21)*, 2021.
20. M. Maier, et al., "Transforming underwriting in the life insurance industry," *The Thirty-First AAAI Conference on Innovative Applications of Artificial Intelligence (IAAI–19)*, pp. 9373–9380.

21. E. Adamopoulou and L. Moussiades, “Chatbots: history, technology, and applications,” *Machine Learning with Applications*, 2, 100006, Elsevier (2020).
22. Agent Bot with multiple skills: https://github.com/IBM/watson-assistant-multi-bot-agent.
23. M. Merler et al., “Automatic curation of sports highlights using multimodal excitement features,” *IEEE Transactions on Multimedia*, 21(5), pp. 1147–1160 (2019).
24. G. Neff and P. Nagy, “Talking to bots: symbiotic agency and the case of Tay,” *International Journal of Communication*, 10, pp. 4915–4931 (2016).

CHAPTER 3

It's Not Just the Algorithms, Really!

"It's not clever enough ... we need a better algorithm!"

Whilst delivering a Question Answering system for a major Client, that statement was heard at every project review. The sad reality was that the challenges being experienced were nothing to do with the AI. The system wasn't working as expected because the training set was full of conflicting data. There were many different versions of the same question, each with different answers.

It was the AI equivalent of being an inconsistent parent; giving a child conflicting messages and then telling the child that he or she is stupid for not learning.

However, for the project team, there could be only one explanation for the failure of the AI to learn. It wasn't clever enough, so they needed a more intelligent, better AI. They believed that they only way to succeed was to discover a better algorithm that was capable of making sense of the nonsensical training data. The reality was that they didn't need a better algorithm. They needed to clean up the training data!

James Luke

We now live in a world dominated by algorithms, and it is therefore no surprise that we tend to blame the algorithm when things go wrong.

For those involved in delivering real Artificial Intelligence (AI), success starts by asking yourself a very honest question ... are you addicted to algorithms? Algorithm addiction can attack any member of the team from the sponsoring executive to the business analyst or the most junior developer. Understanding the risks posed by algorithm addiction is essential in successfully delivering AI so, in this chapter, we aim to provide you with the knowledge required to identify and avoid algorithm addiction.

DOI: 10.1201/9781003108498-4

For those with less technical background, we will explain what algorithms are and position their use within a broader AI application. We will examine some important algorithms in the domain of AI and use some simple problems to demonstrate how different algorithms can be applied to solve the same problem. The objective being to demonstrate that there is no single magic algorithm, they are relatively interchangeable with each algorithm having pros and cons. Algorithms are only as good as the data they operate against and so we will introduce the concept of feature extraction and explain its importance within AI applications. Finally, we will pull out some fundamental principles for the safe use of algorithms that will enable you to successfully deliver AI applications.

INTRODUCING ALGORITHMS

Algorithm is a great word! Not long ago, most people would never have even heard of the word. Yet, now it is the subject of dinner party conversations. The media and press are full of articles about the brilliance of AI *Algorithms* in understanding who we are, what we do, what we want and what we plan to do next. What is an algorithm anyway?

An algorithm is just a coherent set of steps to do a task. The purpose is to take some input and create an output that meets the objective of the task. In some sense, a cooking recipe with the necessary ingredients and the steps to cook the perfect *filet mignon* can be thought of as an algorithm. Computing algorithms have a rich history [1]. The accompanying vignette gives the details of three examples of algorithms created over a period of more than 3000 years of human history, including the fact that Babylonians knew Pythagoras' theorem more than 1000 years before he gave us his version.

Before the invention of computers, these algorithms were performed by humans by hand. With the use of computer programs, they become more natural in specifying the required operations to a computer. In that context, there are generally five key properties algorithms must have [2]:

1. Input(s) to the algorithm is defined.
2. Output of the algorithm is defined to have a specified relation to the inputs.
3. Definiteness: Algorithms must specify every step and the order of the steps clearly.
4. Effectiveness: Must be possible to perform each step correctly in a finite amount of time.
5. Finiteness: Must terminate after finite number of steps.

In the algorithm examples described in the vignette, the steps are simple. As the complexity of an algorithm increases, sometimes it is not clear if it can actually do the intended task in time with the available computing resources or not. In fact, there is an established discipline in computer science to assess the efficiency and feasibility of algorithms [3].

ALGORITHMS IN AI

As we discussed above, the purpose of an algorithm is to execute a task. In his 1996 Turing award lecture, Raj Reddy [4] discussed AI algorithms that:

- Exhibit adaptive goal-oriented behaviour.
- Learn from experience.
- Interact with humans using language and speech.
- Use vast amounts of knowledge effectively.
- Tolerate error and ambiguity in communication.
- Have real-time constraints.
- Explain their capabilities (e.g. answer "how-to" and "what-if" questions).

AI algorithms tend to be rich in content and complex in design. He also observed that it was less important if AI had more or less intelligence compared to humans; but it was more important to understand that "AI is about creating artefacts that enhance the mental capabilities of the human being". This brings us to the ideas of "Intelligent Automation" or "Augmented Intelligence", the vision of humans and machines working together to do better than either of them separately. Humans are better at self-directed goals, common sense reasoning, value judgement, etc., whereas machines are better at large-scale mathematics, pattern discovery, statistical reasoning, etc. *Can we work together to do some amazing things, such as to find the cure for cancer or stop climate change?*

In the last decade, the use of an AI component in the business context has had one of three goals:

i. Automate an existing task performed by a human (e.g. Support Bots)

ii. Improve the efficiency of an existing task (e.g. language translation)

iii. Perform a new task (e.g. a recommender system). Break the list here and start a new line at the margin with "Algorithms are typically......". Continue the paragraph with the remaining text below. Algorithms are typically used for classification (i.e. automated labelling), regression (i.e. numerical predictions) or sequence predictions (e.g. predicting a sentence from a few words).

The scope of the AI tasks has been narrow, giving rise to the term 'Narrow AI'. Tasks in Narrow AI deal with one type of data such as text, or image, or speech, with the need for large amounts of training data and result in 'opaque' (aka black box) models that cannot explain their outputs. Recently, the industry has started efforts in 'Broad AI' which bring together knowledge from a variety of sources and formats and perform more complex

ALGORITHMS

The goal of this vignette is to describe three interesting examples of algorithms.

A Babylonian Algorithm (*circa 1700 BC*)

Calculates the diagonal of a rectangle. Donald Knuth [1] has a fascinating discussion of ancient Babylonian mathematical algorithms from the times of the Hammurabi dynasty. Here is an example. For a rectangle of length L, width W and diagonal D, and area A, the algorithm takes two inputs i.e. $L + W + D = 12$ and $A = 12$ and calculates the diagonal D in a step-by-step fashion using the formula:

$$D = \frac{1}{2} \frac{\left[(L+W+D)^2 - 2A\right]}{(L+W+D)}$$

This uses the Pythagoras theorem ($L^2 + W^2 = D^2$) implicitly, more than 1000 years before the time of Pythagoras! As an aside, Babylonians used a 60-based number (sexagesimal) system, where 30 in sexagesimal is 30; 1 in sexagesimal is 60; 2 in sexagesimal is 120, and (2,24) means ($2 \times 60 + 24 = 144$).

Euclid's Algorithm (*circa 300 BC*)

Find the Greatest Common Divisor (GCD) for two numbers [2].

Let us use 12 and 56 as the two numbers to demonstrate the algorithm.

- **Step 1**: Subtract multiples of the smaller number from the larger number until the remainder is less than the smaller number and keep the remainder and the smaller number. Four multiples of 12 subtracted from 56 give the remainder 8. Twelve and 8 are the new numbers to compare.
- **Step 2**: Repeat the process. One multiple of 8 subtracted from 12 gives a remainder of 4. Eight and 4 are the new numbers to compare.
- **Step 3**: Repeat the process. Two multiples of 4 subtracted from 8 gives a remainder of zero. *The algorithm ends. The GCD for 12 and 56 is 4.*

Sorting Algorithms (*circa 1950 AD*)

Put elements of a list in a certain order. Typically, the list consists of numbers to be placed in numerical order or words in an alphabetical order. We will use 'Exchange Sorting' [3] (also known as Bubble Sort) to demonstrate the algorithm since it is intuitive to understand. Let us take the list {6,3,5,2} as the input to the algorithm. We want the algorithm to sort this list from the lowest to the highest number.

The algorithm starts from the beginning and takes each adjacent pair of numbers (**shown in bold**) in the list and put them in the right order. Each pass covers all the pairs in the list in sequence, and the result of each pass is given as input to the next pass.

First Pass	Second Pass	Third Pass	Fourth Pass
{**6,3**,5,2} → {3,6,5,2}; swap 6 and 3 since 6>3	{**3,5**,2,6} → {3,5,2,6}; no change since 5>3	{**3,2**,5,6}→ {2,3,5,6}; swap 3 and 2 since 3>2.	{**2,3**,5,6} → {2,3,5,6}; no change since 3>2.
{3,**6,5**,2} → {3,5,6,2}; swap 6 and 5 since 6>5	{3,**5,2**,6} → {3,2,5,6}; swap 2 and 5 since 5>2	{2,**3,5**,6} → {2,3,5,6}; no change since 5>3.	{2,**3,5**,6} → {2,3,5,6}; no change since 5>3.
{3,5,**6,2**} → {3,5,2,6}; swap 6 and 2 since 6>2	{3,2,**5,6**}→ {3,2,5,6}; no change since 6>5.	{2,3,**5,6**} → {2,3,5,6}; no change since 6>5	{2,3,**5,6**} → {2,3,5,6}; no change since 6>5.

The output of the algorithm is {2,3,5,6}

This sorting is accomplished in first three passes and the fourth pass is just to confirm the result. There are many different algorithms [4] to do the sorting task (Quicksort, Merge sort, etc.). The specific algorithm is selected for a task based on the size of the input list and the time it takes to execute the number of operations needed to complete the algorithm.

REFERENCES

1. D. E. Knuth, "Ancient Babylonian algorithms," *Communications of the ACM*, 15, pp. 671–677 (1972).
2. Euclidean algorithm: https://en.wikipedia.org/wiki/Euclidean_algorithm.
3. E. Friend, "Sorting on electronic computer systems," *Journal of the ACM*, 3, pp. 134–168. (1956).
4. Sorting Algorithms: https://en.wikipedia.org/wiki/Sorting_algorithm.

tasks with less data requirements and some explanations of outputs. It is likely that Broad AI will take many more years to evolve and prove itself. We believe that the age of General AI, when a machine can learn and do any intellectual task that a human being can (i.e. the Skynet and the robots of the Terminator movies), is many decades away, if at all [5].

ALGORITHM ADDICTION

When things go wrong with AI applications, it's often the algorithm that is blamed. One example of this is in the domain of AI bias; we hear reports of *algorithms being biased*. After some investigation of the origins of the bias, one quickly finds out that the data used to train or configure the algorithm was the real culprit. This is the case in most of the instances. This is because the algorithms are just finding the patterns in the data, which may not have been obvious before AI came along. Unfortunately, truth hurts! We use bias as an example of our tendency to blame the algorithm. However, there are many other types of AI failure where the algorithm is invariably held accountable.

The nature of our business contributes to our obsession with algorithms. Despite massive investment, a lot of AI research still happens in relatively small teams with individual engineers desperate to solve the problem of creating intelligence. Many engineers dream of creating a beautiful and elegant algorithm that will, somehow, evolve intelligence when presented with enough data. Algorithms are fascinating, exciting and essential, but it takes much more than an algorithm to successfully deliver an AI solution. This isn't easy to accept because many of us, who work in this domain, need to confess that we are … quite simply … *Algorithm Addicts*. We love algorithms and we spend our lives thinking of them, trying to understand them and dreaming up new ones.

As *Algorithm Addicts*, we must come to terms with our addiction and warn others of the risks. The single greatest threat to the advancement of AI is believing that there is a magic algorithm that will solve the challenge of creating AI. The failure of AI projects is often attributed to the need for a more intelligent algorithm. In our experience, AI projects are more likely to fail because of a fundamental breakdown in systems engineering. Often this breakdown starts with the problem selection and project definition, both of which will be discussed in Chapter 4. However, before looking at problem selection, it's useful to understand some of the fundamentals of the algorithms that underpin current AI. Understanding the fundamentals will enable you to understand that there is no magic in the algorithm. There may be beauty, elegance, wonder and amazing mathematics, *but there is no magic.*

APPLICATIONS VERSUS THE UNDERLYING TECHNOLOGY

Ultimately, what we're interested in doing is adding more intelligence to systems so that they do more, make better decisions and allow us to gain a competitive advantage in whatever operational environment we work in. For those new to the field, understanding what we mean by AI is perhaps made more difficult by the fact that important terms are often used interchangeably. Right now, there is a huge amount of interest in ML and, in particular, Deep Learning (DL). So much so that you could be forgiven for thinking that DL, ML and AI all mean the same thing. That is most definitely not the case (see the vignette on ***AI Discipline***).

To clarify the terminology, let's start with the fundamentals. AI is concerned with the delivery of a broad set of applications ranging from sensory tasks such as speech recognition and visual perception to interpretation and decision-making tasks such as medical diagnosis and resource scheduling. Current applications of AI that are gaining huge amounts of media attention include driverless vehicles, personal assistants such as Siri and Alexa, game playing machines such as Google's AlphaZero, automated trading tools that manage stock market investments and robotic applications such as the robot sorting products in Amazon warehouses.

There seems to be no limits on the range of potential *AI applications*! What matters is that we do not confuse *AI applications* with the underlying technology or algorithms. In selecting the technology and algorithms to use in order to deliver an application, the list of options is endless. Whilst it may be impossible for some options to work for some applications, generally applications can be implemented using a vast range of technologies.

ALGORITHMS AND MODELS

AI Applications use algorithms to perform tasks. An *application* may use a single algorithm to perform a single task such as identifying a malignant tumour in an X-Ray. As tasks become more complex, it is more likely that multiple *algorithms* will be integrated together to deliver a greater level of intelligence. For example, a medical diagnosis system may take the output of the X-Ray classifier and fuse it with other test results to perform a diagnosis. If we add in the patient history, then we may use a third *algorithm* to extract data from the patient's historical notes. (Figure 3.1).

The first point to understand about delivering any AI Application is that there are many different algorithms and it's important to select the right algorithm for each task. In our medical example, we could be using a neural network for the X-Ray classifier, a rules-based entity extract for the patient notes analysis and a conventional mathematical model for the

THE AI DISCIPLINE

Topics in AI have evolved over the years. The classic AI textbook by Russell and Norvig [1] gives the full range of the technology in scope. As is evident, this is very interdisciplinary in nature, requiring concepts in cognitive science, neuroscience, mathematics, computer science, engineering and ethics. Some of the key topics are

- **Perception**: Deriving information from sensory inputs (text, vision, speech, etc.) to build the relevant knowledge.
- **Knowledge Representation**: Accumulating and storing the semantic knowledge as ontologies, graphs, etc., in a knowledge base (a world model) in a specific domain for practical use.
- **Learning**: Ability to learn from data and human inputs to update the knowledge base.
- **Reasoning**: Various process steps that utilise the knowledge base for practical use.
- **Problem Solving by Search**: An agent that helps to find answers to specific tasks by leveraging the knowledge base.
- **Common Sense**: Incorporating various elements in the world model that are assumed to be evident to humans, without explicit training.
- **Rule-Based Systems**: Creation of real-world systems using a collection of rules in a domain, either based on expert knowledge with the use of relevant data.
- **Planning**: The task of designing a sequence of possible actions to achieve a goal.

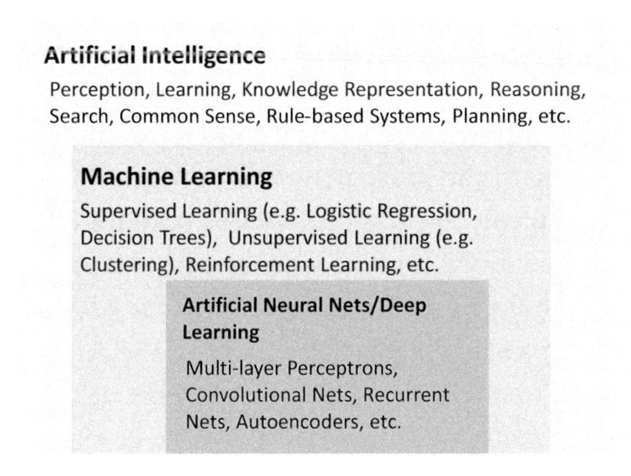

Machine Learning (ML) is a part of the learning activities in an AI system that can be performed with or without human supervision or intervention. An Artificial Neural Network (ANN) represents a particular architecture that is derived from the structure of the brain consisting of neurons, dendrites, etc. [2]. ANNs are made of a network of artificial neurons with an input layer, an output layer and some intermediate ('hidden') layers in the middle and connections across the layers. Deep Neural Networks (DNNs) are ANNs with a large number of hidden layers.

REFERENCES

1. S. J. Russell and P. Norvig, *Artificial Intelligence: A Modern Approach* Pearson, 4th Edn (2020).
2. W. S. McCulloch and W. Pitts, "A logical calculus of the ideas immanent in nervous activity", Bulletin of Mathematical Biophysics, v. 5, pp.115–137 (1943).

diagnosis component. We'll discuss these different approaches later in this chapter. However, remember that there are thousands of different *algorithms* to select from. We've chosen those three just to give you a feel for how different *algorithms* can work in an application.

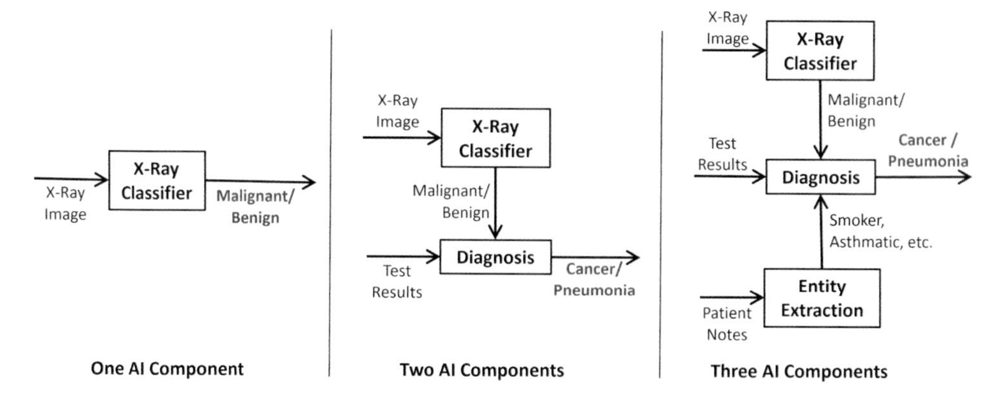

FIGURE 3.1 An example of the use of multiple AI algorithms in an application.

The second point to understand is that algorithms are, in almost every case, generic! Algorithms can be applied to many different problems with the only constraint being the type of data fed into the algorithm and the type of output expected. For example, some *algorithms* can only work on numeric data whilst others are designed to process lists. Similarly, some *algorithms* can only output classifications (e.g. Good/Bad/Ugly) whilst other *algorithms* generate numerical outputs. As long as an *algorithm* can handle the types of input and output expected, it can be applied to most problems.

Algorithms are generally implemented using a chosen programming language (e.g. java, python, etc.) as generic tools. In the brave new world of Cloud-based services, most of the Cloud providers provide implementations of the most popular AI *algorithms* that you can provision as services. So, how do you make a generic *algorithm* work for your particular application? The answer is that you give it an *AI Model*. An *AI Model* is a configuration of an *AI Algorithm* to perform a specific task. The form of the *AI Model* depends on the *AI Algorithm*. In a rules-based system, the *Model* will be a list of rules. In a neural network, the *Model* will describe the configuration of the network in terms of the number of neurons and the connectivity between them. For example, if we use a Neural Network to perform the X-Ray classification, the *AI Model* will comprise the weights and configuration of the neurons required to perform that task.

AI Models can be developed in many different ways that generally fall into one of three categories. Firstly, *Models* can be defined by the developer. A human being can sit down and manually create a *Model* using their human expertise. Secondly, the AI can be trained to develop its own *Model* using training data that has been created by a human expert. In our X-Ray classifier example, this would require a radiographer to manually examine several hundred X-Rays and label each one as either malignant or benign. This form of training is known as 'Supervised Machine Learning' as the human expert has prepared the training data with output labels. Thirdly, the AI can train itself using 'Unsupervised Machine Learning'. In this case, there is no direct expert input (i.e. labels), and the AI examines data automatically searching for interesting patterns. There is a vignette on an example of Unsupervised ML later in this chapter. Much of the discussion in this chapter is on Supervised ML.

What does it mean therefore when you hear, for example, about searching the web for images using neural networks? Quite simply, this means that there is an *AI Application* for image searching that is using an *AI Algorithm*. The algorithm will be a neural network which we'll discuss in more detail later. The neural network algorithm will be implemented somewhere in a conventional computer program. The algorithm will be customised to perform the image search task using an *AI Model* that was developed using ML with a neural network.

It's important to understand that, generally, there are two parts to the *algorithms* used in any *AI Application*: the *algorithm* that uses a *Model* to perform a task and a different *algorithm* that has been used to develop that *Model*. In some cases, these two algorithms are developed using the same technology or approach whereas in other cases it is possible to combine different approaches. For example, a model using a *neural network* will invariably be developed using ML algorithms, whereas the model for a *Rules-based Decision Engine* could be either manually specified or developed using ML.

The key to successfully developing an *AI Application* is understanding which *algorithm* to apply and what it takes to make that *algorithm* work. That also means understanding when the algorithmic approach can't work and not trying to force the impossible. That's not as easy as it sounds in a world of *Algorithm Addicts*.

Now that we understand the basic anatomy of an *AI Application*, let's take a deeper look at what happens under the hood! There is a vast number of books that will explain the multitude of different AI algorithms in detail. Whilst the sheer volume of algorithms may seem daunting, it's really not that bad! In fact, once you understand the fundamental operations at the core of all algorithms, you will find it easier to see through the complexity. In turn, that will allow you to see through the hype and make good decisions about which approach is best for your projects.

OBJECT DROPPING PROBLEM

To develop your understanding of these fundamental operations, let's consider a very simple problem and the different ways in which AI could be used to solve it. Finding a real-world problem that is also simple, is a bit of a challenge, so we need to create a slightly contrived situation. That may go against the spirit of this book, but it will have to do, just to learn the fundamentals.

Let's imagine that we want to help with reducing parcel damage for a mail order company. To optimise space, they are storing items in vertically stacked shelves. When a new order comes in, a robot finds the item on the shelf, takes it and puts it on a conveyor belt for shipping. In this process, sometimes items slip the robot grip and fall and sometimes break. So, our goal is to predict what parcels will break when dropped from different heights of the shelves. This will help us to make the right choices for storing objects in the future to minimise risk of breaking. For simplicity, let us say that the items under consideration fall into three categories: *Fragile*, *Medium* and *Hard*. If the company is already collecting the data, we can use that. If they are not, they could collect the data for the next year; but that may be too long from the business perspective. To get this problem solved more quickly, the company sends a team of minions with a whole truckload of items and asks them to

FOOD ANALOGY TO EXPLAIN AI TERMS

There is a lot of jargon around AI. You will often see terms bandied about in presentations and proposals such as *AI Solution, AI Application, AI Component, AI Model & AI Algorithm*. It can be hard to decipher what an author means (sometimes the author doesn't seem to know either). For the purposes of this book, let us imagine these AI terms in the context of food...

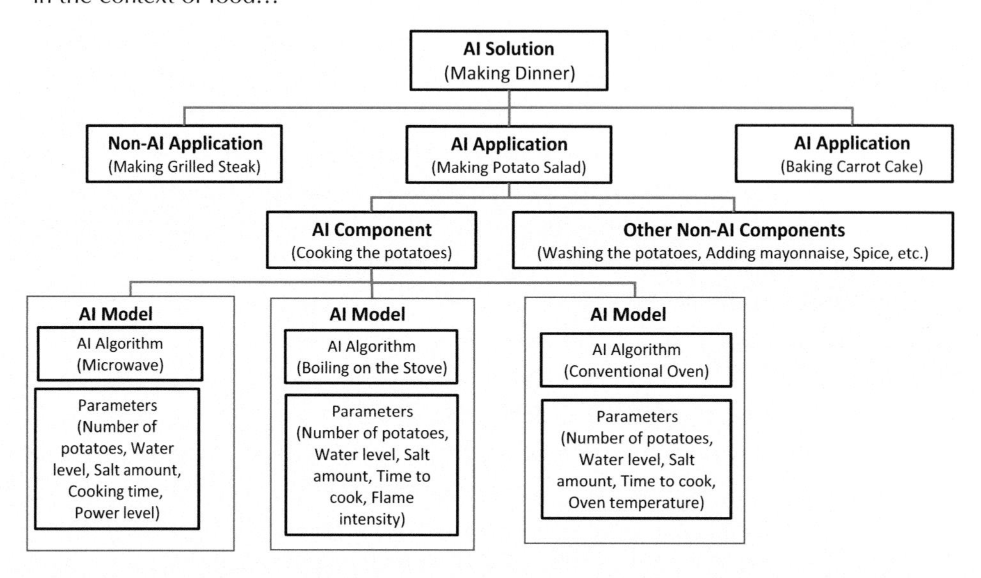

FIGURE A Common AI Terms explained with a food analogy.

It is evening and you are hungry, which is your (business) problem. You could go to a restaurant (which would be outsourcing), but you decide to make some dinner instead, since (i) you are a good cook, (ii) have a kitchen and (iii) your refrigerator has the necessary raw ingredients in it. These represent skills and resources to create the AI Solution. "Making Dinner" is the **AI Solution** to your hunger. The menu consists of Grilled Steak, Potato Salad and Carrot Cake, which are the three **AI Applications** in the solution. We will focus on one application, viz., the Potato Salad in detail.

To make the potato salad, you need the raw ingredients, potatoes, mayonnaise, mustard, etc., which can be thought of as **data**. Then, you have to follow a recipe from washing the potatoes, cooking the potatoes, adding mayonnaise, ... etc., which constitutes the **AI application**. Note that only one step actually involves **cooking the potato**, which is our **AI Component**. Similarly, real AI applications include other processes as non-AI components. The use of one or more **AI components** is what makes it an AI application.

Within this recipe, you have a choice of how you cook the potato. For example, you could use a pot of boiling water over a stove, a microwave oven or a conventional oven. You can think of these three methods as alternative **AI Algorithms**. Not all AI algorithms are interchangeable with one another, you probably won't want to grill your steak in the microwave.

Depending on which cooking device you've chosen for your potato, you are going to need some settings, power and timer settings for the microwave or temperature and timer settings for the conventional oven, etc. These specific settings, optimised to cook the potato to perfection, are the parameters chosen for the specific algorithm and the data. Change the parameters and the outputs will change – your potato may burn! **AI Algorithm** and the related parameters make up the **AI Model**.

Most metaphors can be stretched too far, but this one has one more gift to help us understand AI. Let us consider the role of the humble potato (the "data") in this allegorical story... Could you have succeeded in *any* way to make your own potato salad if you had no potatoes? After all, you're a great cook, you have all the right kit, the perfect recipe and you really really want a potato salad ... The answer of course is no, if you were short of some of the other optional ingredients say capers or even if you had limited kitchen equipment you could probably make a less delicious potato salad but just not without potatoes. I know this seems obvious, but many an AI project, blessed with a clear business problem, equipped with all the latest gear and skilled technicians and a really strong desire to build an AI to meet their demand ... has ended in exasperation because they forgot that all of that the importance of the data completely outweighs the efforts – this is the opposite of a classic IT project. Having the right data, and enough of it, is the immutable core of all AI, there is no way around it.

drop them from different heights in order to see if they break or not. After a day of dropping stuff, the minions return with the data shown in Table 3.1. We know this is a more expensive option, *but remember, we are imagining here! But, the realities of data collection (see Chapter 7) to build AI models for business are not that different.*

If we have this data, can we predict if a particular item of specific type and mass, when dropped from a particular height, will break or not?

UNDERSTANDING THE OBJECT DROPPING DATA

Since we do not have the minions, or the objects, or the resources to perform the real experiment, we wrote a simple simulator to generate data for our worked examples. For each input record, the simulation randomly picks one of five integer values in the range {1,5} for height in metres, one of 50 integer values in the range {1,50} for the mass in kilograms, and one of the three object types {Fragile, Medium, Hard} and calculates the force of impact using the equations in the vignette in this chapter on "Physics Model Behind the Object Dropping Problem". Depending on the breaking thresholds F_b (Fragile) = 750 Newtons,

TABLE 3.1 Example of Data Needed to Train the Models

Height (m)	Mass (kg)	Type	Outcome
3	16	Hard	Unbroken
2	14	Fragile	Broken
5	42	Fragile	Broken
3	34	Fragile	Broken
4	45	Medium	Broken
3	2	Fragile	Unbroken
...	...	...	...
4	50	Hard	Broken

F_b (Medium) = 3000 Newtons and F_b (Hard) = 4000 Newtons, 'Broken' or 'Unbroken' labels are created for each record. We created 1000 records using this approach to serve as a labelled training dataset to create the Supervised ML models. The resulting data consisted of 330 Fragile Objects, 168 Medium Objects and 502 Hard Objects.

In generating simulated data, we are of course revealing the fact that we understand the physics of this problem well enough to be able to define a proper mathematical model. As such, we should ask ourselves why we need to use ML. The answer in this case is simply for educational purposes, and later in this chapter, we will compare conventional, mathematical analysis with the use of ML. As a general rule, however, always be cautious in using ML if it is possible to solve the problem using more conventional scientific means.

Visualising the Data

It helps to visualise the data we just created to get some understanding of what we are asking models to do. Figure 3.2 is a parallel coordinate plot of the three input variables for the 1000 records. For the numerical variables (i.e. Height and Mass), the vertical axes represent the full range of values. Each record is represented by a line connecting the three values on the three axes. The line is in Blue if the object is 'Broken' or in Red if the object is 'Unbroken'.

Even a casual look at Figure 3.2 reveals a few interesting observations: (i) Fragile objects of larger masses dropped from higher distances break at a larger percentage. (ii) Medium objects of large masses break more than the smaller masses. (iii) Hard objects of smaller masses do not break at all.

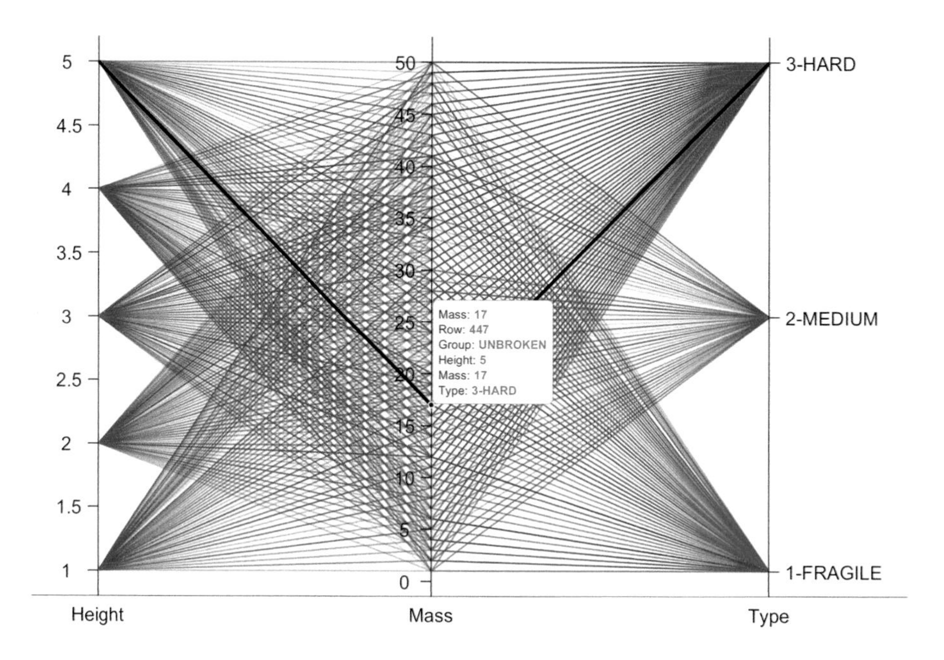

FIGURE 3.2 A parallel coordinate plot to visualise the training data for the object dropping problem. Each record is represented by a line connecting the three values on the three axes. The grey box in the middle gives the data behind the record shown as a black line representing a hard object of mass 17 kg dropped from 5 m that did not break.

FOUR MODELS TO PREDICT OBJECT BREAKAGE

Next, we embark on a discussion of four different models to use the data described above to predict the breakage of falling objects. We will describe the models in detail, discuss the pros and cons and then compare these approaches. There are older references [6,7] that give a broad introduction to various AI techniques. Newer references [8–10, and many other dozens of books and numerous web sites] focus on ANNs as the primary method. Before we delve into the detailed discussion, *remember that all of them are algorithms to automate the task of predicting 'breakage' in our fictitious warehouse. They are all doing the same task of mapping inputs onto outputs!*

Lookup Table

The simplest form of AI would be to create a huge lookup table with every possible combination of inputs to the system together with the desired outputs. The application task will then be to simply take any set of input data and select the corresponding output in the table. The lookup table would be the algorithm used by the AI task.

In practice, generating a lookup table is not a sensible approach for two fundamental reasons. First, we don't always have example data for every possible combination of inputs. Second, even if we *did* have every possible combination of inputs, the table would be massive. Imagine creating such a lookup table for a driverless vehicle and trying to define, obtain and manage every possible combination of input data covering different weather conditions, road configurations and the actions of other drivers, pedestrians and animals.

Human beings do not operate using lookup tables! We learn from our experiences and develop generalised *Models*, which we then apply to infer future decisions under new circumstances. Similarly, in AI, we need to develop general *Models* that perform the specific AI task to make the right decisions even when it has not seen the exact data before. We're going to start by looking at two different ML approaches. We're going to look at how both a Neural Network, and a rules system can each be generated using ML algorithms.

ML via Neural Networks

As we're going to dig a bit deeper into ML, we may as well start with what is perhaps the most inspirational sub domain of AI ... Neural Networks! They are, as the term itself suggests, based on the brain itself. It is, therefore, not surprising that *Algorithm Addicts* are so excited by their potential. Neural networks are made up of a number of artificial neurons that are intended to operate in a manner similar to the neurons in our brains.

Figure 3.3 is the simple neural network used to solve our object dropping problem consisting of an input layer (five neurons), a hidden layer (three neurons) and an output layer (two neurons). Each neuron is connected to other neurons through a series of connections which have weights. When different sets of data are fed into the input layer, the different weighted connections cause different neurons in the hidden layer to fire (or not). The outputs of these neurons in turn cause other neurons in the output layer to fire, generating an output.

Figure 3.4 illustrates an example for the inputs of 3 (Height), 16 (Mass) and "Hard" (Type) have generated an output of 'Unbroken'. This happens to be the right output based on Table 3.1, but the neural network can make mistakes in this process. Since the algorithm

NEURAL NETWORK MODEL

The purpose of this vignette is to explain how a neural network can be constructed to solve the object dropping problem.

The first stage in the process is to define the structure of the neural network. In general, this is not a simple task as understanding the optimum structure for a given problem requires extensive experimentation. For our problem, we need five neurons in the input layer representing the five inputs (see below for an explanation) and two neurons in the output layer representing the two outputs (Broken and Unbroken). To keep the number of parameters low, we chose just one 'hidden' layer but experimented with 3, 6 and 12 neurons in the hidden layer (requiring 26, 50 and 98 parameters, respectively). Even though we have simulated data, we deliberately limited the training data size to 1000 observations. Amazingly, a neural network with just three neurons in the hidden layer (Figure 3.3) was able to provide high accuracy.

You may be wondering how we got to five inputs instead of just three (Height, Mass and Type). This is because Type is a categorical variable with three values, Fragile, Medium and Hard and they need to be converted into numerical variables for the calculation. So, three values for Type are made into three binary variables with 0 or 1 values, making a total of five variables. As an example, a 'Hard' object will be represented in the input by {Hard=1, Fragile=0, Medium=0}.

Here is an explanation for calculating the number of parameters in the model for three neurons in the hidden layer: There are 15 (=5×3) connections between the input layer & the hidden layer and 6 (=3×2) connections between the hidden layer & the output layer. Each of these 21 connections carries a weight parameter. In addition, there is a bias parameter at each of the five summing neurons (three in the hidden layer and two in the output layer), for a total of 26 parameters.

In addition to defining the neural network structure, we need to decide which activation function to use for each neuron. An activation function takes the weighted sum of all inputs and generates an output. In the simplest network, the activation function is a step function that triggers if the weighted sum of the inputs exceeds a threshold. Alternative activation functions include linear or sigmoid functions (see the Figure below). For such functions, it is possible to adjust the steepness of the function. A key point to understand is that the bias parameter mentioned earlier effectively determines the threshold at which the function triggers.

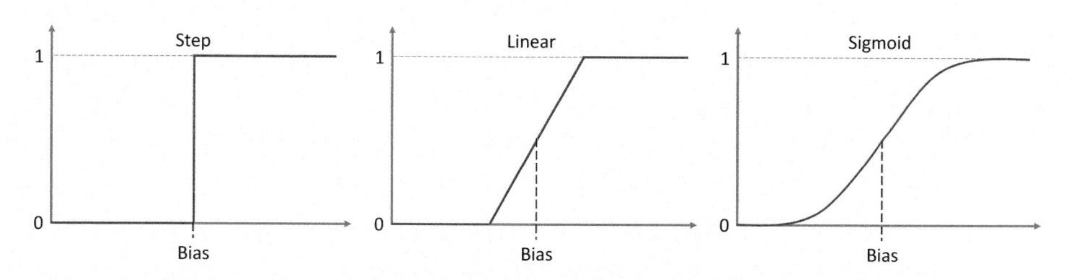

FIGURE A: Examples of neuron activation functions, left to right, Step, Linear and Sigmoid.

When input data is presented to the neural network, each neuron calculates the weighted sum of the inputs, applies the activation function and generates an output. This output is then fed forward into neurons in the subsequent layer. In our example, a 'broken' or 'unbroken' classification occurs when the output of the 'broken' or 'unbroken' nodes exceed 0.5. There is of course the possibility that both nodes could exceed 0.5 and, for complex problems, a strategy to handle such scenarios may be required. Such a strategy could include simply choosing the classification of the neuron that fired with the highest value. For our simple problem, we didn't have to worry about this as it never happened.

At the start of the training process, the weight and bias values are randomised. In order to train the network, each record in the training set is presented to the network in sequence. The output is calculated for each input record and the error between the actual output and the desired input is calculated. This error is often referred to as the loss. A ML algorithm back propagates the error through the network and adjusts the weight and bias values for each neuron.

There are many different strategies for calculating the error and many different ML algorithms. There are further parameters that can be used to define the behaviour of the learning; for example, the proportion by which weights and bias are changed. For experts in the field, we calculated the loss using a binary cross entropy function with a batch size = 10 and we used stochastic gradient descent for the ML. We trained our network with 800 records and use four-fold cross validation. Once trained, we evaluated the model using the test set of 200 records.

The performance of the network is shown in Table 1 below.

TABLE 1 Comparison of neural net model performance for training and test data sets.

Training Set			**Test Set**		
	Model Output			**Model Output**	
Actual	**Broken**	**Unbroken**	**Actual**	**Broken**	**Unbroken**
Broken	236	10	Broken	65	5
Unbroken	3	551	Unbroken	3	127
Overall accuracy	98.4%		Overall accuracy	96.0%	

Kai Mumford

Kai Mumford was born and raised in the south of England. He is currently an IT Degree Apprentice at IBM Client Innovation Centre at the Hursley Park in the UK. He is currently finishing his final year of his bachelor's degree at De Montfort University with plans for graduate school.

has the 'ground truth' (i.e. right answers) because it has labelled data, it can calculate the error (called 'loss') in the model output. One of the key steps in the training process, called 'back propagation', helps to minimise the error systematically by adjusting the underlying parameters of the model in an efficient manner. The key word here is 'minimise', to indicate that the model may still produce errors in the output, given the quality of the data and level of tuning of the neural network structure and the parameters chosen in the end.

We refer to the 'Neural Network' vignette for more details on the model for the object dropping problem. The model was trained with 800 records and evaluated using 200 'hold out' records that were not used in training. The overall accuracy was 98.4% for the training data and 96% for the test data. This is very impressive for such a simple model and shows the predictive power of the technique.

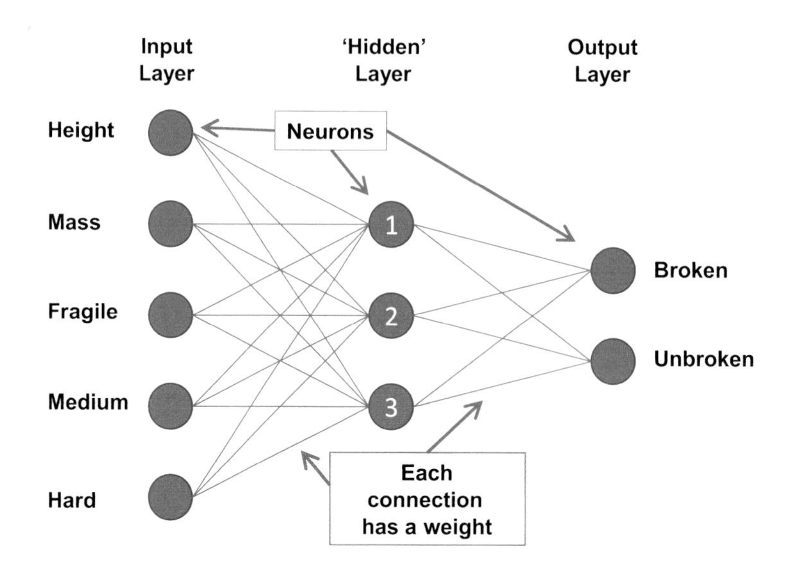

FIGURE 3.3 Description of key concepts for our Artificial Neural Network model.

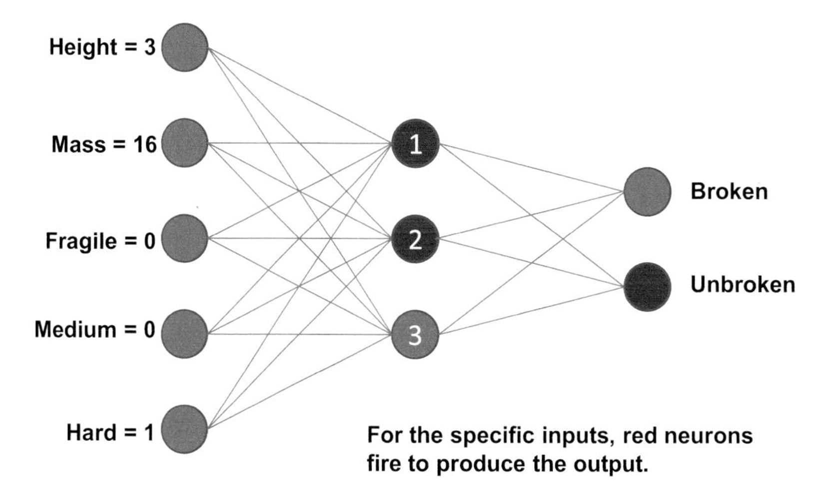

FIGURE 3.4 Illustration of the neural network response to inputs for the case height = 3 m, mass = 16 kg for a hard object to produce an output 'unbroken'.

Overfitting is a common concern with neural network models, when too many model parameters are fitted with too few training data. This is the consequence when the algorithm finds a way to represent all the inputs and the corresponding outputs present in the training data (i.e. 'learning the deck'), rather than creating a generalised model that can learn to respond to input data not previously seen during training. In this scenario, models will yield high accuracy during the training phase leading to a false sense of victory, only to fail badly during the 'test' phase or actual deployment.

In this very simple explanation, the *Model* comprises the configuration of the network (e.g. how many neurons in each layer, number of layers, etc.) and the model parameters (e.g. values of the weights connecting the neurons). Obviously, the model we have described is very simple for the object dropping problem. Many neural networks for image recognition or Natural Language Processing (NLP) use dozens or hundreds of layers with many neurons in each, resulting in millions of parameters. All these properties represent the *Model* and need to be configured and tuned for optimal performance.

ML for Rules Inference

Contrary to popular belief, training a Neural Network is not the only form of ML. Remember the Neural Network itself is just an algorithm that applies an AI Model to a particular task. ML is the technique used to define the *AI Model*. There are many other algorithms that could be applied. One that was extremely popular in days gone past was of course a rules-based system. Rules-based systems generated massive excitement in their day as we naturally assumed that all types of intelligent decision-making could be described in rules. The great thing about a rules-based system is that it can explain its reasoning. When a rule fires, it is easy to understand what caused it to fire!

Rules-based techniques were heavily used to deliver what were known as *Expert Systems* and became unfashionable at the start of the second AI winter in the late 1980s. The main reason for their loss of popularity was that defining and managing large rules-based systems is difficult. This in turn will cause many new to the field of AI to dismiss the use of rules-based systems without really understanding how they work and when they should be applied. It is not uncommon to hear people suggest that ML is an alternative to rules-based systems. This is false, as it is possible to develop a rules-based system using a ML algorithm.

Ross Quinlan [11] developed a series of ML algorithms known as ID3, C4.5 and C5 that enable the construction of decision trees and rules from a training set of data. The first step was to construct a decision tree, by treating the training set as a big bucket of data and then splitting it into smaller buckets according to some parametric test. For each possible parametric test, the algorithm calculated the information gain in terms of classification. Information gain measures how effective a parametric test is in performing the classification. After evaluating all the possible ways of dividing up the data, the algorithm selects the test that was most successful in dividing the data up into the correct output buckets. The algorithm then runs again on each subset of the data until the data has been successfully divided into the right outputs.

Having generated a decision tree, the C5 algorithm converts the decision tree into a set of human-readable rules. The resulting rule set is not identical to the original decision tree, as it is generalised slightly to avoid being too specific (aka overfitting). "C5 Rule-Based Model" vignette provides a more detailed description of how the rules are generated. These rules are of course the *Model* for our rules-based AI task.

The model was trained using the same 800 records and evaluated using the same 200 'hold out' records as in the neural network example above. The overall accuracy was 99.2% for the training data and 97.5% for the test data. This again is very impressive.

Discipline-Specific Model

In general, if the problem we want to solve belongs to a specific discipline, such as physics, chemistry, economics, operations research, control theory and electrical engineering

C5 RULE-BASED MODEL

C5 is a ML system that generates rules from training data. The algorithm was developed by Ross Quinlan [1] and offers the major advantage of generating explainable ML models.

At a very simplistic level, C5 operates by generating a decision tree and then converting that decision tree into a set of rules. The decision tree is constructed by examining each potential parametric test for each of the input features. In our case study, the algorithm would iterate through all the possible tests for *height* (e.g. height>1, height>2, etc.) and then all the possible tests for *mass* and *type*. For each possible test, the Information Gain is calculated and the test with the highest Information Gain is selected to divide the training set into subsets. The algorithm is then applied recursively to the subsets in order to construct a complete decision tree.

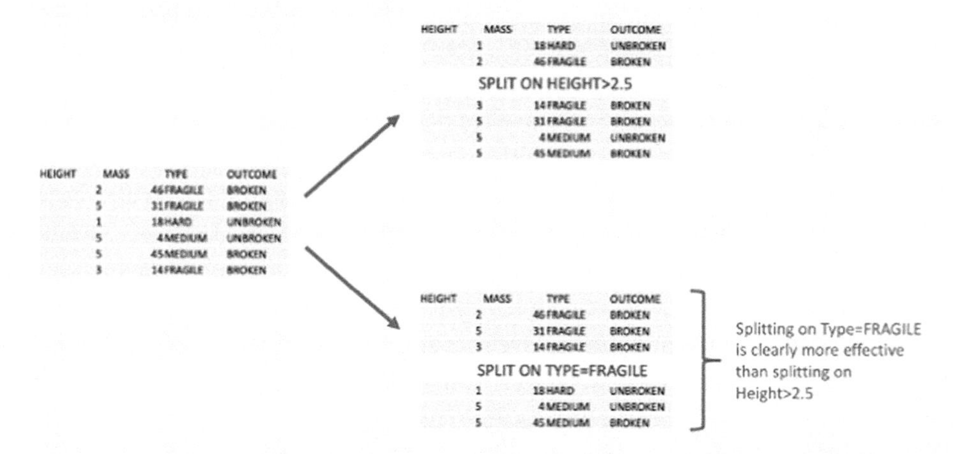

Once a decision tree has been constructed, the C5 algorithm converts the decision tree into a set of rules. In converting the decision tree into rules, there is a small element of generalisation as a pruning algorithm is used to decide if aspects of the tree are too specific. The resulting rule set can then be applied to previously unseen data. A further point to note is that the rules are applied in sequence and the rule that fires first in the sequence is taken as the output decision.

Taking the same 800 record training data used for the neural network deep dive, we used C5 to generate the following rules.

RULE 1
IF Height > 3
AND Mass > 45
THEN BROKEN [0.966]

RULE 2
IF Height > 2
AND Mass > 39
AND Type = MEDIUM
THEN BROKEN [0.950]

RULE 3
IF Height > 4
AND Mass > 39
THEN BROKEN [0.929]

RULE 4
IF Type = FRAGILE
THEN BROKEN [0.754]

RULE 5
IF Mass <= 8
THEN UNBROKEN [0.992]

RULE 6
IF Height <=3
AND Mass <= 11
THEN UNBROKEN [0.989]

RULE 7
IF Height <= 1
AND Mass <= 15
THEN UNBROKEN [0.981]

RULE 8
IF Type IN { MEDIUM, HARD }
THEN UNBROKEN [0.909]

We then tested these rules using the same 200 record test set that was used in developing the neural network model. The performance of the C5 rules against both the training and test sets is shown below.

TABLE 1 Comparison of C5 Rule-Based Model for the Training and Test Data Sets.

Training Set			**Test Set**		
	Rule Output			**Rule Output**	
Actual	**Broken**	**Unbroken**	**Actual**	**Broken**	**Unbroken**
Broken	242	4	Broken	65	0
Unbroken	2	552	Unbroken	5	130
Overall Accuracy	99.2%		Overall Accuracy	97.5%	

- Roy Hepper

Roy Hepper has worked in information technology for over 30 years on wide a variety of projects – including using formal methods, knowledge representation techniques, statistical learning and ML/AI. He continues to work on information architecture projects, applying appropriate mixtures of conventional software architecture, design and programming, augmented with advanced analytics & other ML/AI techniques.

REFERENCE

1. J. R. Quinlan, *C4.5: Programs for Machine Learning* Morgan Kaufmann Series in Machine Learning, (1992).

invariably, the solution involves a good understanding of the discipline to develop the insightful models and do the relevant mathematics. Obviously, our object dropping problem belongs to physics. The key idea is that the knowledge of the discipline helps to formulate the problem better and leads to a better understanding of the problem and the solution than just doing blind data science of finding relationships between elements in the data.

To emphasise this point and having seen how ML can be used to solve a problem, let's go back to basics and try a conventional approach to solving the same problem. Let's try ... don't be scared ... please keep reading ... we'll look after you ... we promise ... some good old-fashioned mathematics and physics.

Why do we need to do this? Well ... it is to help you understand what you are asking ML to do. Such an understanding will help you define problems that are solvable, rather than just blundering into a project and hoping for the best. That is a useful skill to develop as it will help you assess the feasibility of using ML in other potential applications.

So, how would an old-fashioned engineer solve this problem?

First, an engineer would use their extensive knowledge of physics (because all engineers love physics) to understand what was happening. Basic physics tells us that the higher you drop an item from, the faster it will be travelling when it hits the ground. The velocity and the mass will determine the impulse experienced when an item hits a hard surface. The time for the object to come to rest on the ground and the impulse determines the force with which an object hits the ground (Newton's second law). Heavy objects

INFORMATION GAIN

Information Gain is used to determine if dividing a set of data into subsets increases our ability to predict the class of an item in each subset.

Imagine a set of data comprising two output classes referred to as Class A and Class B; for simplicity, let's assume the data set contains 50 items of class A and 50 items of class B.

Now consider two parametric tests that each divide the data into two subsets.

The composition of the subsets resulting from each of the parametric tests is shown below:

TABLE 1 Examples of Partitioning Data. Parametric Test 2 Has More Information than Parameter Test 1.

	Parametric Test 1			Parametric Test 2	
	Subset 1	Subset 2		Subset 1	Subset 2
Class A	25	25	Class A	5	45
Class B	25	25	Class B	45	5

As you can see, Parametric Test 1 has taken a set of data with a 50:50 split in the classes and created two subsets that each still has a 50:50 split. Conversely, Parametric Test 2 has resulted in two subsets comprising a 90:10 split.

In this simple example, the Parametric Test 2 has clearly provided an Information Gain as, by applying the test, we can more easily predict the class of an item in the resulting subsets.

Information Gain can be mathematically calculated for much more complex data sets comprising many more classes. It is a highly valuable tool in comparing different parametric tests to understand which is more effective in dividing a data set.

hit the ground much harder than light objects at the same velocity. Newton's third law explains that the force exerted by the object hitting the ground is equal and opposite to the force exerted by the ground on the object. Finally, we know that the greater the force applied to an object, the more likely it is to break. We can use some structural mechanics and material science to figure out if the observed force is enough to break the object. In its absence, we can use the experimental data to determine the thresholds at which *Fragile, Medium and Hard* objects break. Then, we have the whole story. For those who can't get enough of the physics, we've shown the details in the "Physics Model Behind the Object Dropping Problem" vignette.

Another example of a discipline-specific model is described in the explanation of rise and fall of tides in Chapter 7, which involves the knowledge of astronomy behind the motions of the sun, the moon and the earth.

Summary of the Four Models

Now that we discussed four different approaches to solve our object breaking prediction problem, it is worthwhile spending a little time to discuss their relative merits. Table 3.2 gives a summary in six practical aspects for the comparison. It is clear that discipline-specific

TABLE 3.2 Comparison across the Four Algorithms

Attribute	Lookup Table	ML – Neural Networks	ML – Rule-Based	Discipline-Specific Model
Advanced skill needed	None	Data science	Data science	Knowledge of the discipline
Ability to generalise in the target domain	No	Yes, but needs care to avoid overfitting	Yes, but needs care to avoid overfitting	Yes
Ability to transfer and apply to a similar problem in a different context.	No	Only if the domains are very similar.	Yes, but requires modification.	Yes, if abstracted properly.
Transparency/ explainability	High	Low	High	High
Accuracy	Subjected to errors in the input data	Some errors are inevitable	Some errors are inevitable	Can be exact due to deep knowledge
Maintainability	Needs data	Needs constant monitoring and regular updates	Easier to evaluate the impact of changes in data as we know which rules will be affected.	Easier due to established concepts.

modelling is superior to the other options in terms of the overall assessment, while requiring the deep expertise in the discipline.

COMPARING THE TWO ML APPROACHES

We used two different ML algorithms (i.e. Neural network and Rule-based) to develop two very different types of AI Models. Both are perfectly valid and deciding which one to use will depend on several factors.

First, there is the accuracy of the decision-making. Some approaches are better than others at certain types of problems. In our object dropping example, rule-based algorithm did slightly better than the neural network approach.

Second, there is the type of output. In our example, the output is a class (e.g. *Broken* or *Unbroken*). As we showed, both neural network algorithm and the C5 rule-based algorithm are suitable for this purpose. Instead, if we are looking to predict the share price of a company (i.e. a continuous numerical value), we can use a neural network or a regression tree algorithm [12]. Regression trees use a decision tree similar to C5 algorithm but can yield a numerical output and are also explainable as a consequence. In Chapter 6, we have a vignette with an example of using regression tree for predicting body weight from height.

Finally, we may need to consider other factors such as explainability. The Neural Network cannot explain its reasoning and we have no idea why it made its decision. Conversely, the rules-based system is completely explainable and it's clear why every decision was made. Explainability is a huge issue for Neural Networks as it is a key limitation of their practical application. As a consequence, there is a great deal of research underway into methods of understanding the decision-making of a Neural Network. In addition, there are other

PHYSICS MODEL BEHIND THE OBJECT BREAKING PROBLEM

In this vignette, we explore the physics underlying the data we are going to use for fitting AI models. Turns out this problem relates to the famous legend about Galileo Galilei dropping objects from the top of the leaning tower of Pisa circa 1590 AD and the subsequent development of the three laws of motion by Isaac Newton, described in "Philosophiæ Naturalis Principia Mathematica" (*Mathematical Principles of Natural Philosophy*), first published in 1687 AD. The basic problem is how to include gravity in the motion of an object near the earth's surface. We will use a more intuitive, non-calculus approach to understand the problem.

Background

We are dropping an object of mass, M (in kilograms) from rest at a height, d (in meters) from the ground. The goal is to calculate the force (F) exerted on the object when it hits the ground and relate that to the observation whether or not the object breaks. The composition of the object will decide if the force is adequate to break the object or not, a detailed understanding of which will take us into the disciplines of structural mechanics and material science. For now, we will assume that there are specific breaking thresholds of force F_b for each of object, when exceeded the object will break. In our example, we have assumed F_b(Fragile) = 750 N, F_b(Medium) = 3000 N and F_b(Hard) = 4000 Newtons. We also assume for simplicity that all the objects make contact with the ground for the same time interval, Δt = 0.1 seconds before they come to complete rest. In general, the calculation of Δt also needs a detailed understanding of the momentum transfer between the object and the ground. The acceleration due to gravity at the earth's surface is generally indicated by the letter g and is decided by the mass and radius of the earth and the universal gravitational constant G. Using the parameters for earth, $g = 9.8\,\text{m/s}^2$.

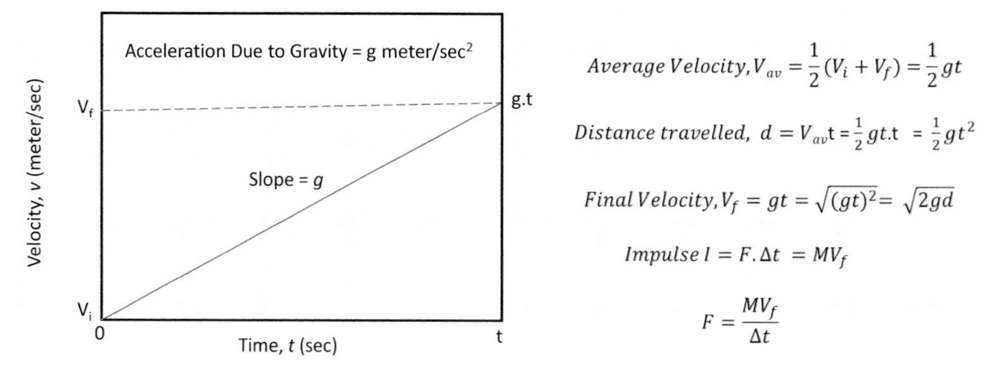

FIGURE A: The mathematics of the object dropping problem. Details are explained in text.

The Physical Model

Since the acceleration g gives the rate of change of velocity and it is a constant in this problem, it gives the slope of the curve (Figure A), relating velocity and time. Thus, starting from rest (V_i=0), velocity of the object increases linearly with time, t, to give the final velocity before impact $V_f = gt$. To calculate the distance travelled, d, we calculate the average velocity V_{av} and multiply by t. Using some basic algebra shown, we arrive at the final velocity, $V_f = \sqrt{2gd}$ as a function of the distance travelled. To calculate the force F exerted on the object by impact, we need two additional concepts: Newton's second law, change in momentum of the object due to the impact is, Impulse $I=MV_f=\Delta F$ and Newton's third law, the force exerted by the object on the ground is equal to the force exerted by the ground on the object, which unfortunately leads to the breakage. If $F>F_b$, the object will break and $F<F_b$, the object will not break. It is worth noting that there are only five parameters in the physics model: acceleration due to gravity, three force thresholds for object breakage and the time for the object to stop after hitting the ground.

Clearly, since this model is based on first principles, it can explain observations anywhere in the universe and more than 330 years of empirical validation of the underlying physics confirms the validity of the model.

forms of explainability that should be considered, and these are discussed in more detail in Chapter 5.

To repeat the same old message ... there are many different AI Algorithms and it's important to select the right one for the right job. To do that, you need to avoid *Algorithm Addiction* and look at the broader aspects of an AI project. You need to consider how accurate the AI needs to be. You need to understand how you will evaluate its accuracy. You need to consider the qualitative issues such as explainability and how important they are to your business. You will usually benefit from testing more than one algorithm against your data, as idiosyncrasies in your data may well favour one algorithm over its competitors. Most importantly, you need to avoid being blinded by the belief that one algorithm alone can magically make sense of your data!

ALGORITHM AWARENESS MATTERS

The importance of a basic understanding of AI algorithms was brought home to me in a Client meeting. We had just demonstrated a highly effective system that delivered everything the Client required. Despite the effectiveness demonstrated, the Client's most senior and experienced engineer said, "Unfortunately we can't use this system. We need a DL system because it can be trained from real world data as opposed to a rules system".

This statement revealed that even the most experienced engineers can be caught up in the media hype surrounding AI. Firstly, DL is not a "system"; it's an algorithm used to define and optimise DNNs. Secondly, and most importantly, it is wrong to say that rules-based systems cannot be generated by ML algorithms. Quinlan's C5 is an excellent example of using ML to generate human-readable rules.

What really frightened me though was not the accuracy of the engineer's position but the strength of his belief in what he just said. This engineer was making an important, multi-million-pound decision based on a lack of knowledge and incorrect assumptions, yet he was convinced of his own expertise. It wasn't that he didn't know that frightened me ... it was the fact that he didn't know that he didn't know!

James Luke

COMPARING PHYSICS MODEL WITH ML

So, what are the advantages of the conventional physics modelling versus ML?

At the basic level, the physics model has only *five parameters* with clear connections to the underlying physics. The rule-based model has more than a dozen parameters for defining the rules applied to the three variables. Neural network model has *a few dozen parameters* to define and train the network. This leads to an interesting observation. Is our understanding of the physical world over the past few hundred years making it possible to describe the natural phenomenon with just a few parameters?

The conventional physics approach is end-to-end complete and backed up by indisputable science. More importantly, we can use it in different situations. We can take our mathematical model to Mars and apply it there even though acceleration under gravity is different on Mars (it is actually 3.71 m/s^2 compared to Earth's 9.81 m/s^2 [13]). If we wished to add new output categories such as *Very Fragile* and *Very Hard*, we could easily do so with minor modifications to the model. We can also develop our mathematical solution with very little new experimental data since many of the underlying principles are established over decades, even centuries. We may want to confirm the expectations by some carefully selected experiments, just to be sure. Ultimately, we could even go to court and defend the behaviour of our solution as it is based on proven scientific analysis and engineering.

What about our ML solutions? The rules-based system has the advantage that it is at least explainable and we can evaluate the rules to understand whether they make sense. The big advantage of the ML solutions is that we did not need to perform the mathematical analysis! ML is most effective in situations where it either isn't possible or isn't practical to perform this type of mathematical analysis. Mathematical analysis may not be possible simply because the system is too complex to understand. Alternatively, it may be possible to conduct the analysis but not efficient to do so. Good examples of this are image classification tasks such as identifying a person from a million photographs or speech to text or speaker identification. These are tasks where the mathematics of the analysis are too complex, so a ML approach offers a solution. Or calculating the daily energy generation levels for a power station, factors such as hundreds of weather parameters, time of year and the previous days' performance figures can all go into such models, they are entirely tractable by traditional mathematical methods except there's not enough time in the day to do them and data scientists are expensive. ML models are much more practical to deploy in this time pressured environment. There is a new evolving area of research that applies ML techniques to solve physics problems more efficiently. As an example, a recent paper by Navratil et al. [14] discusses how ML models can accelerate physics-based simulations of oil reservoirs by orders of magnitude.

In terms of understanding Neural Networks, the mathematical analysis of our object dropping problem gives us a unique opportunity. We now know exactly what we want the configuration of neurons and weights to learn. We know the actual *Model*. So, for our examples, it will be possible to see how different neural network models affect the error rate.

WHAT ARE THE ML ALGORITHMS ACTUALLY LEARNING?

In the case of our rules-based example, the ML is generating a set of rules that describe the underlying model represented by the data in the lookup table. They describe what happens to objects of different types and sizes when dropped from different heights. The rules generated, in this case by the C5 algorithm, are actually generalisation of the lookup table mentioned earlier (Figure 3.5).

So, the C5 ML algorithm generates a set of rules that describe the mapping of inputs onto outputs. But, the rules do not calculate parameters such as acceleration or force. In this respect, the rules describe what is happening, but the rules do not include any understanding of why it happens.

What about the Neural Network? What is it learning? There are at least two possibilities.

The first possibility is that the Neural Network learns exactly what the rules-based approach learnt; a mapping between inputs and outputs.

The second possibility is that the Neural Network actually learnt the underlying mathematics, and that the formulae we used in our mathematical modelling are potentially encoded in the *Model* of the Neural Network. To understand how this is possible, it's important to remember that even the most complex mathematical algorithms are built on the most basic mathematical operations of addition, subtraction, multiplication and division. In fact, multiplication and division are simply extensions of addition and subtraction. As simple Neural Networks can perform these basic operations, then clusters of neurons can perform more complex mathematical operations. It is theoretically possible

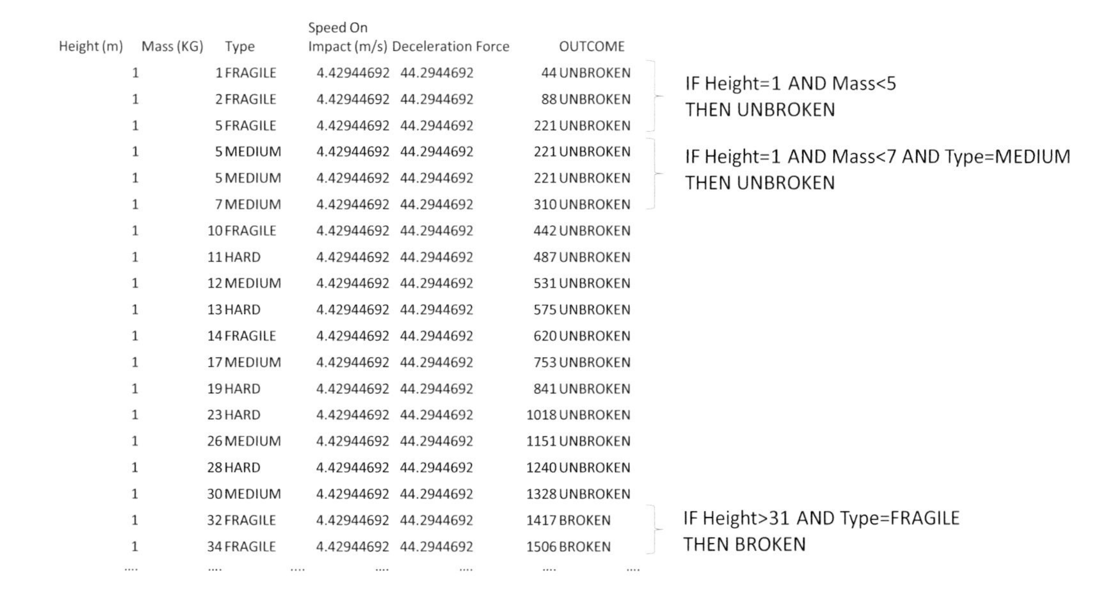

Height (m)	Mass (KG)	Type	Speed On Impact (m/s)	Deceleration Force		OUTCOME
1	1	FRAGILE	4.42944692	44.2944692	44	UNBROKEN
1	2	FRAGILE	4.42944692	44.2944692	88	UNBROKEN
1	5	FRAGILE	4.42944692	44.2944692	221	UNBROKEN
1	5	MEDIUM	4.42944692	44.2944692	221	UNBROKEN
1	5	MEDIUM	4.42944692	44.2944692	221	UNBROKEN
1	7	MEDIUM	4.42944692	44.2944692	310	UNBROKEN
1	10	FRAGILE	4.42944692	44.2944692	442	UNBROKEN
1	11	HARD	4.42944692	44.2944692	487	UNBROKEN
1	12	MEDIUM	4.42944692	44.2944692	531	UNBROKEN
1	13	HARD	4.42944692	44.2944692	575	UNBROKEN
1	14	FRAGILE	4.42944692	44.2944692	620	UNBROKEN
1	17	MEDIUM	4.42944692	44.2944692	753	UNBROKEN
1	19	HARD	4.42944692	44.2944692	841	UNBROKEN
1	23	HARD	4.42944692	44.2944692	1018	UNBROKEN
1	26	MEDIUM	4.42944692	44.2944692	1151	UNBROKEN
1	28	HARD	4.42944692	44.2944692	1240	UNBROKEN
1	30	MEDIUM	4.42944692	44.2944692	1328	UNBROKEN
1	32	FRAGILE	4.42944692	44.2944692	1417	BROKEN
1	34	FRAGILE	4.42944692	44.2944692	1506	BROKEN
....	....	....	....	....	....	....

FIGURE 3.5 Examples of generating rules from data tables.

UNSUPERVISED ML – CLUSTERING ANALYSIS

This is a good opportunity for us to introduce Unsupervised ML to our readers. Clustering is a popular example of unsupervised learning where the machine does not need the output labels (i.e. Broken or Unbroken in our object dropping problem) for the analysis. It provides a way to explore the structure of complex data in an automated manner to get some insights. The output from a clustering algorithm is not technically a prediction, although it is often spoken of as such when people are discussing AI. Let us consider a clustering algorithm in a medical setting to analyze hundreds of X-Ray pictures of cancerous tissues. The algorithm cannot predict cancer but merely places the input pictures into groups of most similar pictures. A cluster is collection of 'similar' records placed in a group. The goal of the algorithm is to segment the data into some number of clusters. Similarity is determined by a mathematical distance measure between the records. The algorithm minimises the distance between the records within a cluster and maximises the distance across clusters. Clustering is rarely used in isolation, because finding clusters is not an end in itself. Often it is used for downstream data mining tasks. We will use it in a similar fashion. We will create the clusters as the first step and then assess the implication of the clusters to predict the object breakage using the label information as the second step.

We will use the object dropping data described in Chapter 3 (just Height, Mass and the Object Type) to illustrate the technique. We choose a popular algorithm called '*k*-means clustering', where the user specifies the number of clusters (k) and the algorithm creates them out of the input data based on the principles outlined above. We use the software IBM SPSS Modeler 18.1.1 for the task with a choice of $k=4$ target clusters. Results of the clustering analysis are shown in rows A–D in the table below:

A Cluster	1	2	3	4
B Type	Fragile	Medium	Hard	Hard
Cluster size/total of type in data	330/330	168/168	212/502	290/502
C Mean height (m)	3.03	3.08	4.5	2.0
D Mean mass (kg)	24.10	24.83	26.42	26.58
E Our description of the cluster	All fragile objects	All medium objects	Hard objects at long distances	Hard objects at shorter distances
F Breakage percentage from the labelled data	77.9 %	17.3%	14.2%	0

Row A identifies the cluster numbers and the next three rows (B, C, D) represent the properties of each of the clusters. Row E is our "human" interpretation of the cluster based on these properties. Row F is the breakage percentage for each cluster **calculated from labeled test data not used in the clustering model creation process**. Columns 3 & 4 (row B) show that out of the 502 Hard objects in the data 212 ended up in cluster 3 and 290 in cluster 4.

Our cluster descriptions in Row E attempt to give a short summary of the data represented in the four clusters. In practice, this means using domain experts to examine the data in each cluster and determine what each *seems* to represent – this is not always straightforward, it's why you need domain experts in your team, especially if the algorithm identifies a previously unknown novel sub-group.

In our example, the SPSS Modeler also outputs a number between 0 and 1 to indicate the Prediction Importance of the input variables in the formation of the clusters.

The prediction importance numbers are 1.0, 0.33 and 0.01 for Type, Height and Mass, respectively, indicating that the Type of the object played the dominant role in creating the four clusters, followed by the Height variable, Mass playing practically no role. It is easy to see how the clusters are related to the breakage percentages in Row F, supporting the intuition observed in the data visualisation discussion in Chapter 3.

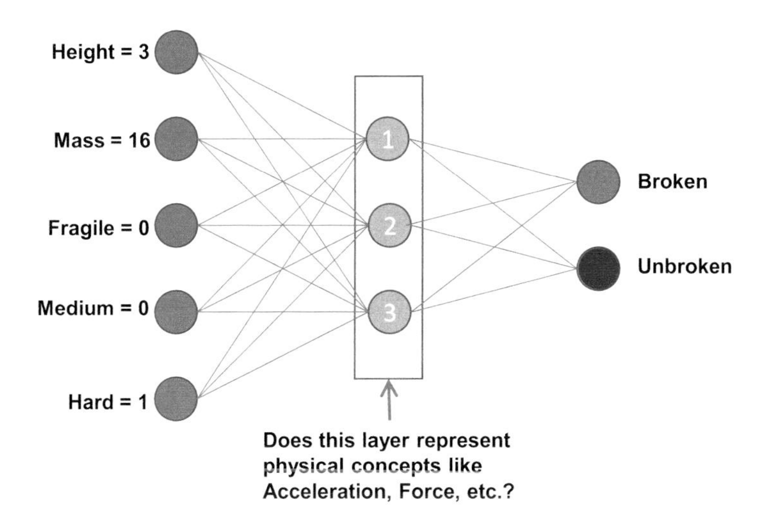

FIGURE 3.6 Imagining what the neural network may be doing; we really do not know.

that a Neural Network may have learnt to emulate the mathematical model that we derived manually (Figure 3.6).

So, which of these possibilities has happened in the Neural Network that we trained? The simple answer is that we don't know and there isn't really, at present, a way of knowing. We do know that parts of neural networks can learn to emulate complex mathematical functions. Unfortunately, the fact that a neural network could theoretically learn to calculate Acceleration and Force doesn't mean it actually has! We just don't know. This is one of the big disadvantages of Neural Networks. We can't, with current technology, fully understand what a neural network has actually learnt.

We can ensure that the neural network is safe to use in an operational scenario by ensuring it is properly tested with data that covers all possible scenarios. Rigorous testing will give us the confidence to use the application operationally.

FEATURE DEFINITION AND EXTRACTION

One thing we can do is to give the neural network a head start! If we know that the force of impact is important in determining whether an object is damaged, then why not change the input data to include force instead of height? The process of analysing a problem and deciding what to feed into an *AI Algorithm* is called *Feature Definition* whilst the process of generating the actual data in an operational solution is called *Feature Extraction*.

Feature Definition is hugely important in the domain of AI … it's also massively challenging.

In our simple example, it is intuitive that any *AI Algorithm* may stand a better chance of success if the input data contained Force instead of Height. Quite simply, the AI will have less to learn because it won't need to figure out that it needs to calculate this new feature called Force using the original feature of Height. The less the *AI Algorithm* has to learn, the better the chance of success.

But what about more complex *AI Applications*? What features are important when predicting the stock market or performing facial recognition?

If we have 50 years of stock market price data, we can *Define* and *Extract* a massive range of features. We can work out the 1-day average, the 7-day average, the 1-year average. We can calculate trends in relative performance between stocks. The list is endless and there are a great number of mathematicians, economists and scientists who invest an awful lot of time analysing stock market data in search of the magic formula that will earn them a fortune.

Hot tip ... if you see a book entitled *How to Predict the Stock Market*, it's highly unlikely the author was successful. The odds are that they've written the book in the hope of recovering their losses.

Face recognition is a really interesting AI challenge because it is a task that humans perform with apparent ease yet it has proved very challenging for AI. In the 1980s and 1990s, face recognition research focused on the use of (conventional) neural networks. The key question was what data to feed into those neural networks. Researchers looked at the features that could be extracted from bitmap images and then evaluated which of those features were most useful in enabling a neural network to learn to recognise a face (Figure 3.7). Typical features included measuring the distance between the subject's eyes, the height of a person's ear lobes relative to their eyes and the angle of a person's nose relative to the line between their eyes. These features were manually defined using conventional image analysis techniques and then fed into neural networks. This effort needed skills and was

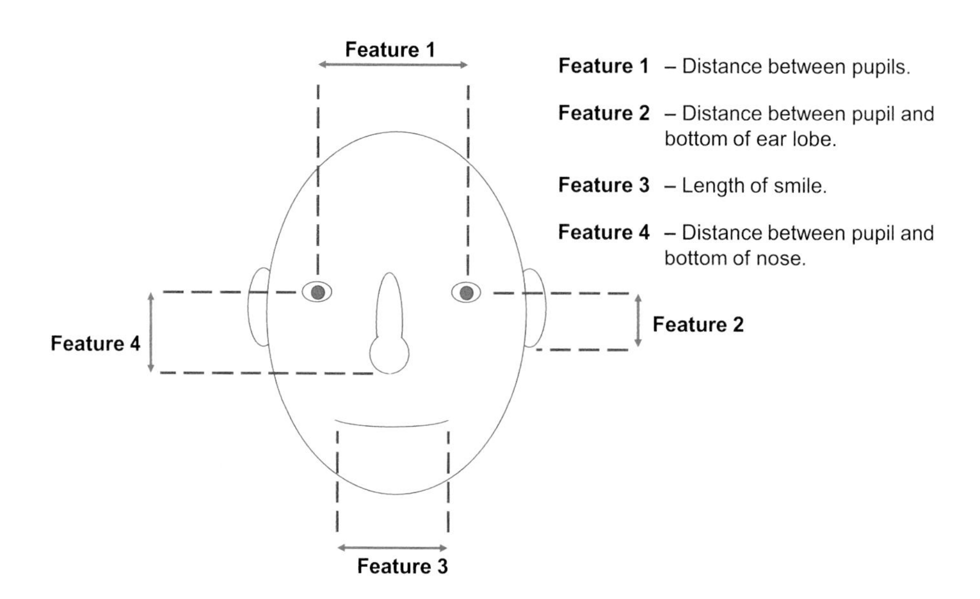

FIGURE 3.7 Examples of manual feature extraction from facial images.

labour intensive. Today, we tend to solve this problem differently using DNNs (see below). In cases when you do not have enough data, manual feature extraction still has a role to play.

REVENGE OF THE ARTIFICIAL NEURAL NETWORKS

ANNs had their origin in 1943 [15]! One of the challenges was the proper design of the neural network to match the task at hand, in terms of the number of neurons, their arrangement in the layers and the connectivity. Over the years, the technology evolved to create different architectures for the ANNs to match different tasks. Two architectures that are immensely popular today are Recurrent Neural Networks (RNNs) and Convolutional Neural Networks (CNN). RNN, initially proposed in 1986 [16], is a neural network for processing sequential data. Natural language, as text or speech, represents an excellent example of sequential data. CNN, invented in 1989 [17], is used for processing data that has a grid-like topology. The data contained in an image can be thought of a two-dimensional grid of pixels. These algorithms were well understood at the time, but their application to practice was limited by the computing power required to run them and lack of adequate training data.

With the advent of the internet age and the maturity of the IT infrastructure, the collection of large amounts of data became possible. The Moore's law advances in computing architecture i.e. Central Processing Units (CPUs), Graphics Processing Units (GPUs), on-board memory and storage, made creation and use of much larger neural networks practical. All these lead to the new age of Deep Neural Networks (DNN). A DNN is a Neural Network with many more hidden layers. In the 1990s, we were building Neural Networks with one, or sometimes two, hidden layers. We are now able to build networks with huge numbers of hidden layers. Consider the two networks in Figure 3.8. The conventional network on the left with just four neurons in the hidden layer has 26 parameters to adjust. One of the popular Deep CNNs for classifying 1000 objects from images had eight intermediate

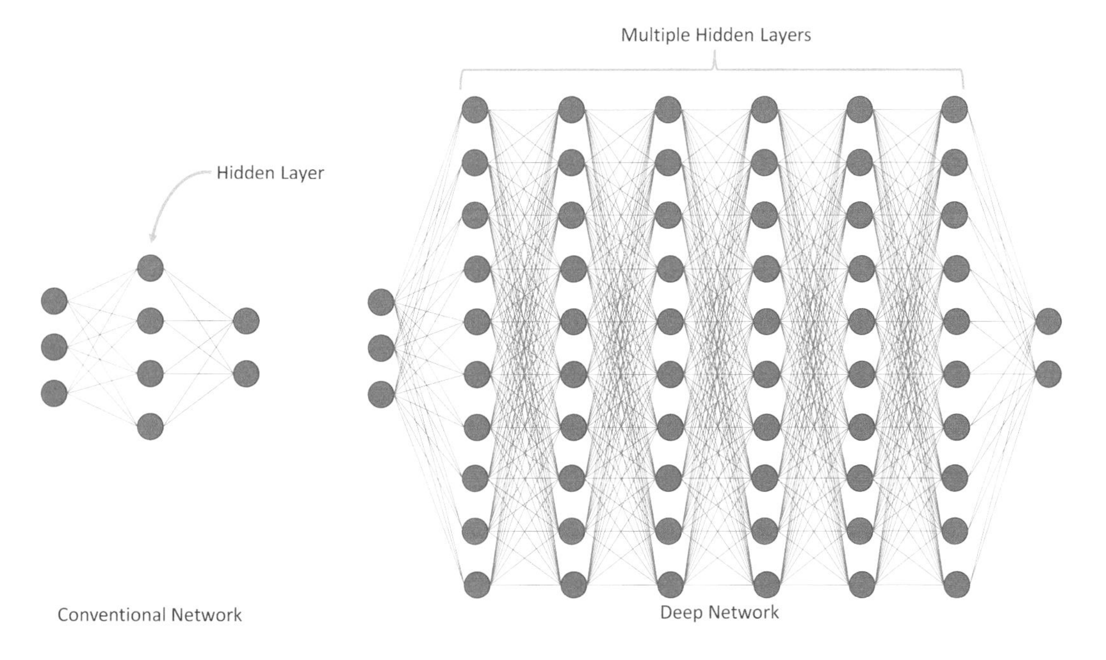

FIGURE 3.8 Comparison of a conventional neural network and a deep neural network.

layers, 650,000 neurons and 60 million parameters to optimise! [18]. This sheer number of parameters that need to be configured means that we need significant processing power and a massive volume of training data.

Whilst the size of DNN causes challenges in terms of processing power and training data volumes, they are considered to have one major advantage. There is less need to undertake manual feature extraction of the type described above and shown in Figure 3.7. In certain applications, the first few hidden layers appear to perform the feature extraction automatically. The extent to which this happens may depend on the type of feature extraction required. Analysis of DNN behaviour suggests that for applications such as facial recognition, the DNN is learning to identify the type of visual features described in Figure 3.7. However, does this mean that is always the case? If we were to apply DNN to a complex radar data challenge, would the DNN learn to perform the type of sophisticated signal processing currently designed by radar experts and embedded into the system design? The simple answer is that we just don't know at present. However, understanding what DNNs are capable of learning and how they can be taught to learn those features with much less data is the subject of considerable research right now.

HUMAN INTERPRETATION OF ARTIFICIAL NEURAL NETWORKS

Researchers at MIT [19] have managed to peer inside a Deep CNN for image recognition to understand what the intermediate layers of a network are actually learning and map it to human interpretations. This work finds some evidence to suggest that, during training, the layers of DNNs evolve to represent a hierarchy of concepts with each layer representing concepts that build on the concepts of the previous layer. The hierarchy begins with colours and works upwards through textures, materials, parts, objects and scenes. The authors conclude by saying that "the units of a deep representation are significantly more interpretable than expected". This is fascinating work and supports the hypothesis that DNNs are effective in discovering and automatically defining features. It is worth noting that in image recognition, the features are directly related to the contents and layout of pixels. Similarly, DNNs are also proving effective in areas such as speech recognition [20] where, again, the features are tightly coupled to the arrangement of the input data. Based on this research alone, we cannot assume that DNNs are automatically deriving complex mathematical features implied in more complex applications.

This is an incredibly exciting aspect of DNNs. In applications such as image classification or speech to text, feature definition was always a major challenge. The ability of a DNN to automatically learn features is massively powerful. However, it's really critical to understand that the key enabler for this process is the volume of data available. These techniques are proving most valuable in situations where the volumes of data are huge and our knowledge of how to define features is limited. Whether or not any of those promises are fulfilled depends on three factors: the complexity of the underlying problem, the volume of data available for training and the efficiency of the DL algorithm.

Let's start with *the complexity of the problem*. The object breakage problem is incredibly simple so it would not be unreasonable to expect a neural network to learn the mathematics of velocity, acceleration and force calculations. In the stock market example, the mathematics is massively more complex and we must question how realistic it is for a ML algorithm to discover the exact weights, and other configuration parameters, required to emulate the mathematics, without access to vast amounts of accurate training data. The same could be said of advanced signal processing or control theory problems. Human beings do not learn this level of mathematics just by example! We are taught mathematics at school and, even then, relatively few human beings are able to solve this type of problem. Is it realistic to expect a ML algorithm to construct such a mathematical model just by observation?

Which brings us to *the second factor, the volume of data*. In cases where there are massive volumes of data, then the chances of a network learning the real underlying model are greatly enhanced. The smaller the amount of available example data for training, the less likely the algorithm is to learn the model.

Finally, it depends on the *efficiency of the deep learning algorithm*. Given enough data, there is no reason why the AI should not learn a real-world model if you test every possible combination of weights and configuration parameters. The ML algorithm is effectively conducting a guided search for the right combination of parameters. Theoretically, with enough processing power and time, it would be possible to simply try every combination of weights and configurations parameters; after all we do joke that if we take an infinite number of monkeys and give each one a typewriter, one of them will type the complete works of Shakespeare. Clearly, trying every combination isn't an efficient strategy so DL algorithms are being designed to perform a guided search. Even so, they are still massively intensive in terms of computational requirements and, of course, the "guided" aspect of the search may mean that the optimum solution is not found.

The reason the search is so challenging is that the search space of a DNN is massive. It is no surprise therefore that the greatest advocates of DNNs and DL are web companies with massive amounts of processing power and access to massive volumes of training data. Applications such as image classification and speech recognition are perfect applications for the use of DL.

In situations where the volumes of data are much lower, these techniques may be more limited. In situations where there is already extensive scientific knowledge, such as signals processing and radar theory, that knowledge should be used. It doesn't make sense to rely on the hope that the ML may discover what we already know.

The object breakage problem we have described above is an incredibly simple problem and the mathematical analysis, whilst frightening for most, should be trivial for most engineers. An engineer, already familiar with gravity, and the likely relationship between higher velocities and propensity to break data, needs few examples to create a useful model. The human has an amazing head start over the algorithm which has literally no prior knowledge and can only learn by examples (lots of examples).

SO WHICH ALGORITHM IS BEST?

There is no best algorithm!

There are thousands of different algorithms and some are better suited to some problems than others. DNNs appear to be very effective at problems such as image analysis and speech recognition when massive volumes of training data are available. Conventional mathematical modelling, whether you consider it AI or not, is more appropriate when deep expertise about the problem exists and can be leveraged to produce a well-engineered solution.

Other factors need to be taken into account besides the accuracy of the actual solution. If explainability is an issue, then you may wish to consider a rules-based approach ... remember rules can be developed using ML as well!

From a strategic perspective, it is also important to ensure that you design your solution in a way that enables continual evaluation of different algorithms and products such that you can swap in and out as appropriate. That means that your programme needs to develop and maintain appropriate test data and the tools to automatically evaluate different capabilities.

The key message is that there is no single best algorithm! At any point in time one particular approach may prove more effective than others; however, it's a continuously changing domain. Solutions should be architected so that the AI algorithms are interchangeable. Above all, it is important not to focus on a single algorithm ... fight your algorithm addiction ... and ensure you are using the right tool for the job!

TRANSFER LEARNING

As discussed earlier, the best scenario for ML is when there is plenty of labelled training data and adequate computing power. In many cases, collecting sufficient training data is expensive or time consuming or simply not possible. This is where Transfer Learning can help. It is a technique for knowledge reuse across related domains. Can we use the data and the models developed for doing tasks in one (i.e. source) domain to help with tasks in another (i.e. target) domain? For example, a photo analysis company could take the image classification model from a large web search company and then use Transfer Learning to develop its own customised version.

Transferring Human Skills

To understand the topic better, let us explore human ability to transfer skills across domains. Someone who is skilled in playing the violin can learn to play the cello very easily, compared to learning to play the trumpet. Violin and cello are both string instruments played with a bow, whereas a trumpet is a different type of instrument. In addition, learning all these instruments needs common knowledge of music (e.g. music scripts, notes, rhythm, tempo, etc.). In contrast, these skills have nothing whatsoever to do with riding a bicycle, which is a different domain that requires completely different skill sets and experience. Thus, having skills in musical instruments does not help at all with riding bicycles. There are also examples where having a skill in one domain can hinder learning another. Even though Spanish and French are romance languages derived from Vulgar Latin, knowing one can actually hinder learning the other due to the significant differences in word formation, phonetic pronunciation, conjugation, etc. This discussion highlights the fact that success of the Transfer Learning across the domains depends on a number of factors and these need to be considered carefully.

Learning across Domains

Five important factors in the transfer of knowledge from a source domain to a target domain in ML are: (i) features in a domain, (ii) feature distributions in a domain, (iii) the nature of the tasks in a domain, (iv) the availability of data in a domain and (v) the choice of the algorithm.

Discussion of the technical details behind Transfer Learning and the various approaches is beyond the scope of this book. We refer to excellent survey papers [21–23] on the topic. We want to discuss a practical example of Transfer Learning to demonstrate the nature of the 'related' domains and the contribution of Transfer Learning to the task at hand.

Consider the problem of classifying customer sentiments as positive or negative based on the customer reviews of products in four domains (Books, Kitchen Appliances, Electronics and DVDs) on Amazon [24]. Each customer review consists of some textual input and an associated rating of 1–5 stars. More than three stars was taken as positive sentiment and less than three stars was taken as negative sentiment. These provide the natural labels/target variables for the classification of the text. Zhuang et al. [23] showed the comparison of Transfer Learning results from every domain to every other domain; that is $4\times3=12$ Transfer Learning experiments, each using ten different algorithms. The baseline was the in-domain classifier. Most algorithms achieved better accuracy than the baseline, when the source domain was electronics or kitchen appliances. This indicates that these two domains may contain more transferable information than books or DVDs. Also, five algorithms performed well in all the 12 tasks while three were relatively unstable. The better performing algorithms were feature-based, which is currently the popular approach to Transfer Learning.

Pretrained Models

Let us get back to our music analogy and pick the piano as our first instrument to learn. Piano is a unique musical instrument that covers the widest range of notes on the musical scale than any other instrument and gives the broadest exposure to music. Besides, it requires the coordination of the brain, hands and feet. After learning to play the piano, learning other instruments is much easier. Analog of this in transfer ML is to use generic data in a domain to pretrain a model and then fine tune it to a specific application domain. This is particularly effective when there is not enough data in the application domain.

Here is an example of this approach in medicine. Nanni et al. [25] studied the classification of patients in the various stages of developing Alzheimer's disease (AD) from 3D-MRI images. There are four groups: (i) cognitively normal patients (CN), (ii) patients with AD, (iii) patients with mild cognitive impairment who will convert to AD (MCIc) and (iv) patients with mild cognitive impairment who will not convert to AD (MCInc). They had labelled MRI data for a total of 773 patients in these groups, and this was not enough to train large traditional image classification neural networks. The authors demonstrated very good classification performance using pretrained standard 2D image recognition architectures (trained on non-medical images) when supplemented by an effective method to decompose 3D MRIs into 2D structures. This suggests that the features learnt from non-medical images in pretrained networks are effectively transferred to medical images.

Another area where pretraining models have become popular is in natural language processing [26]. Compared to the days when NLP methods relied on discrete hand-crafted features, neural network-based approaches now use distributed representations to capture the syntactic and semantic features of the language. Recent work has shown pretrained models can learn universal language representations from large generic data corpuses that can be used by downstream applications such as question answering, sentiment analysis, named entity resolution and machine translation [27].

Now an important footnote to close our discussion on Transfer Learning: Given a target domain, you may have some hunches on what could be a good source domain or the best generic dataset. But till you actually try the Transfer Learning experiment, it is difficult to predict whether it is going to work adequately to meet the level of performance you need for the business. So, you need to put aside some resources to help you make the decision.

REINFORCEMENT LEARNING

There are scenarios where the AI must decide on a set of sequential actions leading to an optimal outcome. Obvious examples are games (i.e. Chess, Go, Atari, etc.) where individual moves/actions lead to a win or loss. As for a business example, you can think of the problem to optimise the supply chain for a widget business for most profit that considers widget pricing, consumer purchase behaviour, manufacturing cost, widget shelf time, etc. To address this type of problems, Reinforcement Learning model [28] using historical (or simulated) data may be the best place to start, and depending on its performance, it can be integrated into the live environment. Figure 3.9 describes the interaction between the environment (i.e. system) and the AI agent. In each step, environment provides the system state, AI agent takes an action which changes the state and the environment provides a reward (positive or negative) to nudge the actions of the AI agent to move in the desired direction. While Reinforcement Learning does not need labels for outputs, setting up an AI agent to respond to the appropriate rewards for each likely action and letting the system figure out the optimal solution is the hard part that can take considerable effort and compute time.

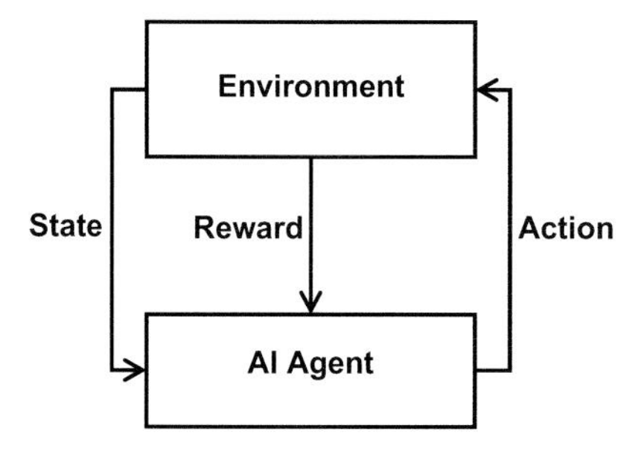

FIGURE 3.9 Interaction model between the environment and the AI agent in Reinforcement Learning.

BRAIN VERSUS ARTIFICIAL NEURAL NETWORKS

As a concept, ANNs really do play to the inherent weakness of the Algorithm Addict! The fact that they are inspired by the human brain cell is always going to be seductive. In addition, the natural character of Algorithm Addicts is attracted to the idea of writing a relatively simple, small and elegant computer program that somehow replicates itself and naturally evolves into a form that is capable of achieving human levels of intelligence … or greater! Before being seduced into believing that we are on the verge of this seismic event, how about taking a step back and comparing an artificial neuron with a real neuron?

Artificial neurons are relatively simple and can be described in just a few sentences. They take a set of inputs, multiply each input by a weight and then apply some form of activation function to generate the output. Networks of these very simple neurons are connected together and the output of each neuron feeds forward into other neurons to generate the final network output. When implemented in a computer program, everything operates in a controlled sequence starting with the first neuron in the first layer and finishing with the last neuron in the final layer.

Sejnowski [29] gives an excellent comparison of today's DNNs to the structure and performance of the human brain. Evolved over 200 million years, human neocortex (called the grey matter) is about 30 cm in diameter and 5 mm thick when flattened. There are about 30 billion cortical neurons in the human brain forming six highly interconnected layers. Each cubic millimetre of the cerebral cortex (about the size of a rice grain) contains a billion synapses. The largest DL networks today are reaching a billion weights. *The cortex has the equivalent power of hundreds of thousands of deep learning networks, each specialised for specific problems!* There is also work [30] in high-performance computing with Blue Gene/P supercomputer that requires 147,456 processors and 144 TB of main memory to simulate the brain of a cat with approximately 800 million neurons.

Real neurons are far more complex with chemical processes enabling a single neuron to recognise and respond to hundreds of different input patterns. The simplicity of the weighted connections between artificial neurons bears no resemblance to the complexity of the synapses connecting real neurons. Each individual synapse comprises a complex system of dendrites, axons and neurotransmitters. Each neuron is connected to other neurons through thousands of synapses. The neurons are not arranged in neat layers that are triggered in a formal sequence, and there are feedback loops far beyond anything we see in any man-made engineering system.

While the idea that ANNs are based on the human brain is attractive and even seductive, it isn't really a fair representation of the truth.

There's a long way to go before an ANN of 86 billion neurons is equivalent to a human brain with the same number of neurons!

Furthermore, the human brain is not just a huge collection of neurons. The brain has an architecture that has evolved over hundreds of thousands of years and contains subsystems responsible for different functions. Even with the huge number of neurons and the sophisticated architecture that we, mere mortals, are only just starting to understand, the human brain does not teach itself even the most simple concepts. Human beings require 2 years of 24-hour care followed by several years of primary education and even more years of secondary education to acquire relatively basic skills of mathematics and literacy.

Neural networks are an exciting and stimulating branch of AI … but they should be used in the right way for the right problem with the right understanding of their capabilities and their limitations [31–33].

FUNDAMENTAL PRINCIPLES AND FUNDAMENTAL MISTAKES

There are two fundamental mistakes in applying AI to real-world problems. The first fundamental mistake is assuming that the AI, usually the ML, will always derive the correct model.

As AI Engineers, we rarely know what the underlying model is. We are only provided with a subset of the data and we need to derive the model either manually, using a mathematical modelling approach, or automatically, using a ML approach.

In a ML approach, the effectiveness of the model will depend entirely on the volume and distribution of the observed data available. If we only have observed data for 3 months of the year, the ML algorithm can derive any number of models that fit the data but will probably bear no resemblance to the actual, real-world model. For a ML approach to work, we must have the right volume and distribution of training data. Later in this book, there is an entire chapter looking at this subject.

In a mathematical modelling approach, the AI Engineer may derive a model from historical data or may use other knowledge to determine the model. Using historical data alone is very similar to the ML approach, but performed by a human, and the effectiveness is therefore also reliant on the volume and distribution of the data.

It's important to remember that in many scientific fields from weather forecasting to control theory, humanity has invested phenomenal resources in developing models and theorems that describe the behaviour of complex systems. We must not assume that a ML algorithm can simply take a few samples of historical data and automatically perform more effectively than thousands of person years' worth of science.

The second fundamental mistake is assuming that the input data available to our system is sufficient for the AI to work effectively. Even the most simple problems become increasingly complex as you dig into the various nuances and edge cases.

To build an effective solution, we need to increase the number of input features and, in doing so, we increase the complexity of the problem. With more input features, we now need to derive a more complex multi-dimensional model. Looping back to the first fundamental principle, we can only do that if we have sufficient data and the amount of data we require increases exponentially with every new feature. We call this the curse of dimensionality and we'll discuss it in more detail later in this book.

Many of the points we have made using our simple example of object dropping problem may seem obvious. However, everything above applies in the real world! Quite often, we turn to AI to solve complex multi-dimensional problems. We believe the AI should be able to derive accurate, complex models where human beings can't. Before suggesting that AI is the answer to any problem, it's really important to ask whether we have the right volume and distribution of data for a ML approach or the right domain knowledge for a defined approach.

Even in cases where we are awash with data, never assume that there is sufficient data in the right form to solve the problem. Many AI projects are doomed to failure at this very early stage, being able to recognise this early will save you money (and tears).

In selecting and defining any AI project, it is important to understand these fundamental principles.

1. AI Algorithms map inputs onto outputs.
2. The mapping can be defined manually or derived using ML.
3. The mapping can only be defined manually if there is sufficient domain knowledge and mathematical skills to do so.
4. The mapping can only be derived using ML only, if there is the right volume and distribution of data.
5. No AI … no matter how clever … can derive insight when the input features and data are insufficient to do so.

If you are ever asked to evaluate an AI approach, remember and apply these fundamental principles.

SO … IT REALLY ISN'T ALL ABOUT THE ALGORITHM

By now, you should understand that there is no magic in AI.

There are those who seem to advocate collecting huge volumes of highly dimensional data and then using *Machine Learning* to discover patterns that are neither explainable to, nor understandable by, human analysts. Such approaches do not represent good engineering.

There are many reasons to consider the application of AI algorithms:

- when intelligence can be achieved by performing simple repetitive steps at massive scale. For example, in chess playing applications, the algorithm will effectively simulate moving a piece and then evaluating whether or not that is a good move. That's no different to a human player … except that the AI can evaluate several million moves per second and look 14 moves ahead.
- when the algorithm can leverage technical capabilities beyond those of human analysts. For example, in the facial recognition example, the AI is able to measure features that a human analyst cannot. The AI can also store millions of biometric patterns and compare them in order to find a match.
- when there are multi-dimensional correlations that a human analyst does not have the mental capacity to compute.

In all three of these reasons to apply AI algorithms, there is a fundamental underpinning of good science and engineering. Understanding these principles are important in selecting, defining and scoping any AI project. Quite simply, the fate of most AI projects is sealed at that early stage!

IN SUMMARY – THERE REALLY IS SO MUCH MORE TO AI THAN THE ALGORITHMS

Algorithms are clearly essential in the delivery of AI; however, we hope this chapter has convinced you that they are just one part of the delivery. In particular, we hope you can overcome your Algorithm Addiction and take away a few key points:

- An algorithm transforms inputs to outputs in a finite number of steps and there are many different types of algorithms to build models (neural networks, rule-based, using knowledge in a discipline, etc.)
- Obsession with any particular set of algorithms can lead a project astray.
- A business application may need different types of algorithms, AI and non-AI.
- Much of the recent success in AI is due to the advances in Supervised ML using ANNs and DNNs are ANNs with many (dozens, hundreds) hidden layers.
- Building good DNN models needs large amount of labelled (i.e. output specified) training data of sufficient quality.
- ANNs are not the only form of ML. There are algorithms that automatically derive rules from data.
- ANN models are opaque (i.e. black box). There is no easy way to understand the model output from the inputs.
- Unsupervised ML helps to understand the data.
- Rule-based algorithms are easily understandable to humans.
- Models built using detailed knowledge in a discipline are richer in describing observed behavior compared to generic ANN models.
- ANNs are particularly useful when the detailed discipline-specific models are hard to build.
- Transfer Learning helps to leverage knowledge captured in models in one domain to create models in a related domain with less data.

For those who are interested in really digging deep into this subject, we recommend the book *Artificial Intelligence: A Modern Approach* by Stewart Russell and Peter Norvig [6]. It's an essential resident of every Algorithm Addict's bookshelf. Please take the time to understand the multitude of different algorithms in this fascinating domain … but, whatever happens, beware of Algorithm Addiction!

THOUGHT EXPERIMENT: COULD AI BE JUST ONE BIG COMPUTER PROGRAM?

To finish this chapter, we thought you'd appreciate a little thought experiment ... there isn't a right or wrong answer ... just something for you to think about and discuss with friends and family.

One of the techniques used in our object dropping problem was straightforward mathematical modelling. There are many who would argue that hard coding an algorithm is not really an example of an AI System.

Increasingly, some argue that AI must include some form of ML.

However, what if a brilliant Engineer presented a system that demonstrated all the features of human intelligence including creativity and empathy. Imagine if we could talk to it and interact with it as if it was a human being. Imagine if, to every observer, the machine behaved exactly as a human being would and was considered by them to be intelligent.

Then one day someone opened the box and found that it has been hard coded just like our simple mathematical algorithm. It was one long computer program with no Neural Network or ML (as we currently define it).

Would we still think it was AI?

REFERENCES

1. D. E. Knuth, "Ancient Babylonian algorithms," *Communications of the ACM*, 15, pp. 671–677 (1972).
2. D. E. Knuth, *The Art of Computer Programming: Volume 1: Fundamental Algorithms* Addison-Wesley, 3rd Edn (1997).
3. M. R. Garey and D. S. Johnson, *Computers and Intractability- A Guide to Theory of NP-Completeness* W. H. Freeman, (1979).
4. R. Reddy, "To dream the possible dream," *Communications of the ACM*, 39, pp. 105–112 (1996).
5. R. Fjelland, "Why general artificial intelligence will not be realized," *Humanities and Social Sciences Communications*, 7, Article 10 (2020).
6. S. J. Russell and P. Norvig, *Artificial Intelligence: A Modern Approach* Pearson, 4th Edn (2020).
7. P. Jackson, *Introduction to Expert Systems* Addison-Wesley, 3rd Edn (1998).
8. A. Ng, "Machine Learning Yearning," https://www.deeplearning.ai/machine-learning-yearning/.
9. A. Geron, *Hands on Machine Learning with Scikit-Learn & TensorFlow* O'Reilly, (2017).
10. I. Goodfellow, Y. Bengio and A. Courville, *Deep Learning* MIT Press, (2016).
11. J. R. Quinlan, *C4.5: Programs for Machine Learning* Morgan Kaufmann Series in Machine Learning, (1992).
12. L. Breiman, et al., *Classification and Regression Trees* Chapman and Hall/CRC, 1st Edn (1984).
13. M. Williams, "How strong is gravity on other planets?" https://phys.org/news/2016-01-strong-gravity-planets.html.

14. J. Navratil et al., "Accelerating physics-based simulations using end-to-end neural network proxies: an application in oil reservoir modeling," *Frontiers in Big Data*, 2, Article 33 (2019).
15. W. S. McCulloch and W. Pitts, "A logical calculus of the ideas immanent in nervous activity," *Bulletin of Mathematical Biophysics*, 5, pp. 115–137 (1943).
16. D. Rumelhart, G. Hinton and R. Williams, "Learning representations by back-propagating errors," *Nature* 323, pp. 533–536 (1986).
17. Y. Le Cun, et al. "Handwritten digit recognition with a back-propagation network," *Neural Information Processing Systems Conference*, pp. 396–404 (1989).
18. A. Krizhevsky, I. Sutskever, and G. E. Hinton, "Imagenet classification with deep convolutional neural networks," *Advances in Neural Information Processing Systems*, 25, pp. 1097–1105 (2012).
19. B. Zhou et al., "Interpreting deep visual representations via network dissection," *IEEE Transactions on Pattern Analysis and Machine Intelligence*, 41, pp. 2131–2145 (2019).
20. D. Yu et al., "Feature learning in deep neural networks - studies on speech recognition tasks," arXiv:1301.3605.
21. S. J. Pan and Q. Yang, "A survey on transfer learning," *IEEE Transactions on Knowledge and Data Engineering*, 22(10), pp. 1345–1359 (2010).
22. K. Weiss et al., "A survey of transfer learning," *Journal of Big Data*, 3, p. 9 (2016).
23. F. Zhuang, et al. "A comprehensive survey on transfer learning," *Proceedings of the IEEE*, 109, pp. 43–76 (2021).
24. J. Blitzer, M. Dredze, and F. Pereira, "Biographies, bollywood, boom-boxes and blenders: Domain adaptation for sentiment classification," *Proceedings of the 45th Annual Meeting of the Association for Computational Linguistics*, Prague, Czech Republic, (Jun. 2007), pp. 440–447.
25. L. Nanni et al., "Comparison of transfer learning and conventional machine learning Applied to structural brain MRI for the early diagnosis and prognosis of Alzheimer's disease," *Frontiers in Neurology*, www.frontiersin.org, 11, Article 576194 (2020).
26. T. Young et al., "Recent trends in deep learning based natural language processing," *IEEE Computational Intelligence Magazine*, 13(3), pp. 55–75 (2018, August).
27. X. Qiu, et al., "Pre-trained models for natural language processing: a survey," *Science China Technological Sciences* 63, 1872–1897 (2020).
28. V. Francois-Lavet, et al., "An introduction to deep reinforcement learning," arXiv:1811.12560.
29. T. J. Sejnowski, "The unreasonable effectiveness of deep learning in artificial intelligence," *Proceedings of the National Academy of Sciences*, Jan 2020, p. 201907373.
30. D. Modha, et al., "Cognitive computing," *Communications of the ACM*, 54, pp. 62–71 (2011).
31. A. Darwiche, "Human-level intelligence or animal-like abilities?" *Communications of the ACM*, 61, pp. 56–67 (2018).
32. J. Pearl, "The seven tools of causal inference, with reflections on machine learning," *Communications of the ACM*, 62, pp. 54–60 (2019).
33. G. Marcus, "Deep learning: a critical appraisal," *arXiv:1801.00631.*

CHAPTER 4

Know Where to Start – Select the Right Project

It was 1994 and I had just given up a secure career with a guaranteed pension! I needed a job and fast. More importantly, I wanted to make my fortune so that I never had to work again.

I wrote to one of the top Formula 1 teams in the history of the sport and told them that I knew how to use AI to revolutionise motor racing. The team responded by inviting me to work with them to test out my ideas. In our initial brainstorm we came up with some great ideas and two seemed to be particularly interesting.

One idea was to use AI to understand how to set up racing cars for races. It was a massively complex problem that needed a really intelligent system and there was a load of data to work with. The other idea was to use AI to determine race strategy. This problem seemed to be less important, a bit mundane and we didn't really have the right data.

I started work immediately on the car setup problem and soon discovered that the data, whilst plentiful, was still insufficient. The problem itself was highly complex and the business process I was trying to align with was led by highly skilled individuals who were using years of experience to make sensitive judgement calls. Conversely, the race strategy problem could be tested easily without the need to modify a proven business process. The lack of data was easily solvable with simulation. By the time I realised that I had chosen the wrong problem, it was too late! The project was shut down and I failed to make my millions.

By selecting the wrong problem, I failed to deliver a working solution and instead of living a life of glamour in Formula 1, I now have to write a book!"

James Luke

DOI: 10.1201/9781003108498-5

In the previous two chapters, we introduced the key ideas behind building artificial intelligence (AI) applications for the enterprise and explained that it takes more than algorithms to deliver those applications. Given that a large percentage of AI projects in the enterprise end as end as proof of concepts (even when successful) and do not move to production [1–4], we want to address the important question, *How do we select AI projects that are most likely to succeed in the enterprise?* We call it the "**The Doability Method**". It is a key contribution of this book. This method was developed in IBM over many years, and we found it to be very useful in prioritising AI projects in an enterprise portfolio and in managing an AI project during the execution. Let us start.

THE DOABILITY METHOD

There are two steps in this method:

Step 1: Assess candidate business ideas and determine which ones are suitable for the current state of practice of AI technology.

Step 2: Evaluate the ones that are suitable for AI in terms of business value and technical feasibility using five themes: Business Problem, Stakeholders, Trust, Data and AI Expectation.

In this chapter, we present the basic ideas behind the first step of the Doability Method ahead of deeper discussions about AI in the subsequent chapters. The second step of the method is in Chapter 9. The intervening chapters provide the rationale to get you ready for the second step. Applying the Doability Method requires an understanding of the practical application of AI to real problems. Whilst the chapters that follow this one aim to provide the understanding required to support the application of the Doability Method, we hope and believe that the knowledge of AI shared in those chapters will be of value all by itself.

Before getting into the detail of the Doability Method, it's worth taking a few minutes to talk about Innovation and Emerging Technologies. Even after 70 years (or more), AI is still very much an emerging technology and, as such, must be approached differently from conventional projects.

INNOVATION AND EMERGING TECHNOLOGIES

Innovation is not just about dreaming up new ideas … it's more about figuring out how to make those ideas real. The challenge of innovation is figuring out which project to back and delivering real operational value. With any emerging technology, most enterprises are awash with ideas of how to exploit and benefit from the new technology; however, only a few are successful in realising those potential benefits.

Successfully exploiting emerging technologies invariably comes down to knowing which projects to back and which to leave alone. This is especially true in AI projects where, in the early stages, it is extremely difficult to predict the path that the project will take. Even a basic definition of project objectives can be challenging without first undertaking a period of experimentation and evaluation.

Considering the uncertainty inherent in AI projects, best practice is to adopt a portfolio-based approach where a programme is defined to develop a set of ideas through project definition into successful delivery.

A PORTFOLIO-BASED APPROACH

Adopting a portfolio-based approach creates the flexibility and agility required to optimise the chances of success. A typical portfolio may be initiated with 10–20 ideas out of which two or three may be immediately deliverable. A handful of projects will be dropped due to the lack of achievable business value or technical feasibility. The remaining projects may require further evaluation, or technical development.

A key element of this approach is not closing out options too early: you will need to accept that your best project ideas might need to be put on ice if blockers are encountered and brought back to life when those blockers are removed. For example, it may be that the application of AI in a control system is not practical due to the lack of a sensor with the accuracy required for the solution to work. Rather than pursue an unachievable goal, it may be wise to shelve the project and focus on another idea. However, it's critical that when projects are pushed onto the back burner, they are not considered failures. In the case of the control sensor, good practice would be to maintain a 'technology watch' such that if an appropriate sensor becomes available, the project can be re-started and the benefits realised.

An undervalued aspect of recognising that a project should stop is that you have probably learnt something profound about your business. It may indicate another opportunity and/or something that you need to better protect. In our sensor example, presumably the current system is working because of the hidden, and presumably undervalued, expertise of some of your workers. Maybe special care should be taken to retain those workers.

DOABILITY METHOD STEP 1 – TO AI OR NOT AI

Anyone who has followed AI in the media could be forgiven for believing that AI is the answer to all problems. Despite all the hype around various approaches, right now, the most proven approach to AI for the enterprise is the use of supervised machine learning (ML), which requires data of sufficient quantity and quality, consisting of application inputs and outputs (i.e. labels) for training. Given that status, like any emerging technology, AI should only really be used when it is appropriate to do so. When presented with an idea for an AI project, the first thing you should do is to ask whether AI is really the answer.

The first step of the Doability Method is to evaluate proposed ideas by following the Decision Diagram shown in Figure 4.1. We should state upfront that any such attempt to simplify the process for deciding on AI adoption involves the risk of oversimplification. Key assumptions behind the diagram are:

- Supervised ML is, by far, the most proven AI technology across a wide range of problems.
- Supervised ML is best (easiest to build, test and maintain) when applied to problems of narrow scope (e.g. image or text classification). Therefore, it is better to break any complicated task into a set of simpler tasks if you can.

- It is necessary that the use of AI fits the appropriate business process to minimise the risk. So, it is critical that typical project team members (with no deep expertise in ML algorithms) can understand and evaluate the task.
- This picture reflects the maturity of the state of practice in the AI technology for real business use now. It doesn't preclude the use of ML techniques such as Reinforcement Learning (see Chapter 3), which relies on interaction with the application environment (e.g. user feedback) for learning instead of labelled data.
- If there are other ways for machines to learn without the reliance on large quantities of labeled data, this decision diagram in the next edition of this book may look different.

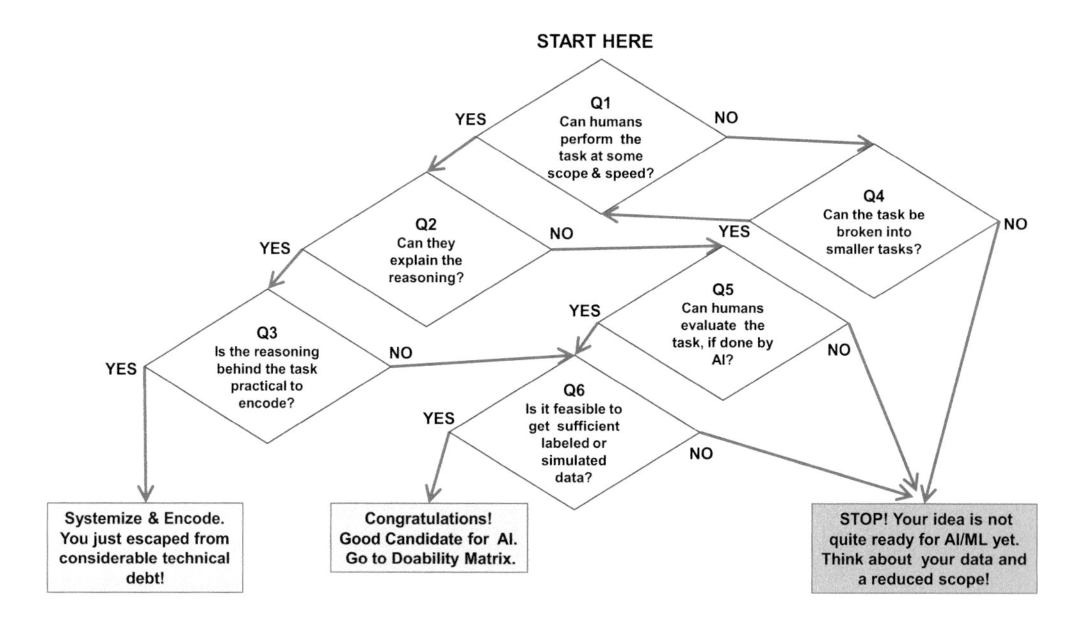

FIGURE 4.1 Is your business idea compatible with using AI technology?

We now discuss the diagram in terms of the questions posed at the decision nodes (i.e. diamonds).

Q1. Can Humans Perform the Task at Some Scope and Speed?

We start with the question of whether human analysts or specialists can already perform the task, albeit at lower speed and may be of some limited scope.

This question has three important implications:

i. Ideally, the best scenario for AI use is when there is an existing business process where a specific task could benefit through automation to improve the productivity and timeliness. In this scenario, clearly someone is already performing the task.

ii. Creation of AI needs business domain knowledge to make sure that the right factors are considered to create the AI.

iii. If there is someone who knows how to do the task, it increases the chance of success of creating a viable solution, whether it is AI or not. Remember it's not just any old human doing the task; it's a human that your solution development team has access to (e.g. don't try to build a medical diagnosis AI without a medical professional on your team).

Q2. Can They Explain the Reasoning?

This question has at least two implications:

i. If there is an existing reasoning/rules explaining the task in an algorithmic way, this provides at least one option for implementing a solution which may not involve AI.

ii. ML-based approaches are notoriously weak in explaining their outputs due to their black box nature and so any other way to explain the results can only be positive and potentially help to corroborate the AI output.

Q3. Is the Reasoning Behind the Task Practical to Encode?

In the early days of AI, there was a view that all tasks performed by humans could be manually encoded into an expert system. In effect, it was believed that human intelligence could effectively be written into a single program, or perhaps a great big "IF… THEN.." statement, for execution by a machine. Unfortunately, it was soon discovered that the challenge of encoding human intelligence is significant. For every use case or rule defined, there are counter examples that render the use case or rule incorrect. This was the principle reason that expert systems were not successful.

However, there are many times when it is possible to encode a carefully scoped task into some form or engineered solution. For example, in the early days of flight, it would have been considered unthinkable for a machine to land an aircraft taking into all the different variables ranging from runway conditions to freak weather and aircraft faults. Dealing with all those variables was a complex task that required human judgement and skill to manage. Autopilots and autoland systems are perfect examples of where a human skill can be systemised and encoded.

The practicality of encoding a task is clearly something that depends on the scope. If you were to ask an insurance underwriter to explain why they accept or refuse an insurance application, they could probably articulate a core set of rules quite quickly. It will take longer for them to articulate the edge cases and more unusual aspects of the judgement. Something that simplifies the task, of course, is that the underwriter bases their decision on the information supplied in the application. The problem is well scoped, and an underwriter may deal with many hundreds of applications a week so there is a good set of general data to consider in encoding the decision-making. Finally, it is possible to determine the effectiveness of the encoded system against historical data. All these factors make it practical to encode the underwriting task.

Now consider, an Oncologist developing a treatment plan for a complex cancer. The number of variables and parameters that the Oncologist takes into account is massive and includes a lot of data not immediately apparent in the notes or test results. The Oncologist may notice the way a person walks or paleness in their appearance. The Oncologist will ask questions about lifestyle and form a judgement based on a very broad set of data. Typically, an Oncologist will see a relatively small number of cases and, unfortunately, we're not able to go back to historical data and test what would have happened with a treatment plan. As you can see, this is a far more challenging task to encode.

Q4. Can the Task Be Broken into Smaller Tasks?

As the technology stands today, AI can do narrow tasks quite well and so it behoves us to pick simple tasks that are more suitable for AI to increase the odds for a successful project. Combining a set of simple narrow AIs, to function as one application, is often better than trying to make one complex AI do everything.

The specific business task to be automated can range in complexity across a spectrum. At one end, the task can be simple for a human but repetitious. Think of classifying thousands of incoming emails every day to a service provider into (say) three bins: billing questions, new service request or customer support. At the other end, it can be a complex task, requiring special skills and aggregation of many subtasks. Example for this is the patient diagnosis by a physician based on multiple sources of information. Such a task may have to be broken into sub tasks, addressing electronic medical records, mining patient history, diagnostic imaging, lab results, etc.

Q5. Can Humans Evaluate the Task, if Done by AI?

This is another very important question with strong implications. It is well known in the field of computational complexity [5] that an algorithm is not useful unless we can validate the correctness of it. The expectation is that validating the algorithm may be easier than creating the algorithm. If the AI comes up with an algorithm to solve a business task, but we have no way of verifying the results, it is a significant business risk. AlphaFold, from Google Deepmind, is an example of an AI that can solve a problem not (practically) solvable by humans but where humans can evaluate the answer; AlphaFold's neural nets exhaustively try millions of combinations of protein folds until a shape is reached that obeys the tens of thousands of folding rules per protein. It is not practical to do this manually, but human researchers can validate the correctness of AlphaFold's candidate predictive shapes using the actual protein's chemical properties.

Q6. Is it Feasible to Get Sufficient Labeled Training Data?

Labelled data is a key enabler for supervised ML technologies and labelling data can be a difficult and laborious task. In cases where you have an existing business process, it is usually possible to obtain labelled data based on historical decisions. In cases where a completely new business process is being defined, then obtaining the data will require resources to be assigned to manually label data. A really important point to remember is the need for the data to be consistently labelled. This may seem obvious, but in practice, it is rarely considered. It is not uncommon to review a failing ML project and to find that the ML is being trained with data that is inconsistent and contradictory. For example, there may be two identical images of a cat where one image is labelled "cat" and the other is labelled "Siamese". Even worse, you may find one of the images labelled "dog".

In terms of labelling data, there are strategies that can be adopted to make the task more practical. For example, crowd sourcing can be a very good way of generating very large sets of training data. An important aspect of crowd sourcing data labelling is of course to send the same data to multiple annotators and then normalise the results to ensure consistency.

Another approach is to use the ML as part of the labelling process. This technique, often referred to as Active Learning, involves interactive analysis of the data. The human annotator is presented with a set of data for manual labelling. As the human labels the data, the ML is being trained in the background and additional algorithms are working to identify which data to present next. The Active Learning algorithms may, in one strategy, pick records similar to those which are achieving the poorest performance in the background training.

Simulating data is often seen as a great panacea in AI projects. Why waste all that expensive human time manually labelling outputs (e.g. pictures of cats, dogs, etc.) when you can just write an algorithm to do that? Please be careful when it comes to simulated, or synthetic, data. We discuss this in greater detail in Chapter 7. The obvious challenge is that if you can write an algorithm to do the job perfectly then why bother with training an AI to do the same thing. If your algorithm gets it right most of the time, your AI will at best do as well but most likely worse than the original algorithm. That being said, it's not quite that simple, and there are cases where the use of synthetic data can work. For example, if developing a decision support system where the AI makes decisions about the environment rather than classifying data within it. In such cases, the use of synthetic data can work as a simulator simply generating all possible permutations and the AI learns what works and what doesn't.

Ultimately, using ML is all about the data and you should consider the practicalities of how you are going to obtain and manage high-quality data. Knowing whether you have sufficient data is the billion-dollar question and there is no simple way of answering (e.g. you may have millions of rows of training data, but does it cover all possible eventualities?). However, you should have the data science experts on your project to at least make an initial statistical evaluation.

THREE RECOMMENDATIONS FROM DOABILITY METHOD STEP 1

From our experience, we know that these six questions can only really be answered qualitatively. It will almost certainly be impossible to give definite answers when all you have is a list of ideas written on a white board. The point of this first step of the Doability Method is not to make a final judgement but to allow you to make an initial pass and perhaps select those projects that are most appropriate for further evaluation. This will depend on the resources you have available to conduct a more thorough evaluation.

Going through the decision diagram in Figure 4.1 results in three possible recommendations:

i. Use traditional programming to capture the business logic, where it is manageable to implement the proposed idea. This interestingly avoids the complexity of maintaining an AI solution and the inevitable related technical debt.
ii. Abort the plan to use AI as the technology to implement the proposed idea since the underlying circumstances do not show promise for success.

iii. AI may be a good option, but you must do Doability – Step 2 to assess the Business Value and the Technical Feasibility of this venture.

To demonstrate its utility, let us walk through the first step of the Doability Method with three examples.

DOABILITY METHOD STEP 1 – WORKED EXAMPLES

As we go through the examples, it will become clear how the choices for implementing AI applications have changed in the last decade. This is because three key technology advances have changed the landscape:

i. With the internet becoming the primary platform for business applications and increase in the digitisation of business transactions, there is an explosion of data captured by an enterprise [7].

ii. Increasing use of public and private clouds is making it possible to acquire 'elastic' computing resources as you need, compared to a traditional IT infrastructure with upfront capital equipment costs.

iii. The field of AI has made phenomenal advances in ML algorithms in the last decade, particularly in supervised ML.

With this in mind, let us evaluate three ideas using Doability Method Step 1 to see where we land in the three recommendations in Figure 4.1.

Example 4.1: AI Playing Chess

This is a good old favourite among AI enthusiasts. For decades, chess was seen as the ultimate test of intelligence, and many believed that it would never be possible for a machine to beat a world chess champion. *Let us imagine living in the early 1990s and AI-based chess does not yet exist. We want to start the Deep Blue Chess project* [8,9]. The goal is to build an AI application to beat the world chess champion. Let us walk through Figure 4.1.

- Q1: Yes. World champions are people (apparently) and they become world champion by beating other people. So, yes … a person can perform the task.
- Q2: Yes. While a Grand Master cannot sit down and write out an end-to-end process for winning at chess due to its sheer complexity, he or she may be able to explain their reasoning for any specific board position. Since chess is a game made up of moves, it is possible to anticipate and evaluate some number of moves on either side in terms of a path to a win (or loss), loss of pieces on either side, the freedom to manoeuvre individual pieces, etc.
- Q3: No. Whilst a Grand Master could explain some number of moves, it would not be practical to write a procedure for every possible "what if" outcome/board position. There are just too many!

- Q6: Yes. Indeed, there were many thousands of historical Grand Master games. Yet, they still DID NOT cover all possible positions, move sequences and game outcomes. BUT, you could construct a simulation – you know how each piece moves, the board is finite and there is a definite goal/end game.

In 1997, the ML algorithms and the infrastructure were not mature enough. The best you could do was to:

i. Create a system that could simulate many moves ahead, sometimes more than 20, consisting of millions of combinations on the fly.
ii. Encode a set of programmatic rules that could use the simulation to search for the optimum choice to win the game at every move, based on game knowledge and history. By doing so, you would have built a system that defeated a World Chess Champion … at least that is how IBM's Deep Blue defeated Garry Kasparov.

An interesting point to note here is that Deep Blue was simply simulating a subset of every possible permutation of the game, evaluating 200 million moves per second. By doing so, Deep Blue was able to perform a task that an ordinary person can perform, BUT at a massive scale. This gave Deep Blue its "intelligence", i.e. AI. It is important to note that the technology behind Deep Blue was very different from the current focus on ML as the primary technique in AI.

However, Google Deep Mind's AlphaZero program can play championship level chess today with NO human input, but just using the game rules augmented by self-play with reinforcement learning [10]. Given the advantage demonstrated recently by AlphaZero over Starfish (the best current chess program in the DeepBlue genre) [11], if we were to create a chess application today, we would be foolish not to consider ML based approaches.

Example 4.2: Analysing Financial Transactions to Spot Money Laundering

Let's consider other examples where scaling a simple function allows an AI to demonstrate intelligence that appears superior to humans. Think about an application for analysing financial transactions to understand possible money laundering.

- Q1: Yes; A human analyst can read through a sequence of financial transactions involving a person or an organisation and identify the potential evidence of money laundering. He may be able to do a few cases in a day but considering the large number of financial transactions in any given day and most of them are perfectly legal, this scenario is begging for an AI application.

- Q2: Yes. Typically, this takes the form of enumerating rules that capture the human understood patterns of money laundering. While this approach may not be complete, it will definitely help to automate the task.

CHESS PROGRAMS: THEN AND NOW

As we were building Deep Blue [1] in the early/mid 1990s, there were limited choices for the system design. Statistical ML techniques, including deep neural networks (DNNs), were not yet up to the task of training a system to play world-class chess (despite the success of TD-Gammon [2] in backgammon). We therefore chose a more classical approach that had been popularised in the 1970s [3], using a state space search algorithm (a variant of alpha-beta game tree search) and a hand-constructed evaluation function. Deep Blue extended this methodology in a few ways:

- Large-scale parallel search using 480 custom-designed chess processors
- Highly non-uniform search algorithms based on our earlier singular extensions work [4]
- Manually chosen evaluation function features, with a combination of manual and automatic tuning

In the 2020s, there are a number of alternatives to building a program to play chess or a similar type of board game.

1. **Classical Approach:** Following the style of Deep Blue and its predecessors, this approach would rely more heavily on automated tuning of the evaluation function parameters and use modern search tree pruning methods that are highly efficient. Most older versions of Stockfish [5], a popular open source chess program, are based on this method.
2. **Deep Reinforcement Learning Approach:** AlphaGo [6] demonstrated that training a Go program with self play, using deep neural networks, reinforcement learning and Monte Carlo tree search [7] could produce very high-level play. AlphaZero [8] showed how an improved version of this approach could train even stronger systems for three different games, including chess.
3. **Hybrid Approach:** Combining the classical approach with neural networks has also resulted in a very high-performance chess program (Stockfish 12 and later versions [9]).

From the point of view of a game developer, either approach 2 or 3 is likely to be an effective for developing a game playing program. Approach 1, the classical approach, while quite effective, requires significant domain knowledge about the game and would likely require a much longer development time. Approach 2, as implemented in AlphaZero and recent open source variants (e.g. Leela Chess [10]), has a relatively large and complex evaluation function, which requires significant computation for training. In addition, the relatively slow execution time of the neural network forces a smaller scale search (based on Monte Carlo tree search). Approach 3 has a much simpler and faster neural network evaluation, which enables a very fast alpha-beta-based tree search that is close to the speed of the classical versions of Stockfish.

Murray Campbell

Murray Campbell is a Distinguished Research Scientist at the IBM T. J. Watson Research Center and a Fellow of the Association for the Advancement of Artificial Intelligence (AAAI). He is best known for his work on Deep Blue, the IBM computer that was the first to defeat the human world chess champion in a match.

REFERENCES

1. M. Campbell, Jr, A. J. Hoane and F. H. Hsu, "Deep blue," *Artificial Intelligence*, 134(1–2), pp. 57–83 (2002).
2. G. Tesauro, "TD-Gammon, a self-teaching backgammon program, achieves master-level play," *Neural Computation*, 6(2), pp. 215–219 (1994).
3. D. J. Slate and L. R. Atkin, "Chess 4.5—the Northwestern University chess program." In *Chess Skill in Man and Machine*. Springer, New York, NY, (pp. 82–118, 1983).
4. T. Anantharaman, M. S. Campbell and F. H. Hsu, "Singular extensions: adding selectivity to brute-force searching." *Artificial Intelligence*, 43(1), pp. 99–109 (1990).
5. https://stockfishchess.org/.
6. D. Silver, A. Huang, C. J. Maddison, A. Guez, L. Sifre, G. Van Den Driessche ... D. Hassabis, "Mastering the game of Go with deep neural networks and tree search," *Nature*, 529(7587), pp. 484–489 (2016).
7. R. Coulom, "Efficient selectivity and backup operators in Monte-Carlo tree search," In *International Conference on Computers and Games*, Springer, Berlin, Heidelberg, (pp. 72–83, 2006).
8. D. Silver, T. Hubert, J. Schrittwieser, I. Antonoglou, M. Lai, A. Guez ...D. Hassabis, "A general reinforcement learning algorithm that masters chess, shogi, and Go through self-play," *Science*, 362(6419), pp. 1140–1144 (2018).
9. https://www.chessprogramming.org/Stockfish_NNUE.
10. https://lczero.org/.

- Q3: No. The reason for this answer is simply that the human created rules cannot keep up with the new ways of money laundering invented every day by other criminally minded clever humans.
- Q6: Yes. With the digitisation of financial transactions in the last decade, it is possible to get the training data with human annotated labels on normal and money laundering scenarios.

Clearly, human analysts cannot scale to perform tens of millions of such tasks every second. By exploiting the scalability of AI, we can deliver solutions that far out-perform human capabilities ... even though the task they are performing is simple. The key point is that AI is not being used to solve a difficult problem that no human can comprehend! AI is solving a very simple problem that any human could easily do. However, the AI can do this task millions and millions of times per second to deliver a capability far greater than that of a human. We refer to the paper by Chen et al. [12] on the evolution of ML techniques for anti-money laundering applications.

Example 4.3: Analysis of Medical Images

Now, let's consider a different problem. One of the most interesting applications of AI is in the analysis of medical images. The ability to look at an X-Ray or an MRI scan and determine whether an image of a lump is potentially malignant would be of huge benefit. Let's push this example through our decision tree.

- Q1: Yes. the task is currently performed by radiographers and other medical imaging specialists. It is a skilled task and does require extensive education, but it is performed by very large numbers of medical professionals around the world.
- Q2: No. The answer to this probably depends on the exact use case, but the general answer is that they can partially explain their reasoning. They can look at an image and point out features that draw their attention to a particular part of the image. However, there will be variance and exception cases so their ability to fully explain their reasoning is limited. Ultimately, they will be using their experience to identify features that "look like" points of interest. So, we're going to say no to this question. They can't fully explain their how they do it.
- Q5: Yes. A trained healthcare professional will be able to evaluate the performance of an AI.
- Q6: Yes. Millions and millions of medical images are captured and analysed all around the world every day. These images are labelled using a standard medical terminology and could form a very valuable set of training data for ML.

Our overall decision therefore is that analysing medical images is potentially a very sound application of ML. In fact, Wu et al. [13] compared the performance of ML algorithms vs. Radiology Residents and concluded that "...it is possible to build AI algorithms that reach and exceed the mean level of performance of third-year radiology residents for full-fledged preliminary read of anteroposterior frontal chest radiographs".

A caveat here is that in the real world, a medical doctor may interpret the image in conjunction with other facts about the patient (e.g. demeanour, skin colour, shortness of breath, difficulty of movement etc.) for a proper diagnosis. Clearly, these are not present in the training data consisting of medical images alone.

DOABILITY METHOD STEP 2- PRIORITISING AI PROJECTS IN THE PORTFOLIO

In Chapter 9, the AI Project Assessment Checklist will take your AI idea and consider it against the five key pillars of successful AI deployments:

- **Business Problem**: a key part of any complex project. Clearly defining the scope is important, however in an AI project we also need to consider how the impact of the AI will be measured, the skills you will require to deliver and how the new capability will be integrated into the business process.
- **Stakeholders**: as mentioned in the introductory chapter, AI solutions will impact society to a much greater extent than previously. In addition to managing internal Stakeholders within an enterprise, the values and beliefs of a whole range of external Stakeholders, from regulators to customers, need to be managed.

- **Trust**: for AI to be successful it really does need to be trusted … unless you are a James Bond villain of course. Trust is not something you can specify; it's up to your consumers to decide whether to trust the AI. However, you can aim to develop Trustworthy AI by considering important factors including accuracy, ethics, bias mitigation, explainability, robustness and transparency.
- **Data**: it's all about the data, so any project evaluation will need to consider privacy, availability, adequacy, operations and access to domain expertise.
- **AI Expectation**: what is the real necessity driving the project and has this type of thing been done before? What is the true scope (again) of the application and is it really feasible (a more thorough version of the step 1 evaluation outlined above)? Finally, what other complexity factors exist and what are your hopes regarding reusability?

As we have said, having just one AI project in mind could be considered inefficient. AI projects can and do fail, most often when the data doesn't quite live up to your expectations. It pays early in your project to have a portfolio of AI ideas, preferably achievable with the same data. Step 2 of the Doability Method in Chapter 9 is designed to compare these ideas and help you to prioritise the ones most likely to succeed. Before you jump straight to Chapter 9, we know you want to, or preferably read the intervening chapters first, why don't you cycle back to the decision flow diagram (Figure 4.1) and run a couple of alternatives to your main idea through it? It may surprise you which one comes out on top when you put them through Step 2 of the Doability Method.

If you are eager test your top ideas in Step 2, then feel free to skip ahead to Chapter 9 and dive straight into the questions. If you're interested in learning more about the reasoning behind the questions, then we recommend that you read the following Chapters. Chapters 5 & 6 will focus on Value, and Chapters 7 & 8 on Doability.

IN SUMMARY – SUCCESS OR FAILURE WILL DEPEND ON SELECTING THE RIGHT PROJECT

Given the excitement about AI, it's very tempting to immediately rush to the most difficult and challenging project that will transform your business. It's a temptation that can lead to disaster, so it's critical that you understand how to select the right project:

- The Doability Method allows you to evaluate if your business idea can be supported by the current AI technology.
- The method consists of a Decision Diagram (Step 1) and a Doability Matrix (Step 2)
- The Decision Diagram helps to validate the suitability of your business idea to the most proven AI technology (i.e. Supervised Machine Learning)

- The Decision Diagram results in three recommendations. (i) Do traditional programing (ii) Not a good match for AI (iii) Probably a good match for AI to be followed with Doability Method Step 2

The Doability Method Step 2 is covered in full detail in Chapter 9 following a series of Chapters that discuss the finer points of Doability and Value.

REFERENCES

1. KPMG 2019 Report: "AI transforming the enterprise."
2. O'Reilly 2019 Report: "AI adoption in the enterprise."
3. Databricks 2018 Report: "Enterprise AI adoption."
4. MIT Sloan-BCG 2019 Research Report "Winning with AI."
5. NP-Completeness, https://en.wikipedia.org/wiki/NP-completeness.
6. "DeepMind says it will release the structure of every protein known to science," *MIT Technology Review*: https://www.technologyreview.com/2021/07/22/1029973/deepmind-alphafold-protein-folding-biology-disease-drugs-proteome.
7. D. Reinsel, J. Gantz and J. Rydning, "The digitization of the world from edge to core," An IDC White Paper – #US44413318, sponsored by Seagate, November 2018.
8. M. Campbell, A.J. Hoane Jr and F. H. Hsu, "Deep blue," *Artificial Intelligence*, 134(1–2), pp. 57–83 (2002).
9. N. Silver, see the Chapter "Race Against the Machines" in *The Signal and the Noise*, Penguin Press, New York (2012); "The man vs. the machine", https://fivethirtyeight.com/features/rage-against-the-machines/.
10. D. Silver, T. Hubert, J. Schrittwieser, I. Antonoglou, M. Lai, A. Guez ...D. Hassabis, "A general reinforcement learning algorithm that masters chess, shogi, and Go through self-play," *Science*, 362(6419), pp. 1140–1144 (2018).
11. AlphaZero Crushes Stockfish in New 1,000-Game Match https://www.chess.com/news/view/updated-alphazero-crushes-stockfish-in-new-1-000-game-match.
12. Z. Chen et al., "Machine learning techniques for anti-money laundering (AML) solutions in suspicious transaction detection: a review," *Knowledge and Information Systems*, 57, pp. 245–285 (2018).
13. J. T. Wu et al., "Comparison of chest radiograph interpretations by artificial intelligence algorithm vs radiology residents," *JAMA Network Open*, 3(10), p. e2022779 (2020).

CHAPTER 5

Business Value and Impact

The story of the Allied code breakers in World War 2 represents one of the greatest engineering achievements in human history. On the eve of war, the British Government established a code breaking group in a country house, called Bletchley Park, 50 miles northwest of London. The team was tasked with de-coding intercepted German messages and providing the Government with intelligence of exceptional value.

Initially a small group of code breakers including mathematicians, linguists and those who had learned their trade in World War 1, gathered at the establishment. However, within a very short time, the group started growing rapidly as the original code breakers were joined by more mathematicians, scientists and engineers. Collectively, they were doing something which many, including the enemy, thought was impossible. They were breaking the enemy's codes and reading their messages.

Whilst code breaking had been practiced for centuries, Bletchley Park represented a massive step change! The group developed the methods, technologies and processes required to transform coded messages into plain text on an industrial scale. In doing so, they gave the Allies a massive military advantage that some argue shortened the war by at least two years and laid the foundations of the computer industry. One of leading mathematicians at Bletchley Park was Alan Turing, one of the founding fathers of both AI and Computing, captured in the 2014 movie "The Imitation Game".

One fascinating aspect of the Bletchley Park story is that, at the start of the war, there was no way of knowing that the codebreakers would be successful.

Try to imagine writing a modern-day business case for Bletchley Park! The executive summary would be, "We would like to take several hundred theoretical academics and put them in a country house where will give them unlimited resources to invent a completely new technology aimed at breaking an unbreakable code".

Would that business case carry much weight with a twenty first century CEO?

DOI: 10.1201/9781003108498-6

Fortunately, the British Government was willing to take a leap of faith and make a massive investment in the industrialisation of code breaking. There was no upfront business case for Bletchley Park ... just total belief in the importance of the enterprise and trust in those empowered to deliver. This was demonstrated very clearly in October 1941 when three of the leading code breakers, including Turing, felt that they needed more resources. They wrote a letter to the Prime Minister, Winston Churchill, expressing their concern about the lack of resources. On reading the letter, Churchill wrote on it a very direct and clear instruction, "Action this day! Make sure they have all that they want on extreme priority and report to me that this has been done" [1].

WHAT IS DIFFERENT ABOUT AI APPLICATIONS?

There are many differences between Artificial Intelligence (AI) Applications and conventional business applications when it comes to business value and impact. You could argue, AI is just a software component to be integrated with other components to build an application. Most traditional IT projects are already difficult to manage for scope and schedule. However, AI business cases are intrinsically more challenging. Here are some critical aspects that need careful consideration:

- Since AI applications are going to make decisions and judgement calls that were traditionally left to humans, topics such as ethics, explainability transparency *are not peripheral topics. They are at the core of an AI project.* We are at the infancy of dealing with these 'human' issues.
- Stakeholder issues are more significant due to the social and moral aspects. For example, you may have to worry about what the media would say or how the employees of the enterprise would react to the choice of the AI being deployed.
- AI projects need more experimentation (e.g. choice of models) compared to traditional IT projects. This introduces uncertainty in the quality of the outcome and the schedule of the project.
- The availability of data of sufficient quality and quantity is critical. Most often, this is not easy to assess in the beginning of a project since some level of iteration with model building is necessary to get a reliable assessment. This adds to the uncertainty.
- There are significant differences in the engineering of machine learning (ML) systems compared to traditional IT systems in terms of resources (i.e. people, engineering environment, etc.) and processes (e.g. DevOps, testing, persistence of training data, etc.). These have to be in place for a successful execution. Also, the engineering environment has to persist for as long as the application itself is deployed.

BUILDING BUSINESS CASES

In the years since the Second World War, management practices have evolved. In today's enterprise, very few projects are approved without a sufficiently detailed, upfront business case! AI projects should be no different. *If the primary motivation to do an AI project is simply to do an AI project because it is the shiny object of the time, you are very likely to fail.*

You should think about the business processes you have and assess where introducing AI will give you the best business value, while being practically doable. If the AI is going to require new business processes, you should be aware of potential challenges lying ahead due to their introduction. Let us talk about factors contributing to building business cases for AI projects and see how they are different.

Examples of AI Business Cases

Let us consider an IT portfolio of applications supporting the various business processes in an enterprise. A typical business case for a new AI project in this context will belong to one of three basic buckets: (i) efficiency, (ii) enhancements and (iii) new capabilities. The benefit to the enterprise is measured typically in terms of increased revenue, reduced cost, increased customer satisfaction, improved market share and, for publicly held companies, increased share value and market capitalisation.

Efficiency

AI is going to automate a task, currently being done by one or more humans. Expected benefit will be less expenses, a shorter time to do the task and potentially much more consistent output, particularly if multiple humans are currently doing the same task. The details of the business case will revolve around net cost savings due to the use of the AI instead of the humans or the financial value of the saved time. An example of this is in Customer Relationship Management where incoming email or text from clients needs to be classified into different bins depending on the type of follow-up action needed, a tedious job for a human, great for AI. A second example of efficiency gains would be the use of "chatbots" in customer support to answer simple questions from the customers. In either case, primary evaluation criterion has to be an objective comparison of AI versus humans in performing the task.

Enhancements

This can have two flavours.

i. Your current technology for doing a task that could be improved by using AI. An example of this can be seen in computer vision where the older technologies relied on human efforts to define distinguishing features in the image, whereas the more recent deep neural networks are able to extract these features automatically using the training samples with labels and also perform image recognition task at a higher accuracy.

ii. AI can do certain narrow tasks better than humans in terms of speed and accuracy. Examples of such tasks include speech to text conversion, text to speech conversion and text translation. Given the advancements in speech processing and text processing in the past decade, this can be done quite well by AI. The business case will involve an assessment of the value of the use of AI for these tasks vs. the current approach in terms of performance quality and effort.

New Capabilities

These are tasks simply not possible for humans to do or complete in the time allotted. For example, recommender systems are really ideal for machines because they can pick out the patterns in the behaviours of individual users and user communities very easily, but this is something that a human or a group of humans cannot do. Reading thousands of pages in a few seconds to look for specific concepts or entities of interest is another example of something that would be impossible for a human to achieve at the required speed but easy for an AI. The business case will contain the new revenue opportunity for the company due to these new AI capabilities that do not exist in the company portfolio today or may be anywhere in the market.

Measurability and Impact

Ideally, each of these benefits will be measurable, you must also remember to record the "as was" performance of your business before you introduce your AI. In many cases, it is possible to extrapolate these measurable features directly into some form of business *value*. For example, it may be possible to link the expected improvement in the quality of a product to business *impact* in terms of the saved money that would be spent in fixing defects and managing complaints in the field. Measurability is, unfortunately, not as easy as you would think and the challenge of measuring success will be discussed in more detail later in this chapter.

There may be an additional business *impact* that is more difficult to quantify and associate with the AI delivery. For example, improving the quality of a product may also improve the reputation and the brand of a company. With customer surveys and detailed market analysis, it may be possible to estimate the *value*, but it is an estimation and it is indirect. You get the message ... some things are directly measurable and some things less so. It is also important to remember that not all *impact* is positive! Consider the *impact* of a media story stating that your AI is behaving in some Machiavellian way. There are many stories in the media warning the public of the dangers of evil AI applications. A negative press report could wipe out the expected *value* of a project and so it is critical that you understand the potential issues.

There are many factors that contribute to creating business value. One of the key factors is understanding your stakeholders ... not only the people critical in enabling the delivery of AI but also those vital to accept its use in practice in the first place. We will discuss them next.

STAKEHOLDERS

Your AI project will have a lot of stakeholders, more than for your typical IT application project, and they won't all be obvious. We are accustomed to thinking of the Head of Marketing and the Finance Director as stakeholders. In AI projects, we generally need to consider many more groups and individuals – in our examples we've underlined a dozen, your project may well have more.

The first scenario (Figure 5.1) is when the enterprise AI application is directly interacting with the end consumers. Examples are internet applications in banking, insurance, e-commerce, etc. There are three groups of stakeholders in this scenario:

- **The Enterprise**: This group includes various people involved in the creation of the AI. *Investors* with the financial stake in the project; *Business sponsors* who support the specific AI project and assign resources; *Domain Specialists* whose knowledge you need to build and verify the AIs efficacy; *Engineers* who build the AI application and

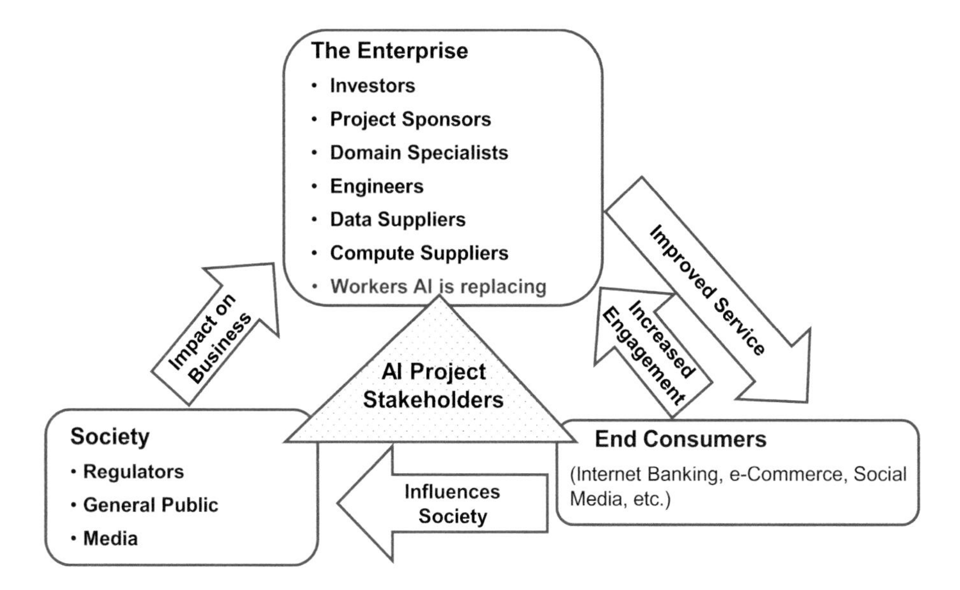

FIGURE 5.1 Scenario where the AI application directly interacts with end consumers.

any *suppliers* of data used by the AI and computing environment (e.g. *cloud provider*). In addition, the introduction of AI either in the internal enterprise processes (e.g. automated email classification) or in the consumer facing aspects (e.g. bots in customer support) may affect the employment of some *workers* in the enterprise while their participation is critically necessary for the creation of the AI. The overall quality of the AI application critically depends on this group. Your *employees* as a whole will have a view on what you are doing. They may be excited that you are moving into this brave new world or they may be worried that this is the beginning of the end as the machines take over.

- **End Consumers/Customers**: These are the intended beneficiaries of the AI outputs. The purpose of the AI is to provide benefits (e.g. a more pleasant user experience) to these end consumers in their interactions with the enterprise. More engagement of the end consumers with the application is generally a positive impact on the enterprise, particularly amplified by social media. Your customers will have a view of AI and will want to know how it impacts their service and what it means for their personal data.

- **Society**: The impact of the AI on the society is really judged by three different sources: The *regulators* who have the job of making sure that the AI conforms to the defined standards, where they exist; the *general public* whose impression of AI matters for its acceptance in the large; and the *media*, who have insatiable appetite for sensational stories either good or bad. The public perception of the AI can have significant impact either positively or negatively on the business success of the enterprise.

  In the second scenario (Figure 5.2) where the purpose of the AI application is to help professionals (e.g. doctors, loan officers, radiologists, etc.) with their decisions affecting their clients (i.e. patients, loan applicants, etc.), you need to add them as "Operators/Decision Makers" to the list of stakeholders.

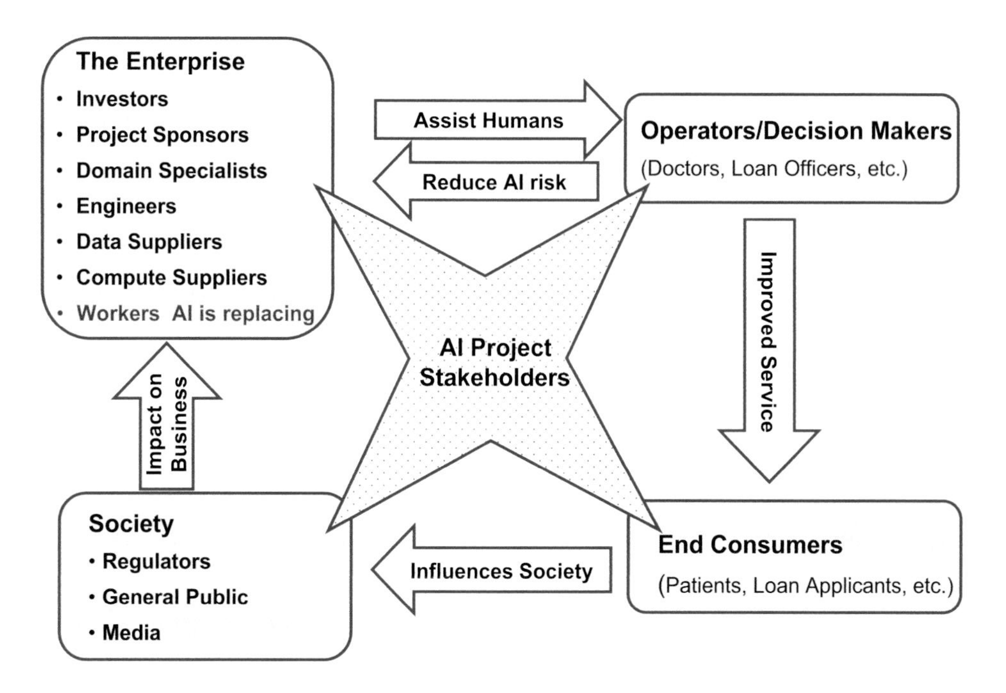

FIGURE 5.2 Scenario where the goal of the AI application is to help in a professional task.

- **Operators/Decision Makers**: You need to make sure that AI functions are really making them more productive and generally be more effective from their clients' perspective. As discussed in Chapter 2, they also provide a way to reduce the risk of AI recommending inappropriate decisions to the end consumers of AI. They play the critical role by adding their professional judgement to apply the AI output for the benefit of their clients (i.e. end consumers of the AI).

Workers Affected by AI Are Also Stakeholders

One particular group of stakeholders that will require particular attention are those employees whose jobs will be directly impacted by the new AI capability. The reason that these employees are so important is that you may need their assistance to develop the new system. Here is a real story of how that played out in one instance. A particularly successful AI project involved the delivery of a classifier to a military organisation. The key stakeholders included a team of four Analysts whose job it was to classify incoming data. Their superiors faced a massive challenge though, in that the volume of data was about to increase dramatically, but the resources available were going to be fixed. In other words, they needed to do more in a shorter timeframe with no increase in resources.

From the outset of the project, it was clear that the Analysts were the only people who could cleanse and validate the training data. While working with them to prepare the training data, we developed a strong understanding of what they did and how they did it. Our understanding of the challenges grew along with our respect for the Analysts and our solutioning evolved as a result. In working closely with the Analysts, we realised that their roles would not disappear but would evolve into the Administrators of the AI.

In addition to managing the training data, they were able to take on additional responsibilities for analyzing more complex data. In effect, *we had developed an augmented system where AI and Analysts worked together to achieve far more than the Analysts had been able to achieve previously.*

It would be nice to think that this success was our intention from the outset, but the reality is that we were lucky! Had we simply approached the project with a view to replacing the Analysts, there is a good chance that they would have been hostile to the project. Instead of helping cleanse the data and teaching us about the challenges, they would have focused on defending their personal value. As it was, they loved their inclusion in the project and they loved the technology even more. The fact that we stumbled into this success should not prevent us from learning from the fact. The key learning point being that AI is most successful when not approached as a replacement for existing resources. *AI will be most effective in augmenting existing resources and growing their roles to be more effective.*

A further factor that contributed to the success of the classifier project was the ease with which it was possible to demonstrate value to the Stakeholders. On the very first day of the project, the Analysts were able to produce a spreadsheet of example data manually labelled as part of the existing business process. The data was in a form that could be immediately used to train and test the classifier. The ease with which this initial evaluation could be undertaken meant that we were able to prove, and establish trust in the capability right at the outset of the project.

Understanding Stakeholders' Perception of Value

Unfortunately, not all situations are so simple. Quite often, the Stakeholders are looking for an AI to enable a completely new business process. This can result in a situation where the Stakeholders demand proof that the AI will work before investing resources to implement the new business process. However, without the new business process in place, there is no data available with which to develop or benchmark the AI to prove that it is doing a good job. With a difficult problem, could a team of humans do better, you just won't know. Ultimately, something has to give!

The greater the resources required and the longer the time needed to prove the AI, the more difficult it will be to secure the support of the Stakeholders. It isn't just time! It's also important that the value of the project can be understood by the Stakeholders. While that sounds obvious, it isn't always so. In particular, in situations where AI delivers an intermediate, as opposed to a direct, benefit, key Stakeholders may struggle to see the benefit.

One example of this is in the domain of text analytics where it is hard to see the value of transforming free text documents into structured data without an end user application. Many organisations have a wealth of valuable data locked away in free text documents. Transforming these documents into structured data is an intermediate step that enables other forms of data analysis. Once structured data has been extracted, it can be used for graph, temporal, geospatial and many other types of analysis. However, very few Stakeholders are willing to sponsor the intermediate, data transformation step without an end application in mind. By insisting on an application with end user value, they change

MAIL ORDER CASE STUDY

I once took on a challenge from a famous mail order company to see if we could analyze their customer data – could we build an AI model to predict their potential to buy new products. Try as we might we could not get any model to more than 5% accuracy (yes that's 95% wrong). The project's allotted month ended, and I remember swallowing hard as I broke the bad news to the customer. This was my third real analysis project, but my first failure, I was still young, I wasn't sure how they were going to react, my team's month of wasted effort had cost them real money. This was going to hurt ...

I dropped the 5% bombshell, the customer's eyes flicked to the right as they did the math ... and then broke into a beaming smile "that's fantastic! The best prediction we're working on now is half that, you've doubled our profit. In saved postage stamps alone, we will have paid for your time in less than a month."

The lesson here is that any method was going to be imperfect, but because this was an existing business process and the customer had a grasp, and a measure, of the problems' complexity, what looked like a worthless AI was deemed a success. If this had been a challenge to support a new business process, it's very unlikely that code would have left the lab, and I might well have taken up a different career well away from AI.

Dave Porter

the scope of the project and quick to deploy "out-of-the-box" solutions are proposed that perform end-to-end analytics in a proprietary manner. Any intermediate data transformation may even be locked away in a proprietary format such that it is not available for use by other potential downstream analytics.

In cases where an open, generic approach is maintained, the initial project is required to cover the cost of the intermediate data transformation. The data transformation may have the potential to support multiple downstream applications; however, the business case is evaluated solely against the first project. At this point, the business case can fail to demonstrate value. The challenge of developing AI capabilities to perform this type of intermediate service is significant. We often refer to it as the "telephone exchange challenge". Every project requires a telephone, but none of the projects can afford, or are willing, to fund the procurement of the exchange.

So, when it comes to managing the Stakeholders of AI projects, it's important to remember three key points. Firstly, the nature of AI projects is such that the range and number of Stakeholders are far greater than for conventional projects. Secondly, it is essential to bring onboard the key employee Stakeholders whose roles will be directly impacted by the new capability, they are your greatest asset. Thirdly, give serious consideration to the "telephone exchange challenge" and develop a strategy that will enable the delivery of high-value intermediate capabilities.

VIRTUAL AGENT ASSISTANT

For a major high street bank, supporting customers during the homebuying process had become increasingly difficult. Regulations, products and processes were continuously evolving, and keeping up with the most up-to-date information can be difficult, especially when that information was spread across multiple data systems.

IBM worked with the bank on the design and implementation of a Virtual Agent platform to support staff who manned the contact centres for the Home Buying and Ownership division in the UK. The Agent's role was to provide contact centre employees with the right information at their fingertips to support homebuyers. Call centre staff could access this information via the platform through either a natural language chat interface or an AI-powered search.

Appreciating how threatening an AI solution may appear to key stakeholders (including staff), a carefully thought-out approach to designing and implementing the Virtual Agent was taken.

- **Co-creation**: Sitting side by-side, the bank and IBM teams jointly co-created the Virtual Agent. Contact centre staff felt bought into the experience and could see the potential value right from the start.
- **Team Member**: The Virtual Agent was intentionally personified as a member of the bank's team and even has her own personality. She was even given a name to humanise the solution and identify her as a colleague.
- **Bespoke**: The Virtual Agent was created and customised to how the bank would get most value from the platform.
- **Impact to Job**: Using the platform meant that the contact centre staff could provide better customer service, without having to spend a long time finding information or getting concerned that the information being provided was correct. They were able to trust and rely on the Virtual Agent.

Since implementing the Virtual Agent, the bank has not only seen a 20% improvement in customer feedback but importantly a large upturn in staff engagement. They are engaged in the process and do not see the Virtual Agent as a threat, rather a colleague they can't do without.

Richard Hairsine

Richard Hairsine is an Associate Partner and AI & Automation Lead in Financial Services at IBM Consulting in the UK.

MEASURABILITY AND UNDERSTANDABILITY

The management of Stakeholders is always going to be challenging. Stakeholder management will always be easier, however, if the AI application is intended to do something that is both measurable and understandable. Perhaps the reason the British Government was prepared to invest so much in Bletchley Park was because the potential return was massive but also that the core enterprise was both understandable and measurable. The idea of taking thousands of encrypted enemy messages and transforming them into thousands of plain text ones is so easily understandable that justifying an investment becomes easy. From a measurability perspective, if you take 1000 messages and decrypt 800 of them, you are doing okay. If you don't decrypt any, then you have a problem.

An interesting aspect of measurability at Bletchley Park is that the code breakers knew if they were being successful. If you successfully decode enemy messages, you can read the plain text whereas, if you fail, it's still gibberish. Knowing when you are performing a task correctly, being able to quantify the number of individual tasks performed and a clear operational return on each task, all add up to a rock-solid business case.

Today's Data Scientists will look at Bletchley Park with envy! No upfront business case and a completely measurable return on investment throughout the entire lifecycle of the project. Most projects have a detailed upfront business case, with milestones and must haves firmly in place often before anyone has seen the data and no measurability once the project starts.

Modern business processes and their supporting IT are complex. It can be hard to determine where to invest in order to generate any form of return. Should AI be applied at the point of data collection to improve the quality of data available or should it be applied in the downstream analytics to make better operational decisions?

Often, we need to consider direct versus indirect return on investment. Improving the quality of data at the point of capture may have no initial impact. However, it may be an essential enabler for future projects that subsequently deliver a massive return on investment. This challenge of understanding indirect and enabling capability versus direct impact is significant, in modern management as we tend to focus on near term and direct business benefit.

Right now, organisations are collecting vast quantities of data and the potential of AI to exploit this data is enticing, seductive and exciting. So, at a strategic level, we hear business leaders talking about the potential of AI to transform their operations. But at a detail level, working out exactly where the value is and how to enable it is not always that simple.

Public sector organisations face the same challenge of trying to figure out where to invest. Few government departments have the resources, or at least the commitment, to invest everywhere! Most are less ambitious and want a more focused strategy that delivers a more specific return in a defined time period.

Knowing where to invest requires an understanding of your business processes, either existing or proposed, together with the accuracy required of any AI within those processes. This can only be achieved if you have a measurement strategy for business processes and the ability to measure the impact of AI on the effectiveness of those processes. Sometimes this will mean identifying and measuring distant upstream processes. The next chapter of this book will look in detail at Accuracy and Measurability.

In textbook exercises, the ability to forecast return on investment is relatively simple. In the real world, any forecasts are invariably based on a catalogue of assumptions and judgement calls. When it comes to AI, any forecast needs to take into account a range of additional factors beyond the financial return. AI offers huge potential benefits that can only be realised if the technology is trusted and applied in an ethical manner. It is also critical that consideration is given to the impact of AI on the key stakeholders, especially those whose assistance is needed in order to develop the capability.

Imagine a health application that fuses together data from multiple sources to allow the identification of new treatments and best practice. Imagine if the application improved the quality of life for tens of thousands of people and, at the same time, reduced the cost of their care such that there was a measurable financial return to the treasury. Such an application would clearly offer massive value to society … but what if, for the application to work, the privacy of individuals needed to be compromised?

What if, for the application to work, we need to collect data about peoples' lifestyles? What if the application needs to track exactly how much alcohol an individual consumes, how much exercise they actually take and how many calories they consume? What if that data will be held indefinitely? What if a future government decides that they will take into account that lifestyle data in determining entitlement to medical care and social welfare benefits?

The point here is simple! In determining value, we need to take into account many factors beyond the direct operational or financial return. Often these, knock on effects, create issues that will negate any immediate benefit if not properly considered and managed.

Having considered our Stakeholders' view of the project in terms of measurability and understandability, it's time to consider the topic of AI Ethics from a business point of view in more detail.

IMPORTANCE OF ETHICS IN AI DEVELOPMENT

AI Ethics

There have been many recent books on this topic [2–5]. Ethics is a human judgement of right and wrong, and therefore, it requires moral agency. *Since AI systems are not human, we cannot expect them to be* 'ethical' or 'unethical'. Somehow, we do expect them to act in a way that we recognise as ethical, have been designed in an ethical way and be used for ethical purposes. Adding to the complexity is the fact that there are no universal standards in ethics. What is not ethical in one part of the world may be completely normal in another part of the world. It is the responsibility of the creators of the AI to understand the ethical norms of the community where AI is to be deployed and to make sure that the AI system they are building is 'Trustworthy' for the intended users of the system. This gets us to the multi-disciplinary field of 'AI Ethics' whose main goal is to maximise the benefits of AI *for all stakeholders* while minimising its risks and negative outcomes. The core principle behind AI ethics is the expectation that human agency and wellbeing will prevail over any choice made by the machine. The original three laws of robots proposed by Isaac Asimov (see the Robotics Vignette in Chapter 2) were an attempt to get the ethics ingrained in the robots (at least in science fiction) for them to behave in a way consistent with human agency and wellbeing. In addition, the processes in the creation of AI have to be sensitive to the concerns about data responsibility and privacy, fairness and inclusion, human value alignment, accountability, transparency, technology misuse, etc. The implementation of AI ethics in practice will require technical and non-technical solutions. This is what is expected of the creators of the AI, i.e. to build a 'Trustworthy AI'. Ultimately the users of the AI system will decide if it is to be 'Trusted' or not.

DECISION VS. JUDGEMENT

Why is ethics a recurrent theme in AI discussions and largely absent from traditional IT system discussions? It's all about "DECISION vs. JUDGEMENT".

There are many differences between AI Applications and conventional programs. Perhaps the most significant difference is that we expect AI Applications to make decisions that normally require human judgement. What do we mean by judgement and how does it differ from normal IT decision-making? Let's consider a really simple example!

A police officer is on patrol and spots a car travelling at 36 miles per hour in a 30 miles per hour zone.

The police officer doesn't really have to worry about making a decision. The rules are clear and all the police officer needs to do is apply a simple rule and issue a speeding ticket. However, the police officer may also be allowed to exercise her judgement. She may be able to take into account the road conditions, the visibility, the risk to others and any mitigating circumstances such as "was the driver rushing a person to hospital?" or "would the driver likely get off on a technicality, as the speed limit sign was broken?" In short, she may be able to exercise her judgement to override the typical rule and *not* issue a ticket.

In computer science terms, we can think of decisions as being explicit and deterministic whereas judgements aren't! Two different police officers may make completely different decisions based on the same input. We implicitly expect that from humans but are uneasy if an AI seems to exhibit that behaviour. The exact circumstances considered in making a judgement may not have been previously experienced by the police officer, we expect her implicitly to call on her experience (often called common sense) to form a judgement. Generally, more experienced police officers exercise better judgement and therefore make better decisions.

Judgement involves taking into account a broad range of contextual, often ethical factors, rushing home to keep a take-out pizza warm is not the same as rushing a woman in labour to hospital. The officer needs to weigh up the pros and cons of each particular decision in context. These types of complex, often unbounded, decision process are hard to define using traditional computer programming rules but are a natural fit for AI approaches that learn by example. It should be no surprise that AI is being applied to many situations that would have traditionally required human judgement.

Whilst not all AI Applications will be required to exercise judgements, and those that do will rarely have high consequences associated with those judgements. Some AI will, and any AI tasked with judgement calls that have a human consequence will require extra care (i.e. resources and investment) to build. AI systems, more than any other IT system or indeed human, will need to be seen as explicitly trustworthy. Building trustworthy AI systems introduces four specific challenges.

Firstly, if AI Applications are going to exercise judgement, then the public, at least, will expect them to be held to the same ethical standards as a human making the decision. Baking in every possible scenario, no matter how far-fetched (pregnant mother good, cooling pizza bad, etc.) may be an impossible task for your development team, a sensible solution is to incorporate a human in the loop for any case that is ambiguous or high risk.

Secondly, in exercising judgement, AI Applications will only be as good as their experience. We call that training data, and you can always increase the scope of the training data. In our traffic cop example imagine there was a subsequent enquiry into the officer's decision not to stop the driver .... in this case the fleeing driver was an escaped convict! If the officer could prove her training taught her to show discretion so long as the conditions were safe and it wasn't above 40 mph, she might escape a conviction of her own. The corollary to this, in a high-consequence AI, is you may need to store your training data for future legal cross examination!

Thirdly, like human beings, AI Applications may be unfairly influenced by their experience. If the training data contains bias, much as rookie cop may have been taught badly to apply all the rules rigidly, then they may show bias in their future decisions. Time spent analysing your training data to spot and fix bias, e.g. making sure any gender or race is not under or over represented, is time well spent.

Fourthly, if a decision is questioned by others, then AI Applications will be expected to explain their decision-making. ML models that are used to classify images can be exhaustively tested, using partial/feint images, to reveal the boundary features between classifications. This can help a lot when explaining their decisions, NLP-based AIs do not have this same luxury. In practice, many useful AI methods are poor at explaining their decision-making. AI solutions that need safe decisions, such as driverless cars, often are built to incorporate multiple overlapping decision models (using different techniques) to make the actual decision by triangulation (all models agree equals good, models disagree escalate to a human). Overlapping model systems cost more to build and maintain, but they do offer better safety/surety to AI decisions and can help pinpoint the logic used by the AI. In addition, the cases where the overlapping models disagree can be used to find potential faults and troubleshoot areas in the AI solution that may need model improvement/extra training.

Autonomous Judgements by AI

Perhaps the most significant difference is in situations where we expect AI Applications to make decisions that normally would require human judgement. What do we mean by judgement and how does it differ from normal IT decision-making? Typically, when we think of decisions in business applications (e.g. whether a person qualifies for payment of insurance claims), we consider various variables and the relevant criteria and decide in a deterministic fashion. This is different from cases when we need human judgement.

Think of a robocop of the future (our archetype for a role that needs judgement) who stops a speeding human driver on the road. The creators of the robocop have used a ML system to 'decide' if the driver should get a ticket or not, based on historical data. How hard can it be? The key observation here is that historically any two human police officers looking at the same speeding situation may decide differently, based on their individual judgements guided by contextual data (bad weather, emergency visit to the hospital, pizza delivery, etc.). Our AI designers could not have captured every aspect of every speeding incident. In reality, all they could expect to have for training data is driver speed versus speed limit, and no records of when police officers ignored the speeders because they know it would be a marginal call. Many factors outside the basic information (i.e. posted speed limit and speed of the driver) influence the judgement. In short, our robocop will not be capable of contextual judgment, because the relevant information often involves considerations and societal understanding beyond what is contained in the training data. No doubt, the decisions of our robocop will be scrutinised more than any other IT system or indeed humans for proper social behaviour. In addition, the inherent *biases in the training data* are adopted by the AI and hence needs a careful evaluation. AI builders have to be sensitive to the underlying ethical questions and use the technology appropriately. *Please visit the Decision vs. Judgement vignette for more discussion.*

Data Concerns

AI needs training data for building models (more on this in Chapter 7). This immediately brings concerns about data privacy, governance and inherent bias.

Explainability

As we discussed in Chapter 3, ML models learn complex relationships between inputs and outputs. From the perspective of the users of the models, they are opaque (i.e. black box). Consequently, it becomes critical to explain to the users how the model derives its output from the inputs and provide transparency to the modelling process.

Making Mistakes

AI algorithms can use statistical models which inherently contain some percentage of mistakes, even if it can be designed to be small. This raises two practical questions immediately: (i) Who decides how much of mistakes is acceptable? (ii) Who is accountable for the mistakes when they happen?

Misuse

Just like any other technology, AI is susceptible for misuse by individuals, organisations or governments. The sheer availability of the technology and its increasing number of applications also make it inevitable that there will be large negative impacts.

Social Change

In contrast to other technology revolutions in history, AI is evolving at a very fast rate. This can result in a rapid transformation in the job markets without the necessary period for social adjustment in public education and training.

The discussion above shows the importance of proper ethics in building AI systems, and in the following sections, we discuss the key factors to do just that.

DELIVERING TRUSTWORTHY AI

Of the topics mentioned above, technology misuse and social change can only be addressed by careful human societal judgement. There are many technology examples (e.g. social media) where misuse resulted in significant negative social impact (e.g. fake news). We cannot blame AI for that. In the following sections, we will discuss three topics that are ethical consequences of ML, viz., the need for Fairness, Explainability and Transparency in AI systems. We will also discuss two other topics that expose the weakness of the ML technology: its propensity to make mistakes and its vulnerability to adversarial attacks.

The motivation of an enterprise should be to deliver trustworthy AI applications for their clients, and consequently ethical behaviour has to be at the core of your project. Your project could have the potential to deliver a phenomenal *value* in terms of financial return ... but only if it is delivered in an ethical manner.

By its very nature, AI will be asked to make decisions that previously required human judgement. In doing so, however, AI will be expected to perform to a much higher standard than the humans it replaces. We know, for example, that in the rollout of driverless vehicles, the public is far more tolerant of mistakes by human beings than they are of mistakes by AI applications. We all believe that any system which is going to emulate human behaviour should emulate the best aspects of humanity.

It's important to remember that every system, in any form, can only exist if society wants it to exist. Whilst it is possible for an unethical business to use AI in a Machiavellian way, there is a very good chance that, at some point powered by social media, there will be a customer backlash and the business will suffer as a result. When it comes to understanding the impact of ethics on the value of your project, there are some specific points to consider.

Firstly, it is important to remember that the ethical acceptance of a project is not determined solely by those who initiate the project. Projects will be judged by many stakeholders including your customers, your suppliers, the press, regulatory authorities and the general public.

Secondly, society's perception of what is and isn't ethically acceptable will change with both context and time. The idea of tracking members of the public using their mobile phones would have been considered highly unethical just two decades ago, yet nowadays we all accept that our movements are routinely tracked, profiled and used to target advertising at us. The thought of Governments tracking our movements was still abhorrent to many ... until a global pandemic caused many to think differently.

Thirdly, it's important to drill down into the detail in order to understand ethical concerns such that the right actions can be taken to address them. There are occasions when projects are dismissed for ethical reasons that can be addressed. For example, privacy concerns can be addressed through anonymisation of data and strict data deletion policies. At this point, however, it's important to be very clear! We are not advocating workarounds to avoid ethical concerns! We are advocating a genuine commitment to delivering trustworthy AI by understanding the concerns and ensuring that they are properly addressed. Now, let us discuss the important topic of bias in AI systems.

FAIRNESS AND BIAS

Interestingly, the term bias had been used in AI for many years! It was not a provocative term and had no connotations relating to equal rights or political correctness. Bias was a technical term that was used to describe the statistical qualities of the available data. For example, whilst working with medical data, we may have collected demographic data for everyone diagnosed with a particular medical condition but have no equivalent data for people without the medical condition. Clearly, there is a statistical bias in the data that would invalidate some forms of analysis.

BIAS IN DEVELOPMENT OF AI

You have just walked into a shop, reaching to get your credit card out and realise it's gone. You search your coat, bag and pockets finally coming to the conclusion you've been pickpocketed. You contact the bank, cancel the cards and then try to find out if you can claim any of the lost cash back through your insurance.

Luckily for you the insurance company has developed a chatbot to answer standard customer queries using natural language processing. You open the dialogue box to speak with the chatbot and ask "I've lost my wallet, how do I claim for its contents?" to which the chatbot directs you to a link in order to begin processing a new claim. Fantastic. Easy. Job done.

Let's try again. Same scenario, same problem, same question "Hey Chatbot, I've lost my purse, how do I claim for its contents?" to which the chatbot responds "I'm not sure how to deal with your request. Please see FAQs for lost or stolen items". Why? We asked the same question?

Here is a clear example of the subtle nuances between male and female entity values. The use of "wallet" and "purse" both elicited different responses, because the chatbot was unable to recognise purse, since this variable was unknowingly overlooked from the conversational development process.

The example above is not unusual, bias towards males and majority groups is pervasive throughout numerous applications of AI, from Amazon's sexist hiring algorithm [1], to Google disproportionately displaying high-paying job ads to men [2] and fintech's assigning extra mortgage interest to borrowers who are members of protected classes [3].

But these systems were not 'born biased'. Bias, by nature, is a human construct. We use cognitive biases to make mental shortcuts (heuristics). Since there have been over 180 different types of bias defined and classified by the Cognitive Bias Codex [4], how can we as humans, developing systems to 'mimic the human brain' not expect some of our biases to become embedded within these systems unless we put significant governance measures in place? In the example above, it was clear that the chatbot development team was mostly, if not entirely, male, lacking diverse and representative variables from wider demographics. Therefore, a subtle nuance like purse and wallet is easily missed.

As a community, we need to make an active effort to ensure diversity in our development teams and, thus, diversity in our thinking. Why? So, we don't unintentionally exclude vast amounts of the population from having access to our AI systems. If we continue on this trajectory, it then begs a great question posed by IBM Chief Watson Scientist Grady Booch: "whose values are we using?" since the "AI community at large has a self-selecting bias simply because the people who are building such systems are still largely white, young and male" [5].

Here is a reference to a maturity framework based on ethical principles and best practices, which can be used to evaluate an organisation's capability to govern bias [6].

Daphne Coates

Daphne Coates is a Senior Intelligent Connected Operations Consultant at IBM. She is the UK & Ireland AI Ethics Community Lead driving new research, Intellectual Property and offerings whilst also becoming the UK & Ireland Subject Matter Expert for fair and explainable AI through delivering IBM's first UKI Explainable AI project for public sector.

REFERENCES

1. Business Insider, "Why it's totally unsurprising that Amazon's recruitment AI was biased against women," https://www.businessinsider.com/amazon-ai-biased-against-women-no-surprise-sandra-wachter-2018-10.
2. The Guardian, "Women less likely to be shown ads for high-paid jobs on Google, study shows," https://www.theguardian.com/technology/2015/jul/08/women-less-likely-ads-high-paid-jobs-google-study.
3. Harvard Business Review, "AI can make bank loans more fair," https://hbr.org/2020/11/ai-can-make-bank-loans-more-fair.
4. J. Manoogian and B. Benson (2017) Cognitive bias codex. https://betterhumans.coach.me/cognitive-bias-cheat-sheet-55a472476b18
5. IBM, "Building trust in AI," https://www.ibm.com/watson/advantage-reports/future-of-artificial-intelligence/building-trust-in-ai.html.
6. D. L. Coates and A. Martin, "An instrument to evaluate the maturity of bias governance capability in artificial intelligence projects," *IBM Journal of Research and Development*, 63(4/5), pp. 7:1–7:15, (1 July-Sept. 2019).

In the media, the term bias is now being used in the context of the very serious, and more sinister, concern that AI systems will actually exhibit bias. This is a very legitimate concern and we need to ensure that AI is a part of the solution to historical unfairness. As with other aspects of AI, bias is one of those subjects where it's important to dig through the hype and the headlines in order to understand the reality.

In this section, we discuss the business value of building fair AI systems. We introduce some famous bias examples, relevant definitions and discuss the various factors that can contribute to bias in AI and how to anticipate these during the various stages of a project.

Some Famous Examples of Bias

One of the reasons we are so concerned about fairness in AI is that we have seen some very high-profile examples of bias in AI systems. These systems generated huge media attention and created a very powerful image of the dangers of AI becoming biased. Here are some examples, in case you are not aware:

- Microsoft Tay chatbot that was taught racist language by its social media users. This is a case when the AI designers did not anticipate the type of data provided by the actual users of the application [6].

- Amazon stops the use of an AI recruiting tool after the system taught itself that male candidates were preferable and penalised women candidates [7].

- Google searches involving black-sounding names were more likely to serve up ads suggestive of a criminal record than white-sounding names [8].

- Bias in Word Embeddings "Man is to Computer Programmer as Woman is to Homemaker" [9].
- Google photo classification software grouped people of African and Haitian descent under the heading 'Gorillas' [10].
- Recidivism Assessment tool in the US criminal justice system identified black defendants were twice as risky as white defendants [11].
- Error rate in facial recognition software was 0.8% for light-skinned men and 34.7% for dark-skinned women [12].

AI Is Not the Problem; it Can Be the Solution

Ensuring that AI systems operate in a fair manner is fundamental to the delivery of trusted AI solutions. As AI systems start to make decisions previously requiring human intelligence (remember the basic definition), the question of fairness is almost certainly going to arise! In fact, we should expect the level of scrutiny experienced by AI decision makers to be far greater than that experienced by their human predecessors. Hmmm … isn't that unfair on the AI? Seriously though, just as we insist on a higher standard of safety for driverless vehicles when compared to human drivers, we will expect a far higher standard of fairness to be exhibited by AI than by the humans currently making those decisions. This is often seen as an inhibitor to AI! Something that will cause society to block AI adoption until a time when fairness can be demonstrated. However, in reality, it is potentially a massive enabler for AI. The reason being that AI is not the cause of unfairness, but it can be the solution. In many of our existing, people-centric processes, unfairness already exists unseen, possibly suspected but often unprovable. People reviewing life insurance applications or determining prison sentences make judgements based on their own life experiences. Whilst we all try to be fair, the fact is that our personal biases and different perspectives do lead to inconsistencies and therefore questions of fairness. When considering this issue, it's fascinating that AI is often perceived as the problem. In fact, AI offers us hope for a better society by giving us the opportunity to surface and address these historical biases. *By applying AI to these decisions, we will create systems that are consistent and measurable.* These two factors will enable us to continue to identify unfairness and address it should it re-emerge.

Some Definitions

Let us imagine building an AI to recommend loan approvals in a bank. Based on a set of attributes about each applicant (e.g. annual income, FICO score, prevailing debts, assets owned, etc.), the bank decides to approve or reject the loan application. *Fairness is the desired behaviour dictated by an underlying standard that can be statistical, social, moral, etc.* A protected attribute (such as gender, race, etc.) is used to separate the population into groups for fairness assessment. In our example, 'fairness' criterion can be that the percentage of the loans approved by the bank should be the same for males and females. *Bias represents the deviation from the standard.* If only 40% of the females are approved, whereas 70% of the males are approved, then we potentially have a bias concern. Here we are evaluating the fairness at the group level (i.e. gender). Group fairness expects that groups defined by the protected attributes receive similar outcomes. A fairness metric quantifies the unwanted bias in the specific AI application such as the approval percentage above. There are many dozens of metrics for evaluating fairness [13]. A bias mitigation algorithm is a procedure for reducing unwanted bias in the application. Depending on how much control the enterprise has over the application development, bias mitigation can happen in the preprocessing stage (i.e. by modifying the training data), in-processing stage (i.e. by changing the learning algorithm) or the postprocessing stage (i.e. by detecting the bias during deployment and changing the predictions) [13].

Reasons for Bias

Often you hear of biased algorithms. It is worthwhile asking what makes an algorithm biased [14,15]. Here are the two primary reasons:

- **Training Data Bias**: Since the statistical AI models learn from training data, if the training data is biased, not surprisingly the model will also be biased. In our loan approval example, if the data from the last 10 years contains implicit bias in favour of males, which may not even be known to the bank, AI will learn that. Now that we are looking for bias in the AI, we cannot escape from the biases in the past practices. It is like looking at ourselves in the mirror and realising we have been biased all along. Once we are aware, this problem can be addressed systematically.
- **Biased Algorithm**: Different groups are likely to have different distributions over the attributes, and those attributes have different relationships to the decision we are trying to predict. Since most algorithms are designed to minimise average error, there is more data corresponding to the majority group (by definition), the algorithm gives more weight to the majority group. This effect can be partially addressed by careful data gathering. Another reason for biased algorithm may be the deliberate choice of a simpler model due to the limited data available, while the actual problem may need a more sophisticated model.

To understand how bias in data can occur, and more importantly what to do about it, let's consider three different examples.

Recidivism Prediction

Our first scenario is based on the real-world recidivism incident mentioned above [11]. In this scenario, data was collected regarding decisions taken by human operatives over a period of many years. These historical decisions are used to train an AI system. Unfortunately, because the decisions have been made by ordinary people, they include the types of prejudice that humans have, sadly, demonstrated throughout history. This prejudice is learnt by an ML-based AI and, consequently, the AI exhibits the historical prejudice. In this scenario, it is imperative that the data is cleaned up! The prejudice embedded in the historical training data must be identified and removed. There is no place for it. This goes beyond protecting obvious fields such as "gender" from being used in the model, you need to consider other data markers that might identify a candidate's ethnicity or gender, such as "zip code", "preferred language" or a Social Security Number that is not consistent with the date of birth, possibly indicating an immigrant status.

Insurance Risk Prediction

Now consider a second scenario. In the UK, we know that there is a statistical correlation between young male drivers and road traffic accidents. However, whilst the correlation is irrefutable [16], it is considered discriminatory and unfair to use this information in pricing car insurance. In fact, it is illegal to do so. If we trained an ML using the raw data, and relied solely on it to price insurance policies, then there is a high likelihood that the AI would demonstrate an illegal bias. In this scenario, we should be careful about cleaning up the data. The source data is accurate and does not contain any form of historical human prejudice. The data reflects a statistical reality, and that reality is not in any way illegal or unethical. It is illegal and unethical to base policy decisions based on the statistical correlation so the data should not be used for training an ML. However, we should not try to massage the data to hide a statistical correlation. The upshot of this is a win for the UK insurers in that they get to charge equally high rates to young female drivers, who are at statistically lower risk than their male counterparts.

Predicting Fraudulent Transactions

In our third scenario, let us consider a completely different type of bias. Let's consider a fraud detection solution where financial transaction data has been captured over a long period of time. Within the data are one or two examples of a particularly rare, but very costly, type of fraud. Alongside these one or two examples of fraud are thousands of examples of legitimate transactions. In this scenario, there is a clear bias in the data in that it is skewed towards legitimate transactions. Depending on the ML algorithm in use, the AI can learn to achieve a very high overall accuracy just by classifying all transactions as legitimate. The situation is complicated though by the cost of making an error. The cost of failing to identify a fraudulent transaction may be many times higher than the cost of falsely blocking a legitimate transaction.

So, what should you do in this scenario? It depends! The best approach will depend on the exact scenario. However, a number of options should be considered. One option is to consider the inclusion of a cost function in the ML such that the small number of fraudulent cases carry more weight in the training. Another option is to manipulate the composition of the training data to artificially create more fraudulent examples ... this is not a very scientific approach.

There are many reasons as to why bias happens in real life. These three examples demonstrate the importance of understanding what the actual issue is and adopting an appropriate strategy to deal with it.

EXPLAINABILITY

Until the recent popularity of ML, we did not have a reason to invoke the word 'Explainability' in a real conversation or in an article. So, what is this all about? As we have explained in previous chapters, the problem with statistical ML systems (e.g. neural networks) is that they map inputs to expected outputs using complex algorithms, sometimes with millions of parameters, with no explicit form that you can see. As a result, you do not know why the model gives you the output it does, because there is no real explanation coming with the output [17,18].

Let us take the simplest of example, where a social media tool recommends a friend or a connection for you. It is common for the tool to identify the common friends you and the recommended person have, in order to justify the suggestion. That was an easy one. Every now and then, you scratch your head, "Why this person?" The same thing about movie selections; since you like science fiction movies, here is another science fiction recommendation for you! These are easy recommendations that you could accept or ignore.

Explaining Business Decisions

Let us think of a business application, where the output of the AI model is a decision that has a serious consequence. Here are some examples: treatment suggestion for a patient to a doctor, recidivism recommendation for a prisoner to a judge or loan approval decision for an applicant to a bank officer. In these cases, it is not enough to know just the decision recommendation from the AI; we need to know why. This was not a problem with erstwhile rules-based systems that applied a set of defined rules to make decisions. These rules were easy to understand for the people working in the domain, and as a result, no special explanation was necessary. Unfortunately, a neural network cannot do that. It is ironic that the AI technique that is driving the global excitement cannot explain itself!

The ability to explain the reasoning of an AI has always been a very sensitive issue in AI. If we are to trust AI applications with sensitive and important decisions, then it is not unreasonable to expect the AI to explain its reasoning. What is important is to think about systems and not algorithms (have we said that before?).

Who Needs Explanations?

In our earlier discussion, we identified various stakeholders that play critical roles in the successful implementation of an AI application. We now revisit them to see who will benefit from an explanation. For the kinds of decisions in business applications we discussed, we see four distinct groups.

i. Engineers who build, deploy and support the system need to know the technical details of how it works for the purpose of debugging and identifying improvements.

ii. Operators of the AI system (i.e. Professional Decision Makers) will need explanations in the domain of relevance (medicine, law, etc.).

iii. Government agencies and Regulatory bodies that monitor or audit the relevant business processes want to make sure that proper guidelines on data privacy, fairness, etc., are followed. The explanation has to be in the context of the relevant regulations.

iv. End consumers need to hear explanations that make sense to their situations and help with exploring options in the outcomes. Clearly, they will have the lowest threshold for complexity and domain information.

EXAM SCANDAL IN THE UK

A phrase commonly heard amongst AI practitioners is 'All models are wrong; some models are useful'. This truism neatly summarises the challenges we face in many AI projects. It is almost always technically possible to build some kind of model; however, this technical feasibility must always be traded off against the usefulness of the model that results. Often, in our desire to create 'something that works', we neglect the second half of the equation; making sure that the model that 'works' does something 'that's useful'.

A good case study of this effect is the 2020 exam results in England & Wales. Due to the social distancing restrictions resulting from the global COVID-19 pandemic, exams could not take place as usual; however, the government agency the Office of Qualifications and Examinations Regulation (OFQUAL), responsible for awarding grades needed to award qualifications to students. OFQUAL decided to use an algorithm for this purpose, based on a combination of teachers' grade predictions, historical class performance and historical school performance. The grades awarded by the algorithm ended up being far lower than the grades predicted by the teachers and, therefore, the grades expected by the students. This caused a public outcry, which ultimately led to results assigned by the algorithm being discarded in favour of the original predictions set by the teachers.

Initially the UK Prime Minister, Boris Johnson, blamed a 'mutant algorithm' for this fiasco [1]; however, whilst the deployed algorithm had several technical flaws which have been outlined in some detail [2], the circumstances of the situation meant that any algorithm would have struggled to achieve a useful result. Factors against the useful deployment of the algorithm were numerous.

First there was large expectations gap between the pupils, teachers and OFQUAL. Teachers' predictions were significantly higher than historical achievements, meaning that assigning grades as predicted by teachers would result (and ultimately did result) in significant grade inflation. The algorithm's main purpose was to bridge this gap. However, we now run into the second, more fundamental, problem; a shortage of individual pupil level data upon which a model can be built.

This led to the model being based on aggregated data from schools, and classes, rather than the pupils themselves. Essentially an individual pupil's performance became of function of their school's historical performance, which on, an individual level, is grossly unfair. It meant that it would be practically impossible for a pupil to break historical norms (at either end of the scale – i.e. no pupil would be the first ever pupil in their school to achieve an A*, or a U).

So, it's clear that although OFQUAL *can* build model to assign individual scores, they can't do so effectively. The data to do so just isn't there. Now the question turns to the consequences of applying such a model. In some cases, usually where misclassification costs are low, it's acceptable to proceed. For example, when building a model to predict which sewers to clean, the cost of cleaning a sewer that doesn't need cleaning isn't that high, and cleaning an already clean sewer doesn't adversely impact its performance. This is patently not the case here. The impact school grades have on a person's life chances is enormous. It's not something that's acceptable to get 'knowingly wrong'.

Unfortunately, OFQUAL did not see things that way and applied their model, with predictable results. The adjustments made by the algorithm were seen to be biased in a number of ways [3], and it was this bias that caused the public outcry and subsequent government U-Turn. Of course, the U-Turn has just replaced one problem with another; the fact that the class of 2019–2020 have inflated grades [4].

Ultimately, this situation could have been avoided if OFQUAL had asked the right question. Instead of 'Can we build a model to assign grades', the question should have been 'Should we'. And given the situation, the available data and the consequences, the answer should have been an emphatic 'No'.

Michael Nicholson

Michael Nicolson is IBM's Chief Data Scientist for Energy, Environment and Utilities. In this role he helps clients drive improved performance from their data, be that through energy efficiency, leakage reduction, pollution prevention or otherwise. He holds a First Class degree in Physics from Imperial College, and before focussing on water and utilities he developed real time analytics applications for England Rugby Union.

REFERENCES

1. S. Coughlan, "A-levels and GCSEs: Boris Johnson blames 'mutant algorithm' for exam fiasco," https://www.bbc.co.uk/news/education-53923279.
2. T. S. F. Haines, "A-levels: the model is not the student," http://thaines.com/post/alevels2020.
3. Financial Times, "Were this year's A-level results fair?" https://www.ft.com/content/425a4112-a14c-4b88-8e0c-e049b4b6e099.
4. S. Hubble and P. Bolton, "A level results in England and the impact on university admissions in 2020–21," UK House of Commons Briefing Paper 8989, (September 2, 2020).

In each case, acceptable explainability is dictated by the audience, and a single application may need to explain itself multiple times in different ways.

The approach [17–19] to provide the appropriate explanations depends on the specific context. Some relevant considerations are:

- One shot static output from the system or an interactive explanation with the user.
- Explanation of a single output instance or a global behaviour of the model.
- Explanation is from the actual model or a surrogate model that tries to mimic the actual model.
- Explanation is based on specific samples or features (i.e. attributes) or data distributions.

There are examples of open-source packages [20] for implementing explainability in business applications.

Explaining the Behaviour of a Complex System

Consider, for example, a driverless car comprising multiple AI subsystems. There may be a vision system that identifies and classifies objects. There may be a laser range finder that calculates the distance to objects around the vehicle. There may be a GPS system that identifies the location of the vehicle, and this system will be supported by mapping data. There may be a noise classification system that identifies the sound of sirens. All this data needs to be fused together and fed into some form of Command & Control subsystem that decides the next course of action.

What level of explainability would we expect in a driverless vehicle of this type? It would not be unreasonable to expect the system to state,

> *I made the decision to pull over to the side of the road because the noise classifier detected a siren approaching from behind, the vision system identified an emergency vehicle approaching from behind, the road was straight and it was a safe place to stop without causing an obstruction.*

Within this explanation, there are "classifications" that may have come from an AI subsystem that cannot fully explain its reasoning. For example, the vision system may not be able to explain how it identified an emergency vehicle approaching. Whilst we may feel that is unacceptable, it is important to accept that there are many situations when we as human beings cannot explain our own image or sound recognition. However, even when the classifier cannot explain its reasoning, it should be possible to design a ML system so that it retrieves the training data that influenced its classification decisions. The vision system may not be able to explain its classification, but it should be able to say, "here are the closest matching images in my training data".

It may one day be a legal requirement that all critical system AIs retain their original training data, so that a forensic examination is possible should the AI go wrong.

So, whilst certain algorithms cannot explain their decision-making, it doesn't mean an overall system cannot be engineered to be explainable. ML systems should be able to support their decisions by showing the training evidence that most closely resembles their decision. In any system, we should be able to trace the functionality of the AI and reproduce decisions in a test environment. We must also ensure that the AI has been developed using best practices and in a manner that would be deemed competent in any investigation. Make sure you allocate resources in your AI project plan/business case to cover these system requirements.

TRANSPARENCY

Transparency, in the development and operation of AI systems, is intrinsically linked to trust! We can't build trust without transparency.

If we are expecting AI applications to make important decisions, then the Stakeholders of such systems are going to want to understand how the AI was developed and how it works. This means being clear on which algorithms are used in the application and what data was used in the development (either directly as training data or as use case and test data).

Once again, this is an area where AI needs to build on and extend the best practice of conventional complex systems. As with all software projects, it is imperative to ensure proper configuration control. It is also important to ensure the software implementation is properly tested. However, in cases where ML is applied, then the data aspect of AI is again very significant. In developing an application, we need to be able to demonstrate that the training data was correctly managed and that the system was trained and tested in a competent manner.

This includes traceability of the development process. For example, when an AI application makes a decision, we must be able to trace the configuration of the AI including understanding which data was used for training and how the training process was configured. This may sound obvious; however, be careful! In conventional software development, we may have released a new version of the code every few weeks. With ML, we could be developing and deploying new AI Models on a daily basis.

The concept of FactSheets [20] is an appealing idea, much like a food label, to collect the various pieces of information related to the AI development. This information can then be used to satisfy the needs of the various stakeholders. There are ways [22] to automate the capture of most of the information from the development artefacts to make it less painful.

One particular area to think about is the impact of privacy regulations on traceability. If our ML is trained using real data, there is the possibility that we could be asked to delete some of the data records. An individual could, for example, make a request under GDPR for their personal records to be deleted. If we delete data records from the training data, then hypothetically, we may also be required to delete or retrain the AI Model associated with those redacted records. Once the data has been deleted, we may be unable to reproduce the behaviour of the system if subsequently challenged to validate, or explain, a historical decision. This is currently a topic of debate.

TACKLING THE WEAKNESS OF ML SYSTEMS

In spite of all the exciting advances seen in the evolution of ML systems over the past decade, it is important to understand some fundamental weaknesses in their application to mission/business critical tasks. We discuss two specific aspects of this below, the inevitability of their mistakes and their susceptibility to adversarial attacks.

Making Mistakes

As described in Chapter 3, the underlying concept behind most of the current ML techniques is called 'Statistical Learning' [23]. Just as it sounds, the *model is not created by explicit programming, but 'learnt' from the training data*. While it is very powerful in finding patterns in data not easily found by humans, it suffers from the known vagaries of data. The first observation is that due to the statistical nature of the data, the mapping from the inputs to the output may not be unique and have some scatter. The other is the complexity of the statistical models. More complex the model, the more data you need. Simpler models failed to explain the detailed differences in the data. Complex models are so optimised for the training data that they have trouble 'generalising' to new data previously not 'seen' by the model. So, any model that does well on a broad set of data, will inherently contain some errors. Added complication comes from the observation that a ML model can often fail on specific instances, despite being very confident in its prediction [24]. This leads to the need for quantifying uncertainties in the model prediction in addition to the model output, so that the user can use the information accordingly [25].

The question is how to deal with these errors, however small they may be, in business/mission critical systems. While 95% accuracy is amazing for most tasks of no serious consequence, 5% error in consequential systems can lead to significant loss of life or value. This is the conundrum we are in. The complicating factor is that in many of these situations, we do not know the right answer. Think of the use of AI in a doctor's office, where AI recommends treatment option A for a patient, when the doctor prefers option B. If the patient got option B and did not recover, what can we say? May be, option A would have been better, we do not know!

The practical business consequence of these inevitable errors is the need to decide what percentage of errors is tolerable for the specific business application and how to manage the consequence of errors when they do happen.

Susceptibility to Adversarial Attacks

Although Deep Neural Network models are extremely popular, they can be subjected to relatively simple adversarial attacks to cause them to misbehave. The extent of the attack depends on the level of access to the model building process (e.g. type of model used, training set used, etc.) and the goal of the adversary as to misguide the output for one sample or influence overall performance of the model [26]. If an adversary has an exact copy of your training data, they can experiment with it to discover its limitations and design attacks they know will beat it – so if you are building a mission critical AI, keep your training data secure. Also, certain models are more susceptible to attacks. In most business environments, we hope that there is adequate cybersecurity that prevents access to internal networks where the models are trained. Then, the most likely scenario is when the adversaries interact with the deployed model as a black box, just like any other user of the AI application. The common attack in this scenario is called "Evasion Attack" where a small modification of the input to the model can change the model output. The first example [27] of this is how an AI system can be made to perceive a stop sign as a speed limit by just physically adding small stickers to the stop sign. The second example [28] is when a small amount of noise, not noticeable to the human eye, is added to an image of a panda, the vision system classifies the modified image as a 'gibbon'. This problem is not unique to digital images and equally applicable to other data formats such as texts. Good news here is that there are tool kits [29] available to detect and mitigate adversarial robustness attacks in AI systems. As you develop AI applications using Deep Neural Networks for classification problems, you need to think about the potential adversary attacks and their consequences.

IN SUMMARY – THERE'S MORE TO VALUE THAN MONETARY RETURN

If proving the value of a conventional software application is challenging, then AI applications are even more challenging:

- To repeat again, AI applications are different from traditional IT applications.
- Their business cases involve more uncertainty and less obvious and longer lists of stakeholders.
- The monetary value of an AI project is highly vulnerable to ethical problems that can rapidly cause any perceived value to disappear.
- The goal of the enterprise must be to create trustworthy AI applications using proper ethical principles.

- Whether they become 'trusted' or not, is left to the assessment of the users of the applications.
- 'Trust' has elements beyond direct revenue to the enterprise.
- It is up to the enterprise to make the best choice on where to invest in the AI technology for long-term business value and social benefit.

REFERENCES

1. *Action This Day*, International Churchill Society, https://winstonchurchill.org/the-life-of-churchill/life/man-of-action/action-this-day-27/.
2. S. Russell, *Human Compatible: Artificial Intelligence and the Problem of Control*, Penguin (2020).
3. G. Marcus and E. Davis, *Rebooting AI: Building Artificial Intelligence We Can Trust*, Pantheon (2019).
4. M. Liao, *Ethics of Artificial Intelligence*, Oxford University Press (2020).
5. M. Coeckelbergh, *AI Ethics*, MIT Press (2020).
6. G. Neff and P. Nagy, "Talking to bots: symbiotic agency and the case of Tay," *International Journal of Communication*, 10, pp. 4915–4931 (2016).
7. J. Dastin, "Amazon scraps secret AI recruiting tool that showed bias against women" (October 2018) https://www.reuters.com/article/us-amazon-com-jobs-automation-insight/amazon-scraps-secret-ai-recruiting-tool-that-showed-bias-against-women-idUSKCN1MK08G.
8. L. Sweeney, "Discrimination in online ad delivery," (January 28, 2013). Available at SSRN: https://ssrn.com/abstract=2208240 or http://dx.doi.org/10.2139/ssrn.2208240.
9. O. Papakyriakopoulos, et al., "Bias in word embeddings," *Proceedings of the 2020 Conference on Fairness, Accountability, and Transparency*, FAT*'20, pp. 446–457 (2020).
10. M. Zhang, "Google photos tags two African-Americans as gorillas through facial recognition software," https://www.forbes.com/sites/mzhang/2015/07/01/google-photos-tags-two-african-americans-as-gorillas-through-facial-recognition-software/.
11. J. Angwin, et al., "Machine bias: there's software used across the country to predict future criminals. And it's biased against blacks," ProPublica, (23 May 2016). www.propublica.org/article/machine-bias-risk-assessments-in-criminal-sentencing.
12. J. Buolamwini and T. Gebru, "Gender shades: intersectional accuracy disparities in commercial gender classification," *Proceedings of the 1st Conference on Fairness, Accountability and Transparency*, PMLR 81:77–91 (2018).
13. R. K. E. Bellamy et al., "AI Fairness 360: an extensible toolkit for detecting and mitigating algorithmic bias," *IBM Journal of Research and Development*, 63(4/5) (2019).
14. D. Danks and A. J. London, "Algorithmic bias in autonomous systems," *Proceedings of the 26th International Joint Conference on Artificial Intelligence (IJCAI-17)*.
15. A. Chouldechova and A. Roth, "The frontiers of fairness in machine learning," arXiv:1810.08810.
16. P. Jenkins, "Why gender-based car insurance is preferable to a black box spy," https://www.ft.com/content/0e54a5da-8148-11e8-bc55-50daf11b720d (July 9, 2018).
17. M. Hind, "Explaining explainable AI," *XRDS: Crossroads, The ACM Magazine for Students*, 25(3), pp. 16–19 (2019).
18. R. Guidotti, et al., "A survey of methods for explaining black box models," *ACM Computing Surveys*, 51, pp. 1–42 (2018). Article no. 93.
19. V. Arya et al., "AI explainability 360: an extensible toolkit for understanding data and machine learning models," *Journal of Machine Learning Research*, 21 (2020).
20. AI 360 Explainability Toolkit: https://aix360.mybluemix.net/.

21. M. Arnold, et al., "FactSheets: increasing trust in AI services through supplier's declarations of conformity," *IBM Journal of Research and Development*, 63(4/5) (2019).
22. AI Lifecycle Governance, https://aifs360.mybluemix.net/governance.
23. T. Hastie, R. Tibshirani and J. Friedman, *The Elements of Statistical Learning*, Springer (2017).
24. E. Hüllermeier and W. Waegeman, "Aleatoric and epistemic uncertainty in machine learning: an introduction to concepts and methods," *Machine Learning*, 110, pp. 457–506 (2021).
25. U. Bhatt, et al., "Uncertainty as a form of transparency: measuring, communicating, and using uncertainty," *Proceedings of the 2021 AAAI/ACM Conference on AI, Ethics, and Society, AIES'21*, pp. 401–413 (2021).
26. H. Xu, et al. "Adversarial attacks and defenses in images, graphs and text: a review," *The International Journal of Automation and Computing*, 17, pp. 151–178 (2020).
27. K. Eykholt, et al. "Robust physical-world attacks on deep learning visual classification," *Proceedings of the IEEE Conference on Computer Vision and Pattern Recognition (CVPR)*, pp. 1625–1634 (2018).
28. I. J. Goodfellow, J. Shlens and C. Szegedy, "Explaining and harnessing adversarial examples," arXiv preprint arXiv:1412.6572 (2014).
29. Adversarial Robustness Toolkit, http://art360.mybluemix.net/.

CHAPTER 6

Ensuring It Works – How Do You Know?

If one of your children arrives home with a report card stating that he or she has achieved 95% in their mathematics exam, you should be justifiably proud.

Our perception of accuracy and what it means is perhaps heavily influenced by our childhood experience of tests and exams. When we hear that an AI has achieved an accuracy of 95%, our immediate reaction is that the performance is excellent.

Now, let's consider a different perspective ... imagine you board a flight and the Captain announces, "Don't worry folks, I only crash one in twenty!" Will you still stay on the plane? That is 95% accuracy, but it doesn't seem so good now.

The concept of accuracy is something that all those involved in artificial intelligence (AI) projects need to understand. It may at first appear relatively simple. However, as our simple example above demonstrates, our perception and understanding of accuracy may need to be challenged if we are to successfully deliver an AI application. Furthermore, accuracy is just one of the metrics that should be considered in measuring the quality of an AI application. In this chapter, we will look at how the performance of an AI application can be defined and measured. We start with a short description of how application quality is managed in traditional software development and compare it to what is needed for AI applications using Machine Learning (ML).

MANAGING QUALITY OF TRADITIONAL SOFTWARE

In traditional software development [1,2], the expected behaviour of an application is captured in explicit or implicit specifications of outputs versus inputs. Quality management revolves around managing defects (i.e. bugs) that document the unexpected behaviour of the application relative to the expected behaviour, during development or deployment. The purpose of testing [3] at various levels such as unit testing, functional testing, system testing, etc., is to manage the risk by finding and fixing significant bugs before the application

DOI: 10.1201/9781003108498-7

IS THE AI WORKING?

If we are to trust AI, we need to be sure that it's working!

In conventional systems, we do that through extensive testing that covers relevant input scenarios and key edge cases. We ensure the system is developed "to specification". However, in AI systems, we are expecting the AI to make decisions much more akin to those of a human specialist.

In conventional systems, we ask one simple question ...

... is the application working to specification?

We answer that question by testing, testing and testing again! We perform white box test, black box test, factory acceptance test, user acceptance test, site acceptance test, security test and performance test (to name just a few)!

What about AI applications ... they're just software so surely we just test them in the same way? Well, unfortunately, it's not quite that simple. As mentioned previously, AI applications are expected to perform tasks normally requiring human intelligence. As such, they are often expected to exercise judgement and make decisions in circumstances that have are not always expected. We don't expect human beings to behave like robots ... we just want our robots to behave like human beings.

When it comes to assessing the performance of a human decision maker, we use a much broader set of criteria. With AI applications, we will need to do the same. We will also need to ask ...

... is the application being applied in a situation for which it was trained?

... has the application been correctly trained?

... could the application realistically be expected to cope with the situation it was presented with?

... was the application deployed with appropriate governance and the right safeguards to ensure it was making good decisions.

So, can we test AI systems in the way we used to test conventional systems or should we think about AI in a different way and ask questions such as ...

... were the correct development processes used?

... could the developers have anticipated the circumstances in which the AI was applied?

... was the correct training data used to develop the application?

... were the developers qualified to build and deploy an AI application?

is put into production. Once deployed, there is a defined process to collect incident reports during operations and provide a mechanism for support teams to diagnose and resolve them expediently. This may involve code fixes by the original engineers who created the relevant code. There are two key points to remember here:

1. Due to the existence of (written or understood) specifications, it is possible to define a bug unambiguously any time during the application development and deployment.
2. The diagnosis and resolution of a bug can be localised to incorrect requirements or specific incorrect or missing line(s) of code in the application or specific choices in configuration parameters in the application environment.

MANAGING QUALITY OF AI APPLICATIONS

Given that we expect AI applications to exhibit complex behaviour, we should not be surprised to learn that testing these applications is also complex. Complexity exists in all forms of AI. For example, an expert system may be constructed through the application of tens of thousands of rules. Whilst it is possible to test the application functions correctly in interpreting rules, it is not practical to predict the behaviour when tens of thousands of rules interact. Similarly, in ML, it is possible to functionally test the implementation of the learning algorithms but impossible to predict the behaviour of the system when it is exposed to data it has not seen before.

As a reminder, a ML component is really making prediction of the output for the given input, based on what it learnt from the training data. The prediction can be numerical (e.g. temperature or a stock price for tomorrow) or a classification (e.g. identification of an object in a picture, prediction of the next word in a sentence, decision to approve a mortgage or not, recommendation of a specific treatment to a patient, etc.). The underlying knowledge to make the prediction is captured in a complex statistical model with thousands (or even many millions) of parameters not comprehensible to humans. As a result, there are two key differences from traditional software development.

1. Due to the statistical nature of ML algorithms, certain fraction of errors in model outputs is inevitable. We can only minimise the errors by following careful development processes in the management of data and modelling activities.
2. In production under the actual business workload, except in a few situations where there is explicit human feedback (e.g. acceptance of the recommendation by the user), there is no way to tell what the right answer is. Sometimes knowing the right answer may take months or even years!

Debugging individual instances of 'erroneous' model outputs requires a detailed understanding of the training data and model behaviour, requiring significant skills and effort. Even if you can fix one instance of 'erroneous' behaviour, it does not mean that there will not be a new 'erroneous' behaviour created due to the complex learning process. As a result, the most common practice to characterise the quality of ML models is through a metric representing accuracy of the model output against carefully chosen test data representing the actual workload. Of course, there are many other aspects to the quality of AI applications (e.g. trust worthiness) besides accuracy. We will return to them later in this chapter.

AI APPLICATIONS – THEY'RE ONLY HUMAN!

"Cognitive systems are not deterministic. In fact, they're going to make mistakes and I can prove mathematically why they have to behave that way."

That statement was made by a very senior executive in a meeting and, as a professional engineer, I have to say I was shocked. With my entire career focussed on delivering AI applications, I could not accept the executive's point of view. My response was cutting, "The pilot who is flying me home in 6 hours is Cognitive ... but deterministic when it comes to landing a 777 at Heathrow."

Several years later, I am starting to look at this conversation with a very different perspective. If we expect an AI application to perform a function normally requiring human intelligence, then perhaps we do have to accept that mistakes will be made. After all, human beings make a lot of mistakes.

The very nature of AI problem spaces is such that it should be impossible to test the function of the AI for every possible permutation of input data. If it was possible to test every permutation, then why use AI?

So perhaps we need to think differently about how we engineer AI applications. Rather than viewing an AI application as a black box that should be tested exhaustively, we should view the application within the context of a much broader system.

Consider the pilot for my flight home. The pilot does not operate alone! There is a second cognitive entity, or co-pilot, also sitting in the cockpit. Both pilots are surrounded by deterministic equipment that alerts them to issues. The pilot is supported by other professionals in air traffic control who also monitor what is happening and intervene if required. Pilots are trained not just to deal with common circumstances but also given strategies to deal with rare/unusual situations. For example, pilots are trained to adapt to the different handling characteristics of a damaged aircraft. This training has evolved over many decades of aviation in what must be considered the ultimate continuous improvement operation. Every time a safety incident occurs in aviation, it is investigated and the lessons learnt are fed back into the design of aircraft, operating procedures, pilot training and so on.

In short, the cognitive entity, that is, a pilot sits in the middle of a massive system of systems all focussed on ensuring mistakes are not fatal. Perhaps, we need such an approach for our AI applications?

James Luke

We spend the next few sections on explaining accuracy in its various flavours. Readers with prior experience of statistical accuracy may wish to skip over the next few sections. For those, who haven't had the pleasure, we aim to provide a basic introduction to the art!

STATISTICAL ACCURACY

For a start, it is very rare that the effectiveness of an AI application can actually be reduced down to a single number. Understanding what accuracy is, how to measure it and how to define what is required are essential aspects of any AI program.

So, what do we mean by accuracy? Based on the discussion above, there are two common types of AI predictions: (i) Numerical prediction by regression and (ii) prediction by classification. Metrics for estimating accuracy is different in these two cases.

Numerical Prediction by Regression

To make the discussion concrete, let us consider the case of predicting a person's weight from his/her height using regression models. Input is the height and output is the weight. We obviously need data to create the model, and we are going to use the body measures data from the US Center for Disease Control [4] that contains heights and weights of over 8500 people, 2 years and older. The details are described in the accompanying vignette on "Predicting Weight from Height". We created two ML regression models, i.e. Linear Regression and Regression Tree. Table 6.1 is the summary of metrics for the two models.

Three popular metrics that characterise errors in regression models are:

- **Mean Square Error (MSE)** is the error in the model captured as the sum of the squares of the difference between the prediction and the actual value in the training data, divided by the number of predictions. The goal of the modelling process is to find specific values of the model parameters that minimise this error. Obviously, smaller MSE is better.
- A related metric is **Root Mean Square Error (RMSE)** which is just the square root of the MSE. A simple interpretation of the RMSE for the linear regression model is that the prediction of the model is within ±17.839 kg of the actual for 68% of the data, when the error is normally distributed.
- R^2 (***R*-Square**) represents the percentage variation of the output that is explained by the input variable(s). It ranges between 0 and 1. In our example, $R^2 = 0.64$ means, 64% of the variation in the weight is explained by the height. During the model creation activity, this metric also helps to do better feature engineering by identifying input variables (i.e. features) that do not contribute much to quality of the model and hence can be removed from the analysis.

The regression metrics are very close for the two models and the Regression Tree model appears to be only slightly better than the Linear Regression model. But, when you look at the predicted values in the figures (see the accompanying vignette), you can see that for lower heights (i.e. for children), the predictions from the Linear Regression model are basically useless (including negative weights!). While Linear Regression is simpler to understand, the Regression Tree model is doing better predictions across the entire population and therefore a better model between the two. This is just an example of the assessment needed before any model is put into practice.

Now we move on to discuss the accuracy metrics in classification problems.

TABLE 6.1 Regression Metrics for the Two Machine Learning Models from the Vignette, "Predicting Weight from Height"

Model	MSE	RMSE	R^2
Linear regression	318.24	17.839	0.63
Decision tree	311.29	17.643	0.64

PREDICTING WEIGHT FROM HEIGHT

The purpose of this vignette is to show the quality metrics from two different ML regression models. We used the 2007/08 Body Measures data from the US Center for Disease Control [1] that had 8861 observations of both heights and weights for persons 2 years and older. Only 80% of the data was used to train the model, with the remaining 20% (i.e. 1770 points) held back to test the model. Since the actual weights are known for all the data, the quality metrics can be calculated relative to the model predictions.

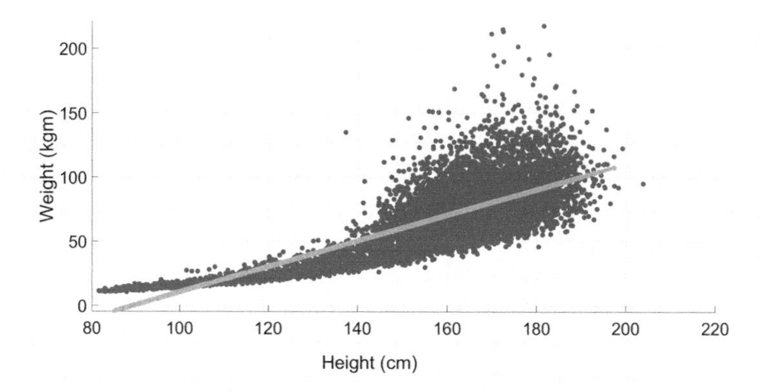

FIGURE A: A Linear Regression Model finds a straight line that best fits the training data and uses that to predict. The blue points are the training data and the orange line represents the predictions for the hold-out data set.

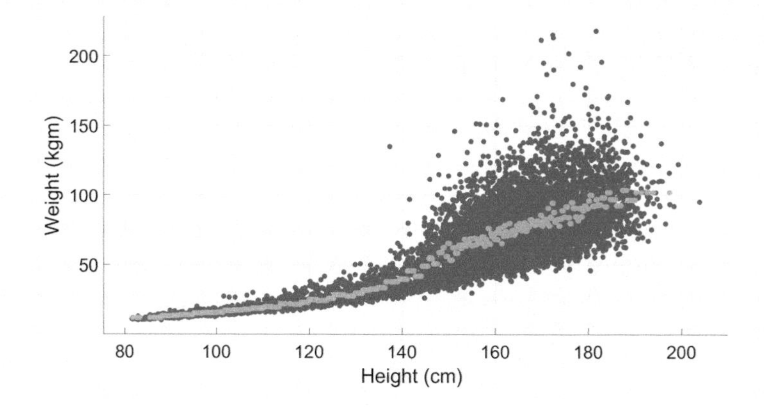

FIGURE B: A Regression Tree Model [2] partitions the input data into bins (called leaves) to minimize the error and estimates the output values based on the bin averages. The blue points are the training data, the orange points are the predictions for the hold-out data set.

The relevant metrics and relative performances of the two models are discussed in Chapter 6.

REFERENCES

1. Body Measures Data, Center for Disease Control, https://wwwn.cdc.gov/Nchs/Nhanes/2007-2008/BMX_E.XPT.
2. L. Breiman, et al., Classification and Regression Trees Chapman and Hall/CRC, 1st Edn (1984).

Prediction by Classification

Let's start with a simple binary classification example where the AI has just two choices. Many of us now use biometrics to logon to our laptops and phones. When you present your thumb to your device, the classifier can either correctly identify you or fail to do so. What about if someone else picks up your device and tries to access it? Does the classifier correctly identify the attempted intrusion and block access? So, we have two separate metrics to consider: the number of times a person who should have access is granted access and the number of times a person who should not have access is blocked. In this case, we can have a simple definition of accuracy as the percentage of times the classifier correctly identifies the user as valid or not, out of the total attempts.

$$\text{Accuracy} = \frac{\text{Correct Response}}{\text{Total Attempts}}$$

Obviously, we need to consider the opposite scenarios: the number of times a person who should have access is blocked and the number of times a person who should not have access is allowed in.

In the world of AI, we have a wonderful terminology that describes these four statistics. You will hear accuracy described in terms of True Positives, False Positives, True Negatives and False Negatives. This basic four box model covers all possible scenarios when considering the accuracy of a simple, binary classifier. What does all of this really mean? Well, quite simply, it means that even a simple binary decision requires four separate metrics to fully describe its accuracy. In practical terms, most applications can probably just rely on a single metric to describe performance. For example, if we were writing a review of commodity biometric classifiers for a consumer magazine, we could simply use Accuracy as defined above. The Correct Response is the sum of True Positives and True Negatives. So, the ratio of the Correct Response to the Total Attempts is a single simple measure of the product effectiveness.

However, if we are formally evaluating a classifier for a more serious application (e.g. biometric security system for a prison), then we need to conduct a more thorough evaluation that looks at all four metrics. We should also be very clear on what is required in terms of performance for each metric.

Figure 6.1 explains the relevant ideas. The box illustration in Figure 6.1a is called a "confusion matrix" (ironic, huh!) in the literature, where the textual descriptions in the four boxes are for our biometric authentication application. An example of an actual confusion matrix is shown in Figure 6.1b based on the results of a test run of 100 attempts with 83 legitimate users and 17 fraudulent users. Eighty results were true positives, seven were false positives, and three were false negatives and ten were true negatives.

It would be nice if some of these metrics could be combined and the good news is that they can be! Ratios of these numbers are used to describe the model's efficacy. The two terms you need to listen out for are *Precision* and *Recall*. *Precision* is the fraction of the decisions by the AI to grant access that are correct. *Precision* is defined in general as:

PRECISION VERSUS RECALL

It is often said that compromise is an essential element of engineering. It's relatively easy to build something that is strong, and it's easy to build something that is light ... the challenge is building something that is both strong and light.

In AI, it's no different as achieving the right level of quality in terms of decision-making is often a compromise. To really understand the challenges of measuring and specifying accuracy in AI, here is a simple, every-day situation for you to ponder.

Two competing research teams each develop an AI classifier aimed at identifying a specific form of cancer. The results achieved by the teams are shown below:

Team A		**Team B**	
True positives 26	False positives 6	True positives 26	False positives 10
False negatives 10	True negatives 60	False negatives 6	True negatives 60
F-measure = 0.87		*F*-measure = 0.87	

The first point to note in examining these results is that the *F*-Measure for both teams is the same. However, they achieve this result in quite different ways with Team A having a higher precision but lower recall and vice versa.

Which is best?

Now, let's add a couple of complicating factors.

Anyone identified by the AI as having the cancer must undergo a long and expensive medical procedure. The procedure is uncomfortable and invasive; in short, it's not a pleasant treatment.

What does this mean for the application of our classifiers?

If we selected Team A's classifier, for every 100 people classified, six will unnecessarily undergo this difficult treatment. Conversely, using Team B's classifier would result in ten people unnecessarily receiving a painful treatment. Clearly, therefore, we should go with Team A.

However, what if the cancer is terminal?

Team A's classifier misses ten people who require treatment and may therefore die. The classifier developed by Team B only misses six cases that require treatment.

Now, which is best? Team B, of course.

This type of dilemma has existed throughout history and, as a society, we are accustomed to hearing cost benefit and moral arguments about similar situations.

The key point to understand is that it's simply not possible to reduce the assessment of two competing AI applications to a single metric. In this example, both applications achieved the same *F*-Measure but in different ways and with different implications for those affected. In real-world applications, you may need to factor in the consequence and or liability cost of the error before you choose.

(a)

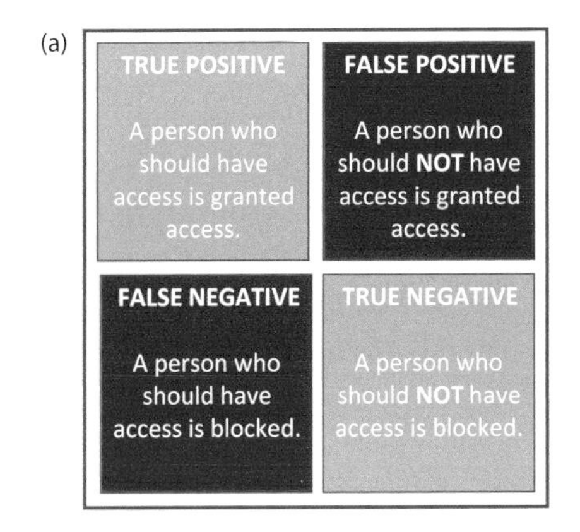

(b)

Ground Truth → Prediction ↓	Should have Access	Should be Blocked
Access Granted	80	7
Access Blocked	3	10

FIGURE 6.1 (a) gives the description of the confusion matrix for a binary decision of grant access or block access. (b) shows the confusion matrix filled in for the example described in the text.

$$\text{Precision} = \frac{\text{True Positives}}{(\text{True Positives+False Positives})}$$

Recall is the fraction of the legitimate users that are correctly identified by the AI. *Recall* is defined in general as:

$$\text{Recall} = \frac{\text{True Positives}}{(\text{True Positives+False Negatives})}$$

These metrics can in turn be combined into a single metric. One particular metric is referred to as the F-Score (or the F_1 score or just plain F) is the harmonic mean of Precision and Recall and is defined mathematically as:

$$\text{F} = \frac{2\times(\text{Precision}\times\text{Recall})}{(\text{Precision}+\text{Recall})}$$

A perfect model will have all three metrics: Precision, Recall and F-Score equal to 1. In our example for the algorithm to give network access to people, Accuracy = 0.9; Precision = 80/(80 + 7) = 0.92; Recall = 80/(80 + 3) = 0.96 and F = 0.94. That is very good.

There are other metrics available [5] to describe accuracy of AI models for specific contexts, such as Area Under ROC Curve for binary classification problems and Cross Entropy Loss (or log loss) for evaluating the predictions of probabilities of membership to a given prediction class.

Combining raw accuracy data together into a single metric, or a small group of metrics, is useful when performing a high-level evaluation. These metrics are useful when conducting a scientific evaluation of different algorithms or when, for example, selecting a general search tool such as an intranet search engine. However, these combined metrics should be used cautiously when looking at more sophisticated and mission-critical applications.

COST FUNCTIONS

In many enterprise applications, the cost of a mistake varies with the type of entity being classified. Consider a fraud detection solution such as the one used by your bank. The solution monitors transactions and blocks your attempt to pay for skiing lessons in the Alps because the spending pattern has deviated from your normal behaviour.

In such a solution, the cost of incorrectly blocking a transaction is very low. The worst-case scenario is that you call the number on the card, verify the transaction and continue your holiday. You may feel a bit irritated about the inconvenience, but you recognise and respect the fact that your bank is being cautious.

In the same solution, the cost of missing a genuinely fraudulent transaction can be very high. A company could lose thousands of pounds by allowing the transaction to proceed.

As you can see, the cost of a *False Positive* is far higher than the cost of a *False Negative*. It is therefore understandable that banks err on the side of caution and tune their systems to be quite sensitive; they tend to shout "fraud" at the first hint of trouble.

It's particularly important in this scenario to look at the performance of the AI against each possible outcome and not to rely on a single, overall accuracy figure. If the bank normally experiences fraud in 5% of transactions, then an overall accuracy of 95% can be achieved just by classifying all transactions as **not** fraudulent. Clearly, that is a very misleading figure and clearly demonstrates the need to drill down into the actual figures for each possible outcome (Figure 6.2).

Now let's consider a different and far more serious scenario. Let's imagine our system is a medical application that examines medical images to decide whether a lump is cancerous. The cost of classifying a malignant lump as benign is huge ... possibly a human life! Just as with the fraud solution, the safe thing to do would be to tune the solution so that it was skewed towards *False Positives*.

What if, however, classifying a lump to be cancerous meant that the patient had to undergo a very expensive and, more importantly, a very unpleasant and painful procedure? In such a scenario, there is an ethical responsibility to minimise unnecessary interventions and a financial incentive not to incur unnecessary cost.

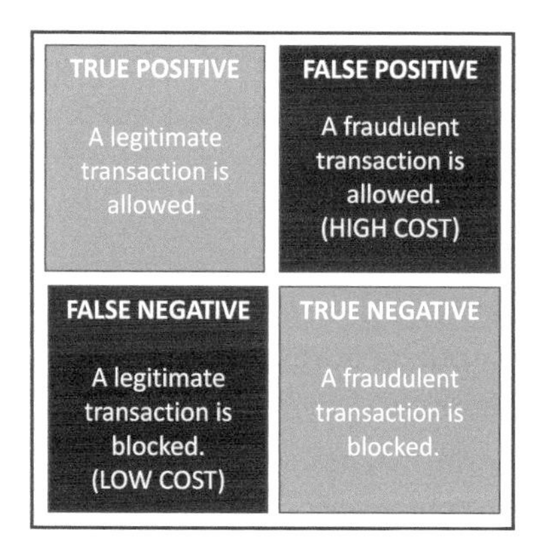

FIGURE 6.2 Description of the confusion matrix for the fraudulent transaction case.

Product is Good & classified as PASS.	Product is fixable, but classified as PASS.	Product is defective, but classified as PASS.
Product is Good, but classified as FIX.	Product is fixable & classified as FIX.	Product is defective, but classified as FIX.
Product is Good, but classified as FAIL.	Product is fixable, but classified as FAIL.	Product is defective & classified as FAIL.

FIGURE 6.3 Confusion matrix for the three decision example of classifying products as PASS, FIX or FAIL. The size of the matrix gives $3 \times 3 = 9$ metrics.

MULTIPLE OUTCOMES

As always, enterprise applications are almost certainly going to be a little more complicated. Imagine your enterprise wishes to apply AI to perform a quality assessment on products. As products roll off the production line, they will pass through a booth where photographs are taken and then passed to an AI to determine if there are any faults. The simplest possible scenario is that the AI has to make a pass-fail decision for each item that passes through the booth.

To make matters more complex, quality assessment is rarely a simple binary decision. Products can have different types of faults and some of them may be fixable, whilst others may be more terminal. If we require our Classifier to make a PASS-FIX-FAIL decision, the number of metrics we need just increased from 4 to 9 (see Figure 6.3). Ideally, we would like the red (off-diagonal) cells in the matrix to be zeroes, implying perfect classification. The decisions in the three red cells on the left (below the green diagonal) are wasteful since good and fixable products are being rejected as "FAIL" or unnecessary additional work is recommended on good products. In contrast, the three red cells on the right (above the green diagonal) are potential sources of business risk by passing defective products for use or unproductive 'fix' on defective products.

As the number of decisions (n) increase, the confusion matrix cells grow as n^2 and (n^2-n) represents off-diagonal elements that need careful evaluation.

QUALITY METRICS FOR NATURAL LANGUAGE UNDERSTANDING

An important area of the application of AI in enterprise is the domain of Natural Language Understanding (NLU). Examples of common applications of NLU in practice are summarised in the vignette "Common NLU Applications" in Chapter 8. To illustrate the variety of quality metrics and challenges, we select three popular NLU applications in this section.

Search/Information Retrieval

You are very familiar with doing searches on your favourite search engine looking for something. You already know that as you type the words for your query, the search engine can suggest potential search terms you can use. One measure of quality of the search engine is how well it can anticipate your intended search term based on what it learnt from you and other users like you. The second measure of quality is when you get the search results back, whether the recommended ranked sources of information meet your need. If the top recommendation is what you wanted, that is pretty awesome. If you get the best information in the top five recommendations, you may still be happy. If you have to do a lot more work than that, you will be unhappy with the search engine.

Information Extraction

Unstructured text documents are often packed full of valuable information that needs to be transformed into a structured form. This process is known as entity and relationship extraction. Extracting this information into a structured form is of huge value to downstream analytics.

Let us consider the use of NLU in law enforcement (Figure 6.4). A typical police report or witness statement may describe a whole raft of different entities. There could be references to names, addresses, vehicle registration numbers, vehicle makes, vehicle models, vehicle covers, phone numbers, email addresses, passport numbers, driving license numbers, immigration visas, credit card numbers, cash, drugs, guns, cigarettes, property of all sorts of types, organisation names, guns, knives, dates of death, dates of birth, dates of arrest, dates of immigration entry, dates of other events and just about anything else you can imagine!

If the police department applied NLU to their entire content, they would have an incredibly valuable resource against which they could perform data mining, time-series, predictive, geospatial and many other forms of analysis. The effectiveness of these downstream analytics will be directly related to the accuracy of the entity and relationship extraction … but, what do we mean by accuracy?

As explained earlier, it is possible to use a single metric such as the *F-Measure* to describe the overall accuracy of the AI. However, to really understand the effectiveness of the AI, we will need to look into the detail.

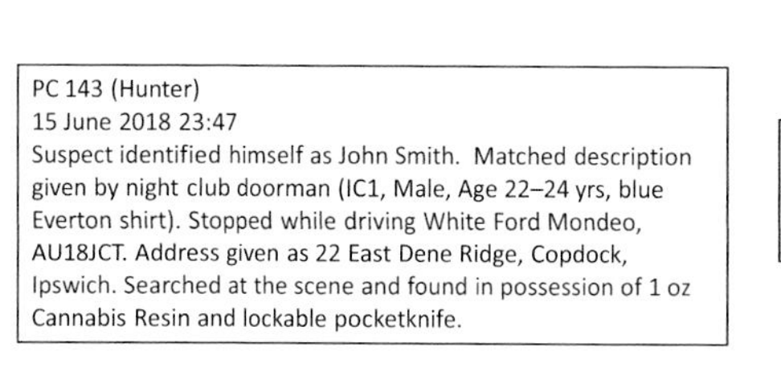
PC 143 (Hunter)
15 June 2018 23:47
Suspect identified himself as John Smith. Matched description given by night club doorman (IC1, Male, Age 22–24 yrs, blue Everton shirt). Stopped while driving White Ford Mondeo, AU18JCT. Address given as 22 East Dene Ridge, Copdock, Ipswich. Searched at the scene and found in possession of 1 oz Cannabis Resin and lockable pocketknife.

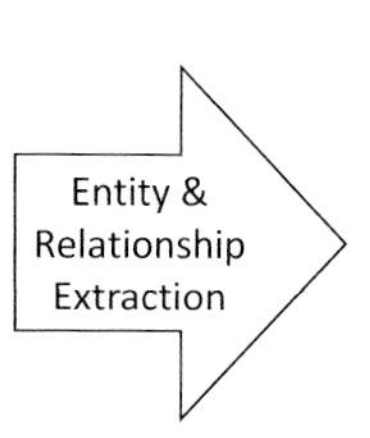

Entity	Value
Arresting_Officer	PC 143
Arrest_Date_Time	15/06/2018: 23:47
Suspect_Forename	John
Suspect_Surname	Smith
Suspect_VRN	AU18JCT
Suspect_Vehicle_Colour	White
Suspect_Vehicle_Make	Ford Mondeo
Suspect_Address_Street	22 East Dene Ridge
Suspect_Address_Town	Ipswich
Evidence_1_Description	1 oz Cannabis Resin
Evidence_2_Description	Lockable Pocketknife

FIGURE 6.4 Example of entity extraction from a police report.

First, some entities may be of higher value than others and we need to take that into account. Examples are name of the suspect and the current address. For those key variables you may have to build extra tests (sanity checks). Nothing is perfect, but you can often use simple rules to isolate possible anomalies. For example, you could flag for manual checking any document that does not seem to have a suspect name, and or extract those key suspect names out as a list and compare them with known criminals, and or against staff names or the report's author name (just in case the AI has misidentified the author of a report, or their fellow police officers as suspects). Extracted addresses can be matched using geocoding software, which in turn will give you (typically) a clean version of the address including zip/post code plus a latitude and longitude for geolocation. If you can get geolocation data, you can quickly plot the Euclidean distance, from an expected location e.g. your police station, the centre of your operating area, etc., and the suspects address, bad guys often perform their crimes close to home, so you can flag any seeming long-distance behaviours. If the distance is small (use averages), your address extraction and geo-matching is probably correct, if lots of suspects seem to live 1000s of miles away, chances are your extraction process is conflating street name with city name or you really do have some travelling bandits.

Second, it is important to know which entities are most problematic for the AI so that we can focus any optimisation efforts appropriately.

Finally, we need to consider the relative numbers of different entity types across the corpus during training and in production. Our test corpus may contain a large number of phone numbers and a small number of credit card numbers. If the AI is great at identifying phone numbers and useless at credit card numbers, then the relative distribution during testing will make the AI look better than it is for real use.

Interactive ChatBots

They are everywhere! Typically, ChatBots are deployed in customer support roles specialising in business-relevant tasks in specific domains (called Skills). In contrast to one-shot Question-Answering, the ChatBot tasks need multiple turns with the user in a specific context. Examples are travel reservations, banking transactions, answering COVID-19-related questions [6], etc. The user interaction can be via speech or text. Speech technology involves acoustic models and thus introduces further complications due to accents, dialects, background noise, etc. Once the user speech is transformed into text (typically using a speech-to-text AI component), NLU processes take over and then the problem becomes the same as for textual inputs.

There are different architectures to support the specific goals of ChatBots [7]. The role of NLU is to understand the user intent from the textual input and map it to the available knowledge and come up with an appropriate response, while keeping the context of the user and the topic. Other embellishments like greetings are added to the ChatBot response to make the user experience closer to a human–human interaction. Kuligowska [8] introduced quality metrics to evaluate commercial ChatBots, addressing various aspects such as user interface, general and special knowledge content, personalisation and conversational ability. Sandbank et al. [9] studied distinct attributes of an "egregious" conversation

CROWD SOURCING ENTITY EXTRACTION

The potential value of entity and relationship extraction was obvious to a large law enforcement agency (the name is being withheld to protect both the innocent and the guilty).

The agency had millions of documents packed full of valuable data that, if extracted into a structured form, would enable all sorts of analysis and reveal everything from the personal networks of major crime bosses to the modus operandi of petty criminals.

All that was needed was a highly effective entity and relationship extraction tool. The fastest way to develop such a capability was to create a huge corpus of labelled data and use ML to rapidly, and supposedly easily, create a model.

The agency gathered tens of thousands of documents and created a team of 50 people as Annotators to label the data. With such a large team, it was expected that a large corpus of training and test data would be generated in no time … and it was!

Unfortunately, the accuracy of the solution was terrible! The training data was inconsistent in every respect. To start with, there was no agreement on the entity types. Some Annotators had labelled every date as a DATE whereas others had used DATE_OF_BIRTH, DATE_OF_ARREST and other such labels. The "Squash Club" was an ORGANISATION according to some Annotators and a LOCATION according to others. The output of the crowd sourced approach was inaccurate and inconsistent … and so was the resulting AI.

TOP TIP: Keep your expert labelling team small, and cross reference their efforts (periodically give them the same documents to label and compare the results. WARNING: this will cause some professional arguments). This will add time to your project, but it will ensure that your model is trained with a consistent data set and will be more likely to succeed.

James Luke

between the user and a ChatBot. These included repetition of responses by the ChatBot and its inability to understand the user intent, rephrasing by the users, emotional state of the user, user request to talk to a human, etc.

This discussion shows that creating a ChatBot for business requires a careful assessment of its task against the NLU technology needed to support the specific skill (e.g. product support) from the institutional knowledge and making sure that the user experience contributes positively to further the business goals. Training data for the ChatBot must come from actual user interactions to increase the success of its deployment.

WHAT DOES THIS MEAN IN PRACTICE?

Well, you must not get seduced by a single-headline accuracy statistic and always remember that 95% score is great for an exam result but not necessarily good enough for an AI application. From a practical perspective, there are several other things that you need to think about.

Importance of Tooling

To use some historical analogies, the Industrial Revolution would not have been possible without the lathe, called the "mother of machine tools" [10]. When combined with the power of steam, creating all types of heavy machinery became much more doable. Similarly, it was the building of a wind tunnel [11] by the Wright Brothers that allowed them to design the aerodynamics for their plane. As we stated in Chapter 3, AI is not just about the algorithms. The importance of tooling to create and sustain AI applications cannot be overstated. Based on their experience of deploying production systems at Google, Sculley et al. [12] have pointed out the substantial hidden technical debt in ML systems and observed, *Only a small fraction of real-world ML systems is composed of the ML code …. The required surrounding infrastructure is vast and complex*. We shall discuss some general tooling needs in this section, with more details in Chapter 10.

Provision of training and test data is always a challenge for AI projects. To train an ML system effectively, we require a large volume of consistently labelled data and a significant proportion of that data is going to be needed for testing. This test data needs to be provided in a form that can be fed into the AI, and we are going to need tools to compare the output of the AI with the ground truth data (data where we already know the correct answer).

Let us return to the entity and relationship extraction application in NLU to illustrate some specific needs. Since our data sources are unstructured text, we need a tool that allows a domain expert to annotate (label) the unstructured text data. This is a human-intensive effort, and crowd sourcing is a popular approach to help with this massive labelling job. A part of the tool's function is to resolve any inconsistencies among the labellers to minimise the noise in the data. Recently, due to rapid advances in NLU technology [13], pretrained language models (called transformers) have been built using large corpuses in the public domain. Any NLU project can leverage these models as the first step and then customise the tooling to support the specific NLU task subsequently. Despite such advances, adapting public domain language models to specialised business domains (e.g. manufacturing, finance, etc.) remains a challenge [14] and requires human expert in the domain to create the appropriate labels.

Once we create a large-enough corpus of 'labelled' data, we need tools to divide that into training and test datasets, making sure that their feature contents are statistically equivalent between the two AND with the operational data. In addition to maintaining a reproducible data pipeline, tooling is also needed to create the model and test the model for various properties (i.e. accuracy, bias, etc.). We also need monitoring tools during production (see below) to assess that the application is behaving properly. For the purposes of debugging and auditing the model behaviour at any time during development or deployment, we need versioning of the models and the related data. We refer to Baylor et al. [15] for an example of an end-to-end tools platform.

These requirements apply to all ML applications (not just NLU) so they should be factored into your AI development programme. One option to consider is whether you are able to capture labelled data through your existing business processes. If that's possible, it may significantly simplify your task and is one of the reasons it is preferable to first apply AI within existing business processes. If you do this, it's important to ensure the captured data is consistent. When multiple people label data, they don't always generate the same labels for each situation. It's important to have cross over (e.g. the same data labelled by multiple Analysts) to enable normalisation of the data into a consistent set.

CHOOSING NLP ALGORITHMS

The imagination of Natural Language Processing (NLP) practitioners has been captured by the latest developments in Transformers [1,2]. A rather natural question that many beginner practitioners ask, while seasoned practitioners take the answer for granted, is: *If Transformers are the state-of-the-art in language processing, should I ever bother with a non-Transformer algorithm?*

The NLP arena evolved over time [3], starting in the 1990s with techniques for encoding domain knowledge as rules, and continuing with the advent of techniques for learning from data: classical statistical ML (e.g., SVM for classification, CRF for entity extraction) followed by deep-learning (e.g., CNN, BiLSTM). Since 2018, Transformers are the latest breakthrough and the state-of-the-art in NLP: they outperform prior techniques in terms of quality of result, require less training data compared to deep learning (DL) techniques, and allow training of multilingual models. They are also the most compute-expensive: an order of magnitude slower at both training and inference time compared to DL models, which in turn are an order of magnitude slower than classical statistical ML.

Therefore, different applications may still choose a non-transformer approach in many cases.

Seasoned practitioners try different algorithms and often make practical choices informed by their specific use case, considering the quality of the model that can be achieved with each technique, and any compute constraints of the application. Here are some general guidelines.

Rule-Based Techniques are suitable for use cases with a small number of variations (dates, email addresses), or when it is difficult to create labelled data (PII information such as social security or bank account numbers). Less suitable for complex tasks with lots of variation or needing information from context (e.g., identifying person names).

Classic Statistical ML is suitable in use cases that require fast training and inference time, and where you have sufficient training data to achieve result quality that is reasonable for your application. Incorporating word embeddings as features can be extremely powerful and alleviate the need for human-engineered features.

Deep Learning (DL) is suitable in use cases with sufficient training data where you want high-quality of result as well as reasonable inference times; especially useful in cases when GPUs are cost-prohibitive.

Transformers are suitable in a few different use cases:

- **High-Value Use Cases** where the quality of result is of utmost importance and worth the trade-off in compute resources.
- **Use Cases Where Labelled Data is Scarce** and classic ML/DL techniques do not achieve the quality necessary by the business application.
- **Multilingual Use Cases** where training data exists in multiple languages and you want to simplify operational costs by deploying a single multilingual model, instead of many monolingual models.

- **Bootstrapping Labelled Data** in the same language or a different language; if you have a small amount of labelled data, you may train a Transformer model with reasonable quality and use it to label a large quantity of unlabelled texts (in the same language or another language), followed by human validation of the labelled data. This approach can speed up the process of obtaining labelled data and may allow you to obtain sufficient data to train a classic ML or DL algorithm, and therefore deploy a more cost-effective model.

Laura Chiticariu

Laura Chiticariu is a Distinguished Engineer and Chief Architect for Watson Core Language at IBM. She has more than 10 years experience developing and productizing Natural Language Processing systems and algorithms, and she now leads the development of core Natural Language Processing algorithms that power many IBM products.

REFERENECES

1. T. Young et al., "Recent trends in deep learning based natural language processing," *IEEE Computational Intelligence Magazine*, 13, pp. 55–75 (August, 2018).
2. X. Qiu, et al., "Pre-trained models for natural language processing: A survey," *Science China Technological Sciences*, 63, pp. 1872–1897 (2020).
3. E. Brill and R. J. Mooney, "An overview of empirical natural language processing," *AI Magazine*, 18(4), pp. 13–24 (1997).

As part of this strategy, you should remember that AI capabilities are continually improving. Therefore, it is necessary to review available AI components on a regular basis and ensure your application is using the most effective component available. This is particularly true if you are delivering cloud-based applications where the AI components are available as Cloud services.

Use of Historical Data

Let us consider the case of an insurance underwriting solution. It may be possible to generate a corpus of historical insurance applications from customers and the decisions assigned by the underwriters. This could be used as the basis of training and test data for an ML application. Note, that we refer to this as the "basis" of training and test data. Even when you have access to a large corpus of historical data, there are several points you need to consider.

The fact that data has been labelled as part of a historical business process does not mean that it has been labelled correctly. When reviewing historical data, we often find many incorrect labels and decisions. It is also quite normal to find inconsistent data; two insurance applications that appear identical but resulted in completely different decisions. This may be because two different people contributed to the original data or it could be due to the fact that, in one of the cases, additional information was available from a different source.

Historical decisions also need to be considered in the context of the time. Historically, banks in the UK would not approve mortgage applications where the borrowing exceeded four times the applicant's income. In recent years, borrowing criteria have changed, and so historical data is not necessarily a useful source of training data.

Another concern with historical data is that it may also include prejudice! We have already touched on the topic of AI Bias in Chapter 5 and will discuss data aspects of bias in more detail in Chapter 7. Human-labelled data can often reflect the prejudices, conscious or unconscious, of those who labelled it. This is potentially one of the greatest benefits of AI in that we can systemise these decisions, identify bias and manage it out of the system to deliver ethical and fair decisions.

What this means, of course, is that you are going to need more tooling! Even if you have a huge corpus of labelled data ready for immediate use, you will still need to check the data to ensure it's acceptable to use and doesn't contain any thorny issues such as personal information that should have been redacted under GDPR. You are going to need tools and processes to validate the labels and continuously improve the quality of the training and test data.

What about when you do not have existing data? Then, quite simply, you need to create it. This is one of the reasons why, in evaluating the *Doability* of an AI project, we ask whether the AI will fit into an existing business process. If it does not, then you are going to need to create training and test data from scratch and you need to ensure it is accurate. One method of doing this is to crowd source. Take a large corpus of data and ask your entire workforce, or other volunteers, to each label some of the data. Many hands do make light work; however, it is critical that you ensure the consistency of the data. Once again, you are going to need the tools and processes to ensure accurate and consistent data.

Much of what we are discussing here will be revisited in the following chapter. The key message is that, if you are to measure accuracy, you need accurate training and test data and that means you are going to need tools to develop and maintain this data.

One point to note, however, is that this is not necessarily an upfront process. It may be perfectly reasonable to develop an AI with the aim of starting with whatever data is available and then continually improving its quality. By deploying an AI early, even when its accuracy is dubious, you may be able to gather feedback and identify conflicts such that you can improve the quality of your training and test data more rapidly. Of course, there is a caveat! If the initial performance is too poor, the users may lose faith in the approach and disengage. Remember that careful Stakeholder management is another essential part of making AI work in the real world.

HOW ACCURATE DOES IT NEED TO BE?

The key question is, what level of accuracy is good enough? It really depends on a combination of operational value and more sensitive factors relating to ethics. Let's consider accuracy in two different operational scenarios, underwriting life insurance and driverless vehicles.

In developing an AI to perform life insurance underwriting, understanding the accuracy required is a (relatively) straightforward business analysis task. We know how many life insurance applications are received. We know how long it takes to process and the cost of performing the underwriting. We can develop business models that examine factors such as attrition; how many potential customers give up on the process due to the time taken to consider their applications and how many would be retained if we reduced the processing time by a certain amount? We know what it costs to insure an applicant and we can quantify the risk of approving an application that should have been declined. It may be a complicated task, but assessing and quantifying risk is what insurance companies do! As such, the company should be able to determine what they need from the AI in terms of accuracy to allow them to reap the benefits of augmenting human underwriters with AI tools. There is an example of a real insurance underwriting solution in Chapter 2.

Now let's consider driverless vehicles. Clearly, 95% would not be acceptable … or would it? Well, firstly we need to be clear on what the 95% accuracy is describing! Driverless vehicles are complicated beasts and made up of many subsystems. This raises the question of how the accuracy of an individual AI component contributes to the accuracy of an overall system.

When AI components interact, they can impact the overall system performance in different ways depending on the configuration and architecture. One pattern touched on in Chapter 2 is to federate decisions to multiple AI services and perform some form of arbitration or fusion on the results. In a scenario where we used three separate Classifiers, each with an accuracy of 95%, a simple voting system could improve the overall accuracy. A fusion strategy could be to select a classification if two or more of the Classifiers generate the same result. If all three Classifiers generate a different result, then the classification generated by the Classifier with the highest confidence is chosen. This type of configuration can, depending on the scenario, achieve a far higher accuracy than any of the Classifiers individually.

Now let's consider a pattern where multiple AI components are integrated in a single sequential architecture (Figure 6.5). In this pattern, errors are normally compounded. An error in one of the earlier components causes inaccurate data to flow into a later component and the errors can grow as a consequence.

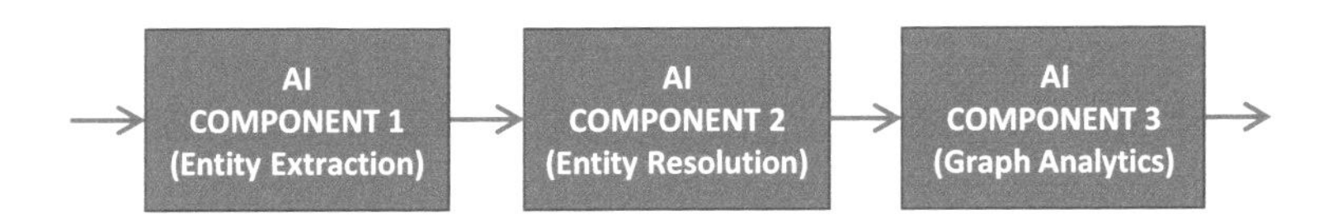

FIGURE 6.5 Three AI components in series.

So, in understanding the accuracy requirement of any individual AI component, it is important that you consider the overall systems view and understand the interaction between the individual components. It is particularly important to understand how improving the accuracy of one component impacts the overall system accuracy. From an optimisation perspective, this is essential in knowing where you should focus your effort in order to achieve the required system accuracy.

With a system engineering perspective, we should be able to determine the overall accuracy of a driverless vehicle (for each of the different scenarios it may encounter). The question remains, however, how accurate does it need to be? Unlike the life insurance scenario, this is not just a question of statistical business analysis. We need to consider more sensitive and qualitative factors. If the driverless vehicle is to replace human drivers in a delivery business, we can calculate both the business cases in terms of cost and the safety case in terms of reduced accidents. However, it will be more difficult to quantify the public perception of the safety. We should expect the public to require a far higher standard of safety than currently delivered by human operators.

As always in engineering, there are few right and wrong answers. The key point, however, is that in order to successfully develop an AI application, you need to understand the accuracy required for the application to be both operationally valuable and ethically acceptable.

WHERE DO YOU ASSESS ACCURACY AND BUSINESS IMPACT?

Understanding the accuracy needed to achieve a specific business impact is essential but not always straightforward [16]. Examples of metrics that matter to business are net promoter score (NPS), sales rate, recommendation to conversion rate, click rate, customer satisfaction, time on page, etc. We need to understand the connection between these metrics and AI model output. This will avoid wasting resources to improve model metrics that may, in fact, have little or no impact on the business. There must be some instrumentation that stores business metrics and the associated model metrics to understand which model operations contributed to significant business results.

In some cases, business impact is directly attributable to the AI application. Consider a customer retention solution that analyses customer spend patterns and detects those that are likely to attrite. In such a clear-cut situation, it is possible to measure customer attrition prior to and after deployment of the solution to determine the business impact.

Sadly, this type of measurability in an enterprise is not always as easy to achieve. At the root of the challenge is the complexity of the data flows within and across departments in the enterprise. Some departments gather and process massive volumes of data from a variety of sources. Due to complex internal organisational structures and processes, it is difficult to understand where to invest in order to get the maximum benefit on the operational effectiveness of the enterprise. To manage this effectively, it is necessary that there is some form of measurability across the enterprise and not just where the responsibility for the AI lies. The complexity of this challenge is compounded, since frequently, the department that invests in the AI may not be the direct beneficiary of the resulting improvements.

OPERATING WITHIN LIMITS

In Chapter 2, we talked about the key differences between AI applications and conventional solutions. One of the fundamental differences is that AI applications are extremely sensitive to the data they are trained with and operate against. Fundamentally, AI applications work well in scenarios that they have been designed, or trained for, but can work poorly once the environment changes around them. In terms of accuracy and business impact, it is therefore important to understand when the system is operating beyond its limits. Lots of emphasis is placed on the accuracy and/or bias when training your model, but the real challenge is continuing this performance measurement for the entirety of the model's productive life!

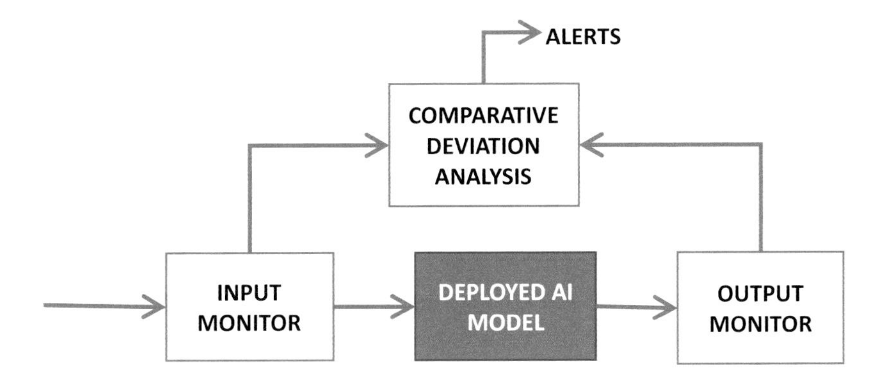

FIGURE 6.6 Monitoring inputs and outputs of an AI component in production for potential deviations from the expected behaviour.

Monitoring Inputs and Outputs

One way of achieving this is to monitor the inputs and the outputs of the production AI model over time (Figure 6.6). Statistical analysis of the input data may reveal situations where the input data has changed significantly, e.g. a larger percentage of male data is being processed compared with the training data which might trigger a gender bias. Similarly, statistical analysis of the output data will reveal when the AI has started behaving differently, e.g. The average input document typically yields between 2 and 20 extracted entities, a recent 500-word document yielded 200 entities (this profile of document wasn't in your training data, it could be an error, or new type of doc, e.g. a telephone contact list, or even a foreign language document where your AI has assumed each unrecognisable (i.e. not in the dictionary) word must be a proper noun. In either or both situations, alerts should be generated such that the developers and administrators know that they need to review the operation of the system. We discuss monitoring in more detail in Chapter 10.

Use of Alternative Models

An established way to architect fault tolerance systems is to build in redundancies [17]. What if you create one or more independently developed "challenger" models running in parallel with your deployed model? (Figure 6.7). There are three distinct benefits to this approach:

i. The challenger models can be used to ratify the deployed model's performance.

ii. Once the relative performance among the models is established and understood, if all the models begin to deviate, there is a good chance that something has changed with the input data assumptions.

iii. Keeping a stable of challenger models (including the ones that are being retrained with the latest data) means that it can be easier to swap out an errant model without major business impact.

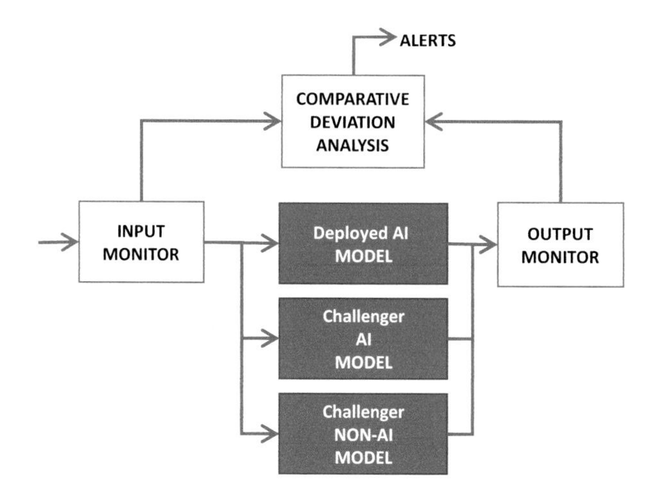

FIGURE 6.7 Use of alternative models to validate the behaviour of the deployed model and provide options in case it fails.

MONITORING TEXT ANALYTICS

Entity and relationship extraction applications are very data sensitive. They are designed for specific languages and particular styles of writing.

In 2003, we designed an entity and relationship extraction tool to analyse short, prose text reports for a police force. The system extracted entities and relationships that were then used to generate network diagrams. It was a great success and was well received by the Analysts who had previously had to manually read the documents and generate the diagrams.

Several months later, we received a phone call saying that the system was behaving strangely. The first sign of trouble was that the reports were taking a long time to generate. Whilst travelling to the customer location to investigate, we were told that the data was corrupted and that network diagrams of criminal organisations were now full of the names of serving police officers and support staff.

On investigating the fault, we discovered that the system was a victim of its own success. The Analysts were so pleased with the results that they'd decided to load more data into the system. They added in reports from a different police force. Those reports were very similar to those for which the system had been designed with one exception. Each report included a long distribution list containing the names of police employees.

We also discovered that the document set also included some very, very large documents. These documents were telephone directories. They were taking a long time to process due to their size; however, the structured nature of these documents was such that they didn't need to be processed by a prose text entity extractor.

Rather than constrain the data poured into the system, we immediately deployed monitors and statistical analysis to spot unexpected changes in the distribution of the output data and flag an alert for further investigation.

James Luke

If there is any ethical concern (e.g. bias) about your deployed model, keeping one of the challenger models as a traditional non-AI model can help to swap them quickly and limit the negative business impact. Even if less accurate, this gives an option of a "safer and more explainable" model while you repair your AI.

Manual Validation of Model Performance

Sometimes there is no completely automated way of monitoring the performance of your AI. Let us consider the example of an authorisation system that allows access to a building based on user data (e.g. facial image). In Figure 6.8, AI is being used to sift incoming cases, automate the decision for the ones in which the model has a high confidence, whilst sending those cases which are borderline or low confidence to a manual evaluation process. In this example, 90% of all cases are deemed straightforward and easy to make an automatic judgement, with only 10% of cases really needing human scrutiny. How do we know that the machine is continuing to perform well? Access to a wrong person can be risky and a retrospective review may be inadequate. To monitor the efficacy of the auto decisions, a random 5% of the decisions with high confidence are added to the manual process. The human decision on this 5% (when compared to the model score) can help to validate the model efficacy.

Each AI is unique and what monitoring/measurement system you employ will need careful consideration. *What is not in question is that you need one. A production AI deployed without monitoring is more of a liability than an asset.*

QUALITY ATTRIBUTES OF TRUSTWORTHY AI SYSTEMS

We have spent nearly all this chapter talking about accuracy-related metrics as the primary evaluation technique for making sure that the deployed AI application is meeting your business goals. While this is a popular story, this is not the whole story. Due to their very nature, a key attribute of AI systems is their trustworthiness. Building Trustworthy business applications is much more than getting the accuracy right. For the sake of

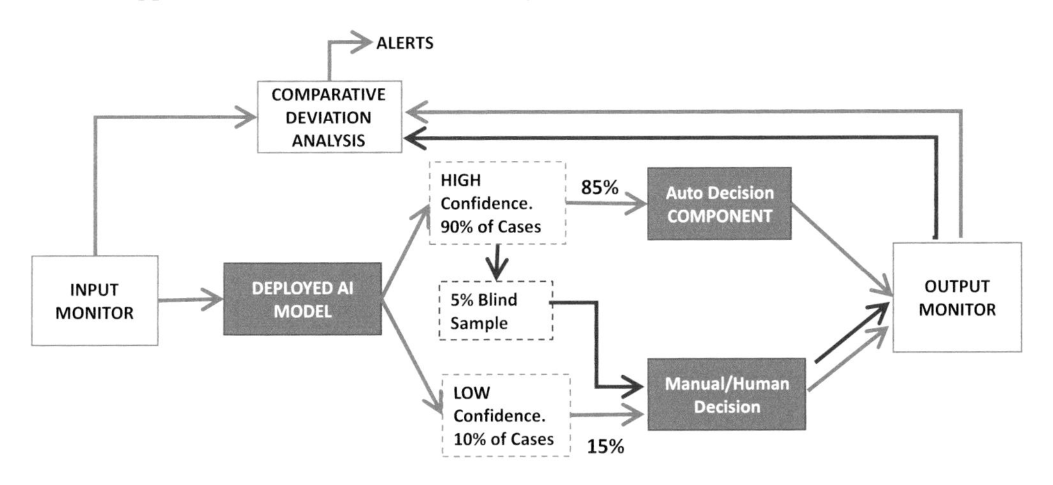

FIGURE 6.8 Manual decision to supplement AI automation. The system has high confidence in 90% of the cases and low confidence in 10% of the cases. Five percent of the high-confidence cases are deliberately routed to manual decision for validation.

completeness, we repeat key components of assessing the Trustworthiness of AI systems here. You will find more details in Chapter 5.

Fairness/Bias

Does your AI application carry implicit biases? Typically biases stem from the imbalances in training data, and it is important to check the data distributions to identify biases even before the AI model is built. As mentioned earlier, blind use of historical data for training is a natural place for biases to creep in. There must also be monitoring during deployment to identify any unexpected biases in the AI application in production.

Explainability

Since ML models are black boxes (i.e. opaque models) created using complex statistical algorithms operating on the training data, there is no simple way to explain the output result from the input data. The various stakeholders (development teams, consumer of the application, auditors, etc.) need appropriate levels of explanation for the model to be trusted. This means, explanations for the relevant stakeholders must be built into the development activities and the application user interfaces.

Robustness

It is well established that the ML models are prone for adversarial attacks. Just small manipulations of an image or text by an adversary can lead to unexpected results. If the adversary has access to training data and the model development process, more virulent attacks are possible. Application owners must design techniques to withstand adversarial attacks as well as to recover from them when they happen. The consequence of adversarial attacks can be severe in mission/life-critical scenarios.

Uncertainty

Due to the underlying statistical learning techniques in ML models, every output from a model has some uncertainty in it. Model owners need to provide an estimation of uncertainty in model outputs and make it visible to the users. This is in addition to the usual confidence levels provided by typical ML algorithms. This will allow the user to opt for human decisions when the model uncertainties are not acceptable in practice.

Transparency/Governance

Due to the extreme sensitivity of AI model behaviour to training data and the AI ethical guidelines from governmental agencies on the use of data (e.g. European Commission's GDPR), there has to be a governance mechanism for built-in transparency in the creation of AI components. Even a technically sound and a very useful application can fail in the society if proper governance is not practiced by the enterprise.

IN SUMMARY – IF THE AI ISN'T TRUSTWORTHY, PEOPLE WON'T TRUST IT

We can only successfully deliver AI if the stakeholders are willing to accept it. For that to be the case, we have to understand how to evaluate the AI and how to ensure it is trustworthy:

- Managing the quality of AI applications in production is very different from traditional software.
- Due to the statistical learning in ML systems, errors are inevitable.
- In many business scenarios where AI is used to make decisions, we may not know the right answer when the decision is made.
- Quality of an ML component cannot be captured in a single metric.
- Cost of the decisions must be included in the evaluation of the quality of the decisions made.
- ML applications need extensive infrastructure and tool support.
- Quality of the application is tied to the quality and quantity of the data used in training.
- Accuracy needed for the application depends on the intended use.
- ML model metrics must be evaluated in the context of business impact.
- Monitoring during production is a 'MUST'.
- There are ways to make the deployed model more fault tolerant.
- Trustworthy AI systems need new quality attributes for business success, viz., bias detection & mitigation, explainability, robustness, uncertainty quantification, transparency & governance.

REFERENCES

1. F. P. Brooks, *The Mythical Man-Month: Essays on Software Engineering*, Anniversary edn. Addison-Wesley Longman, Reading (1995).
2. S. McConnell, *Code Complete: A Practical Handbook of Software Construction*, 2nd Edn. Microsoft Press, Redmond (2004).
3. B. Hailpern and P. Santhanam, "Software debugging, testing and verification," *The IBM Systems Journal*, 41, pp. 4–12 (2002).
4. Body Measures Data, Center for Disease Control, https://wwwn.cdc.gov/Nchs/Nhanes/2007-2008/BMX_E.XPT.
5. J. Brownlee, BLOG: metrics to evaluate machine learning algorithms in python. https://machinelearningmastery.com/metrics-evaluate-machine-learning-algorithms-python/.
6. M. McKillop, et al., "Leveraging conversational technology to answer common COVID-19 questions," Journal of the American Medical Informatics Association, 28(4), pp. 850–855 (2021).
7. E. Adamopoulou and L. Moussiades, "Chatbots: history, technology, and applications," *Machine Learning with Applications*, 2, Elsevier (2020).
8. K. Kuligowska, "Commercial chatbot: performance evaluation, usability metrics and quality standards of embodied conversational agents," *Professionals Center for Business Research* 02, (2015).
9. T. Sandbank, et al., "Detecting egregious conversations between customers and virtual agents," *Proceedings of the Conference of the North American Chapter of the Association for Computational Linguistics: Human Language Technologies*, 1 (2018).

10. History of Lathe, https://en.wikipedia.org/wiki/Lathe.
11. Wright Brothers' 1901 Wind Tunnel, https://wright.nasa.gov/airplane/tunnel.html.
12. D. Sculley, et al., "Hidden technical debt in machine learning systems," *Neural Information Processing Systems Conference* (2015).
13. D. W. Otter, J. R. Medina, and J. K. Kalita, "A survey of the usages of deep learning for natural language processing," *IEEE Transactions on Neural Networks and Learning Systems*, 32(2), pp. 604–624 (2021).
14. A. Ramponi and B. Plank, "Neural unsupervised domain adaptation in NLP--a survey," arXiv:2006.00632 (2020).
15. D. Baylor, et al. "TFX: a tensorflow-based production-scale machine learning platform," *KDD'17: Proceedings of the 23rd ACM SIGKDD International Conference on Knowledge Discovery and Data Mining*, pp. 1387–1395 (2017).
16. M. Arnold, et al., "Towards automating the AI operations lifecycle," *MLOps Workshop at MLSys* (2020).
17. V. P. Nelson, "Fault-tolerant computing: fundamental concepts," *Computer*, 23(7), pp. 19–25, (July 1990).

CHAPTER 7

It's All about the Data

The news was devastating for the young couple.

It was their first child and the 14-week ultrasound scan should have been routine. However, the mid-wife measured the fluid on the baby's kidney and said she was worried. Apparently, the team had been sent a draft scientific paper suggesting such measurements were a soft marker for possible birth defects.

As a PhD Engineer, the Father couldn't fight his instincts and immediately went searching for more information. He discovered that the paper had been written by a team at a nearby hospital, so he called them to ask for more information. During the conversation, he asked how many cases had been considered in preparing the paper?

The answer was that the cases of just 12 children had been considered. In fact, the Father was told that there was insufficient data to form any opinion. The paper had been written as a "call for data" and circulated to ask other hospitals to collect the data. The author of the paper apologised and said that he had never intended for parents to be warned as a result of his work. Quite simply, there was nothing to indicate at this time that there were any issues to worry about.

26 weeks later, my perfectly normal baby girl was born … she is now studying at university, for a first degree in mathematics.

James Luke

"More data … I need more data!" If you are the spouse, parent, partner, colleague or friend of an Analysis Junkie, you'll have heard them saying that at least once (a day). AI Engineers are both Algorithm Addicts and Analysis Junkies so they're going to need feed their habit and will be continuously demanding more data. The reason Algorithm Addicts and Analysis Junkies have such a craving for more data is that it really is the key to success. In fact, it could be argued that availability of data is the most significant factor that is driving the current interest in AI. As a society, we are collecting, or rather, generating more and more data.

DOI: 10.1201/9781003108498-8

DATA TSUNAMI

Let us understand the sources that contribute to the massive data tsunami. Figure 7.1 shows a snapshot of activities on the internet in 1 minute in 2021. Not surprisingly, the activities relate to online searches, shopping, social media, entertainment, mobile applications, etc. Almost everything we do on the internet is recorded. Our financial transactions are increasingly electronic. We also cannot forget the real-time data from a plethora of sources such as stock market trades and Internet of Things (IoT) devices including cell phones, automobiles, refrigerators, Closed-Circuit TV (CCTV) in parking lots, Automatic Number Plate Recognition (ANPR) at the toll booths, etc., that are continually recording all types of data from weather to traffic to power consumption.

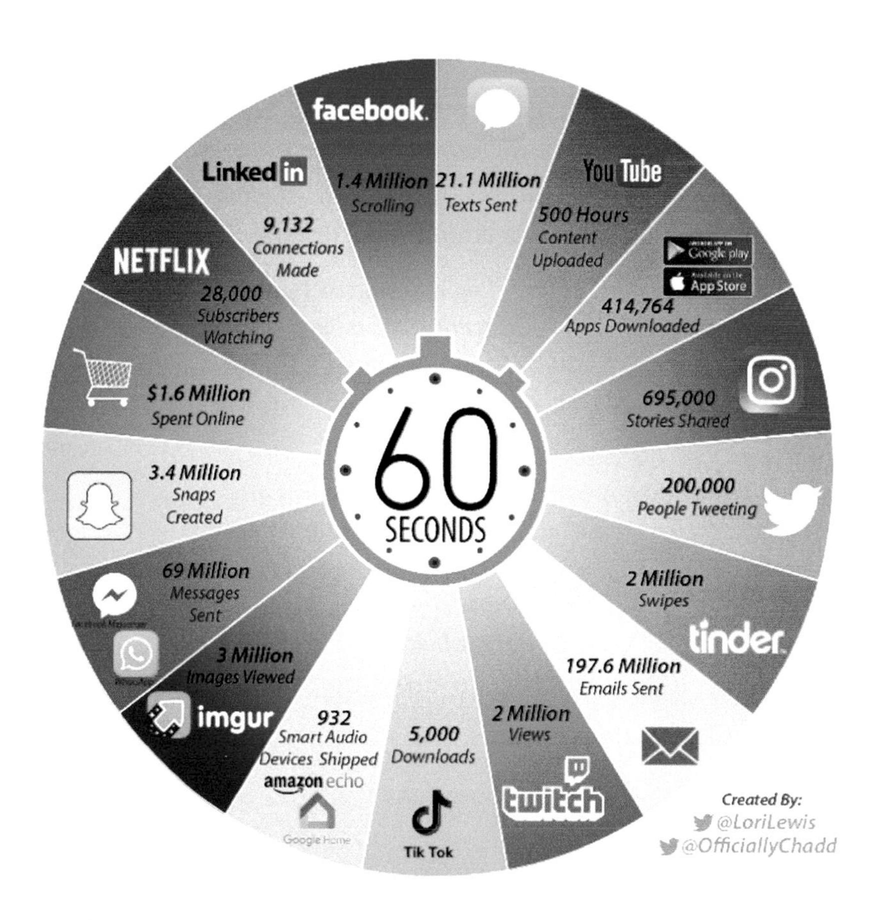

FIGURE 7.1 What happens in an internet minute in 2021 [1]. (Reproduced with permission.)

Beyond the usual suspects on the internet mentioned above, there are digital news media (e.g. CNN, BBC, etc.) and the digital versions of print newspapers (e.g. *The New York Times, The Wall Street Journal, The Guardian*, etc.). To add to this deluge of archival data, most recent books, magazines and journals are currently available in the digital portable document format (pdf). Even publications that were previously published on paper are being rapidly digitised for easier access to the readers, facilitating machine analysis, even if requiring Optical Character Recognition (OCR). Then, there are web sources such as Wikipedia or Data.gov and unlimited number of blogs on almost any topic imaginable.

In addition to the raw transactions, the metadata collected during the transactions (e.g. name, email addresses, IP address, location, etc.) have commercial value to the companies that are collecting them. While these data from these sources may be publicly visible, *using the data for AI requires proper licensing agreements with the owning companies.* In addition, companies (e.g. banks, retail stores, telecommunications, manufacturing, etc.) have their own business-relevant **proprietary** data such as customer profiles, credit card transactions, business processes, product engineering information, personnel records and email. With so much data, it is no surprise that organisations of all shapes and sizes see the opportunity to exploit this data … if they can make sense of it.

A recent IDC study [2] estimates the trend of the worldwide data growth (they call it "Global Datasphere"). It is shown in Figure 7.2, reaching 180 Zeta Bytes in 2025. In case you are wondering how much one Zeta Byte is, it is 10^{21} Bytes. The study is also projecting the data stored in the enterprise data centres to grow from 46 ZB in 2021 to 122 ZB in 2025. That is a lot of data! We can only anticipate with the expanding 5G services across the world by communication service providers, the variety, volume and velocity of the data can only increase even more.

With so much data available, it's all too easy for the Algorithm Addict to fall off the wagon! Surely, all you need to do is gather as much data as possible and the algorithm will do the rest? No! It's time to get back into rehab and remember that there is no magic algorithm and just throwing algorithms at data never delivers value. To successfully deliver AI solutions, you are going to need to learn a little more about data and how to deal with it. It really is all about the data, and in AI project, you can expect to invest up to 80% of the available resources just on accessing, understanding, cleansing, preparing and managing the data. Remember, once your model is trained, you are going to have to include most or all of the data transformations you put your training data through into your live data

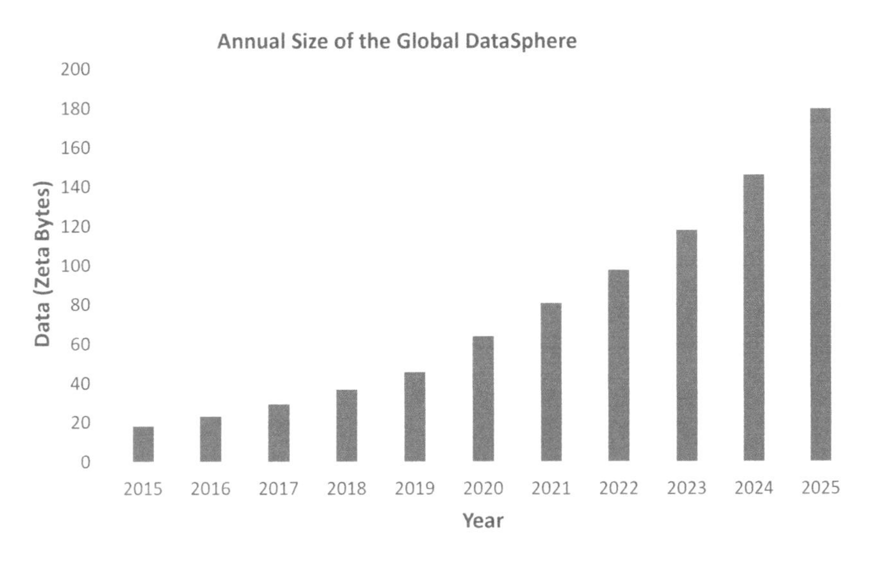

FIGURE 7.2 Estimated trend of world-wide data growth [2]. (Reproduced with permission.)

application before you can use it. Spoiler alert: if your data transformation code needed to look up massive tables of reference data to clean/augment the transaction data, so will your live model. If you need the application that model sits in to have sub second responses, you may want to rethink your business case.

In this chapter, our aim is to help you understand the fundamentals of data science so that you can make good decisions when it comes to selecting, scoping and managing projects. Before digging deeper, it's worth taking a few minutes to remember that enterprise AI is different from consumer AI on the internet. We talked about this earlier in Chapter 2 and the fact that so much of the current interest in AI is driven from the perspective of the big internet companies such as Google, Amazon and Facebook. We make two key observations here:

i. These companies are accessing, generating and storing massive volumes of data that requires an AI to get business value.

ii. In turn, they naturally drive advances in AI to meet their own needs, which are not the same as the needs of a typical enterprise.

In addition, as we discussed above, the content the web companies generate is actually quite narrow in scope and their platforms inherently standardise the data collection so that they can be easily analyzed. Consequently, the internet companies do not share the same concerns about AI applications as typical enterprises, in terms of data volume, types or quality.

DATA TYPES

Structured Data

The format of the data collected can vary widely depending on the source and the intended purpose. Structured data refers to the format of the data that facilitates queries against it in a pre-designed and engineered way. There are three mainstream structured data representations:

i. Traditionally, enterprise data used for business analytics has been in the form of tables (e.g. Microsoft Excel spreadsheets) with columns representing attributes (or features) and rows representing observations. Relational Databases (e.g. IBM DB2, Oracle, etc.) that have schemas and linking tables are used for complex large-scale data and high-performance needs. Structured Query Language (SQL) allows queries against relational databases.

ii. Semantic Web is a knowledge representation promoted by the World Wide Web Consortium and Tim Berners-Lee to make the internet data machine readable. It consists of a Resource Description Framework (RDF), Web Ontology Language (OWL) and Extensible Markup Language (XML). SPARQL is the language to perform 'semantic' queries.

iii. Recently, graph databases (e.g. Neo4j, JanusGraph, etc.) that use nodes and edges to represent relationships between entities have gained wider acceptance. Cypher and Gremlin are examples of query languages for graph databases.

Unstructured Data

Not surprisingly, most of the human-created data is unstructured. Natural language (i.e. English, Spanish, Hindi, etc.) is the most pervasive form of communication among humans and the most natural to capture knowledge. The language can be in the spoken or written textual form. In either case, the sheer number of languages and their dialects pose a major challenge to a machine. In the written text, the syntax of the grammar and the symbols representing the language have to be learnt by a machine. In addition to the grammar, the spoken language has many inherent attributes (e.g. slangs, accents or emotions) present, which are difficult to process. In spite of major advances in Natural Language Understanding (NLU) in narrow contexts, as evidenced by our daily interactions with Alexa or Siri, mastering language in general remains a significant AI challenge.

With the ubiquity of cell phones, data captured as a photograph or a video has become extremely common. Images can be in different formats and/or resolutions. Videos bring in the complexity of actions captured as time sequence of image frames and the fusion of audio & visual information. Computer vision is a very established field over many decades. But, as discussed in Chapter 3, the handcrafted feature extractions of the past are being completely overrun by the capabilities of today's Deep Neural Network (DNN) algorithms.

DATA SOURCES FOR AI

In building enterprise AI applications, the architects have to evaluate if they could use publicly available datasets for training ML algorithms. Depending on the application domain, it may be possible to exploit publicly available datasets. We need to remember three key considerations in selecting training data sets:

- The choice of the training data set depends on the domain of interest. As an example, let us imagine we want to use AI to identify the missiles from aerial images during the Cuban Missile crisis in October 1962. If the reconnaissance flights can happen any time of the day, we need to have training data consisting of aerial images of potential missiles, hopefully taken during day AND night.
- As we discussed in Chapter 3, supervised ML needs labelled data sets. If not already labelled, manual labelling of the data has to be included in the project planning.
- Careful consideration of licensing implications and privacy constraints for the immediate and long-term use in the business application.

There are many public data sets available to support the training of AI models, both unstructured and structured formats [3]. The data sets include images for various purposes (e.g. facial recognition, object recognition, handwriting & character recognition, etc.), videos for action recognition, text data (news articles, messages, tweets, etc.) and structured multivariate data in a tabular form in various domains (financial, weather, census, etc.). The ability of an AI algorithm to extract useful knowledge out of data critically depends on the ability of the data scientist to prepare the data for the purpose. Structured and unstructured data require different challenges in that context. For example, noise in a collection of images manifests itself very differently from noise in a relational database; hence, different techniques are required in order to process the different data types. The processing of unstructured data requires critical steps for extracting 'structured' features in it. An example in natural language text is 'topic modelling' that captures statistical co-occurrences of certain words in a document.

DATA FOR THE ENTERPRISE

Unless all the data needed for the AI development is owned by the organisation, procurement of the necessary data from external sources can be challenging. Even cost-free publicly available data sources such as Wikipedia have restrictions on the intended use of their data under their license. Contracts with data providers typically have restrictions on the duration of the data use and conditions for removing the data after the intended use over a specified period. These are key considerations since the maintenance of the AI models may need the availability of the data in perpetuity. Use of personally identifiable information (PII) without careful consideration can have serious business consequences. As an example, the European Union's General Data Protection Regulation (GDPR) [4] requires that customers have a right to know how their personal data is used, which can include decisions made by AI. Under GDPR, the Data Controller is obliged to follow geographic restrictions on where the data can be hosted and provide the customer timely responses for requests to remove their personal data from the company repositories. Information Technology (IT) infrastructures of most companies are not designed to respond to such surgical responses. The legal framework also provides constraints to what data can be collected in the first place, which has an influence on the algorithmic workings of learning systems. Such data privacy laws are being adopted in other parts of the world such as the California Consumer Privacy Act (CCPA) [5] as of 1 January 2020.

ENTERPRISE REALITY

A typical enterprise (e.g. banks, insurance, retail stores, manufacturing, etc.) faces a different reality from the web technology companies when it comes to data. *To keep this discussion simple, we will use the language of structured data captured in a tabular form to illustrate the basic ideas, but analogous examples in the unstructured data domain are easy to imagine.* Data may have been collected over many years and is often stored in siloed systems. Just accessing the data can be a challenge due to security, privacy and regulatory concerns. Once accessed, data from separate sources needs to be fused together; not easy when there may be no common point of reference between records. Fields may be incomplete, and all too often we see data that has been partially or wholly duplicated across multiple systems. In many cases, the duplication is inconsistent with some records being amended such that it is impossible to determine which data is actually accurate. These are just some of the issues that you will encounter and, in a nutshell, are the reason so much effort needs to be focused on the data.

AI projects really are all about the data! So ... what do you need to know? We're going to start out by considering some key differences in the way human beings and AI algorithms use data when learning and then when making decisions.

HUMANS VERSUS AI – LEARNING AND DECISION-MAKING

Earlier in the book, we talked about the differences between narrow and broad AI. All of the AI solutions currently being delivered are narrow solutions that focus on very specific tasks. A key characteristic of narrow AI systems is that they are best when they operate on closed and constrained sets of data. In human decision-making, we often use far more data from many disparate sources. For example, an AI system that decides whether to buy or sell shares will have access to a massive range of historical share price data. However, the AI will not have read the morning's papers and will not necessarily associate a comment by the Environment Minister with a potential increase or decrease in the value of utility companies.

Consider even the most basic of human and AI operations such as recognising a person on the street. As human beings, we can identify an old school friend by a characteristic of their walk even though we haven't seen them for many years. We recognise a friend by the shirt they are wearing and their hair colour even though they are some distance away and have their back turned towards us. We can discount a possible identity because we know that person is away on holiday. In short, we use an unconstrained set of data sources, extensive human knowledge and context to infer and discount possible identities.

Conversely, a narrow AI system will operate on a very narrow set of data from sources such as CCTV. AI will use sophisticated biometrics to measure specific features that will, hopefully, enable it to uniquely identify individuals. AI benefits from precision in measuring features and the speed with which it can search a massive database for possible matches. However, AI is constrained by its inability to work with incomplete data or to deal with edge cases. AI won't work if the subject is facing the wrong direction or rare match to another person in the database (or perhaps just a twin).

Another key difference between human beings and AI is in the way we learn to make decisions. ML systems require massive volumes of training data. When teaching an AI to recognise animals, we need to provide the ML with massive volumes of data. However, human children learn to identify animals with far greater accuracy using far less data. In fact, a human child can correctly identify an animal such as a tiger after seeing only a drawing or a cartoon image. Understanding how human beings learn with so little data is the subject of extensive research. One of the goals of the AI community is to enable AI to learn with much smaller sets of training data. Clearly, human beings benefit from tens of thousands of years of evolution with core concepts and skills hard wired into the structure of our brains. You will be interested to know that there is a new area of research called "neuro-symbolic AI" [6,7] that hopes to combine broader human knowledge with statistical learning of neural networks.

Finally, it's important to remember that the massive amounts of data being generated today are still in silos! The real value in the data comes when it is fused together and that is not as easy as it sounds. To integrate data, we need to integrate systems and integrating systems requires standards, cooperation and partnerships. That's before we even touch on the sensitive subject of data privacy. Now we enter the zone of Data Wrangling.

DATA WRANGLING

Data wrangling is simply the process of making raw data useful for AI model building, much like taming the wild cattle to behave farm friendly. The objective is to create a coherent useful view of the data so that we can train the AI to learn the underlying patterns easily and efficiently. The process may require manipulating large volumes of data from disparate sources by summarisation, categorisation, aggregation, etc. For example, let us say that you own three stores in the local marketplace, selling fruits, bicycles and fast food, respectively. You record the sales at each store daily using some point-of-sale software. You could choose to keep the three data sets separate or aggregate them by fruit sales versus bicycles versus fast food, or food (fast food plus fruit) versus bicycles, or sum all to create a daily market sales figure. You might want to merge in weather data summarised by day showing the average temperature, rainfall, etc. Depending on what you want to predict, you will need the appropriate data. Your task is to create an input data record that is consistent i.e. can be replicated in the real world and carries enough detail to give your model a chance to spot a pattern that is useful. With the data we have collected this far, here are some examples of questions you can ask and expect an answer from the AI models:

- Should you keep the bicycle store open for longer hours on weekends and nice weather days?
- Do you sell more fast food on rainy days?
- Do you have the same customers coming to all three stores? Perhaps, you can give them special incentives?

Answers to each of these questions is looking for a pattern in the buying behaviour of your customers in each data set or across the combined data set from the three stores.

Data wrangling is data science's dirty little secret, it can be more an art form than science, i.e. it is a creative activity with little formal rules. *Remember to keep a focus on what is practical for the environment you are trying to model, the variables and summaries that you pick for your training HAVE to be available in the time frame for your live AI system.* A common error (more common than you'd think) is to inadvertently bake 'future' data into your training data. For example, when aggregating historical data sets, you may find a field containing actual average temperature per day. You thought it would be useful and chose to merge it by date with your other data. But, if you are hoping to use your model to predict today's sales, what temperature value are you going to use? Today hasn't happened yet and so you can't use today's actual average temperature! Here are some choices:

- Use a predicted average temperature. This is doable in this simple example but possibly inaccurate and only as good as the temperature prediction.
- Rebuild your model with the historical "predicted" not actual temperatures (if they exist).
- Rebuild the model with the previous day's average temperature, etc.

Choosing what variables (aka features) to include or create (by data wrangling) is always messy, do not be frightened to experiment and try multiple approaches. We're now going to dig deep into the question everyone asks ... "how much data do I need?"

HOW MUCH DATA DO WE NEED?

The practical answer is ... *whatever you are doing, you are going to need data that you don't have and you're going to need more of it than you can get!* The slightly more complex answer is ... it depends on the problem and, in order to successfully apply AI to any problem, you are going to need an extensive focus on data science. You are going to need to evaluate the data available, understand it, manipulate it, cleanse it and then figure out what additional data is needed to deliver an effective solution. You will also need to understand that the system and the data it uses will need to be managed throughout its lifecycle. Before discussing the data science required to support an AI project, let's consider the fundamental question of "how much data do we need" and try to understand why it's such a hard question to answer.

The volume of data we need to collect really depends on three factors: the complexity of the underlying pattern you are trying to find, our theoretical understanding of that pattern (aka our domain knowledge) and the accuracy required of our AI model. To understand how these factors impact the amount of data we need to collect, let's work through four practical examples.

Example 1: Modelling Tides (Collecting Enough Data to Spot Trends & Subtle Periodicity)

In the history of computing, one of the first practical requirements was to print tables for use in navigation. In the days when Britannia ruled the waves, the ability to navigate accurately was of huge importance to the Royal Navy. Navigating the seas safely required the ability to calculate a ship's position by observing the positions of stars. In order to avoid running aground, it was necessary to calculate the height of tide at any particular location and time of day. Thanks to Isaac Newton, both of these problems are solvable with mathematics. The relative position of the stars to the earth can be calculated and the position of the moon relative to the earth directly determines the height of tide at any particular location. Whilst the mathematics required to perform these functions was well understood, expecting hardened sailors to perform these calculations whilst at sea was not really practical. Instead, ships were, and still are, provided with tables that they can use to perform these basic navigation functions without the need for a mathematician onboard. One set of tables carried by all ships are tide tables and these allow the navigator to look up the height of tide at a particular location on a specific day at the required time.

You may be wondering what tide and navigation tables have to do with Artificial Intelligence (AI). Due to the importance of these tables for the Empire, organisations including the Royal Society, the Royal Observatory and the Royal Astronomical Society invested considerable resources in developing the technologies and science to enable accurate navigation. The calculation of these tables was the work of human computers and it was laborious task. One mathematician, a Lucasian Professor of Mathematics at Cambridge,

WHY TIDES RISE & FALL

Let us start with some basics. Earth goes around the Sun every 365.25 days. Moon goes around the Earth every 29.5 days. Earth rotates around its axis every 24 hours. As a consequence, even though everything is moving, from the Earth's point of view, the motion of the moon has a bigger impact on its surface than the Sun, because it is much farther away and relatively stationary.

Tidal Impact of the Moon

Moon's gravity produces a bulge of the water level closest to it and on the farther side of the Earth's surface (blue in the Figure A). Due to Earth's rotation, each location on Earth experiences two high tides and two low tides every 24 hours and 50 minutes. The tidal height is given by

$$h_M = A_M \cos\left(\frac{2\pi t}{T_M} + \varphi_M\right)$$

where T_M=12.421 hours is the Principal Semidiurnal Lunar (M_2) period, A_M is the amplitude and φ_M is the phase that depends on the details of the underlying physics. These are the three parameters we need to determine using the measurements discussed in Figure 7.3. If we have this theoretical knowledge, then all we really need is sufficient data to estimate these three parameters at a given location. The generalisation of the model is achieved by the equation itself. We need to be careful at this point, because a lot depends on how we are collecting the data. If we are able to sit on the shoreline, we would only need to record the minimum height and the maximum height of tide together with the times at which they occurred. If the measurements are perfect, we can rely on only three data points because we need as many independent observations as the number of unknown parameters in the model. Due to inevitable measurement errors, we probably need many more data points (say, 30) to build a valid model with appropriate error estimates.

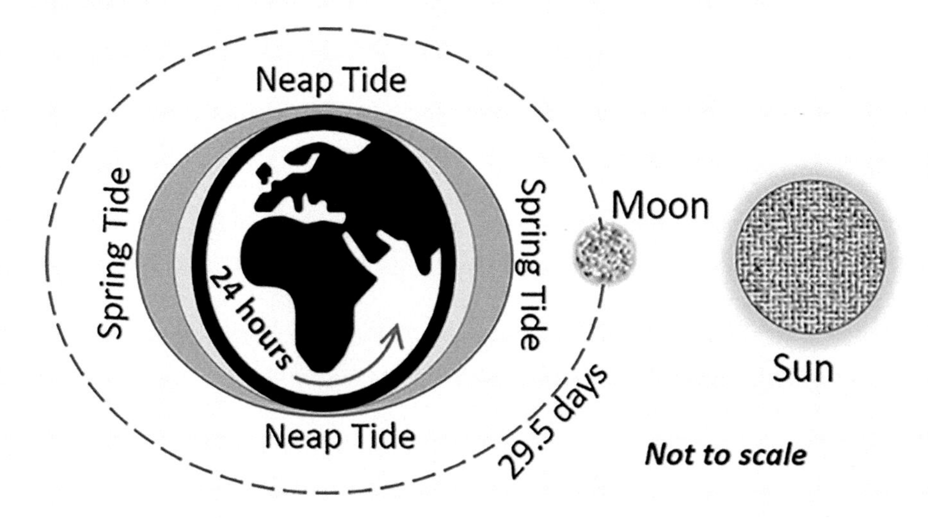

FIGURE A: An illustration of the effect of the sun and the moon on the tides on earth. Not to scale.

Tidal Impact of the Sun

Sun's gravity also produces a bulge of the water level closest to it and on the farther side of the Earth's surface but with a smaller amplitude (orange in the Figure).

$$h_S = A_S \cos\left(\frac{2\pi t}{T_S} + \varphi_S\right)$$

where $T_S = 12$ hours is the Principal Semidiurnal Solar Period (S_2), A_S is the amplitude and φ_S is the phase that depends on the details of the underlying physics.

Combined Tidal Impact of Both Moon (M_2) and the Sun (S_2).

The combined effect of the Moon and the Sun can be written as

$$h = a_c \cos\left(\frac{2\pi t}{T_c}\right) + a_m \cos\left(\frac{2\pi t}{T_m} + \varphi\right) \cos\left(\frac{2\pi t}{T_c} t\right)$$

$$T_c = \frac{2\, T_M T_S}{T_M + T_S};\quad T_m = \frac{2\, T_M T_S}{T_M - T_S}$$

resulting in $T_c = 12.2$ hours and $T_m = 29.5$ days. There are five parameters (T_C, T_m, a_c, a_m and φ) in this equation giving rise to the shape of Figure 7.4. Here, the amplitude of the higher-frequency (i.e. smaller period, 12.2 hours) wave is modulated by a lower-frequency (i.e. higher period, 29.5 days) wave. The larger amplitude (Spring tide) is caused by the Moon and the Sun reinforcing each other while the smaller amplitude (Neap tide) is when they are opposed to each other.

believed it was possible to build a machine to calculate and print these tables. His name was ***Charles Babbage***! He built a small version of a 'Difference Engine' in 1822, but the work on a larger engine was suspended in 1833. It could be argued that the need to understand tides was a key factor in the development of early mechanical computers.

What would happen though if we knew nothing about the physics of tides [8,9] and we wanted to use AI to predict tidal height. How much data would we need? The simple answer is that we don't know how much data we would need. All we could do is start collecting data. What happens if we collect data at different sampling rates (See Figure 7.3)?

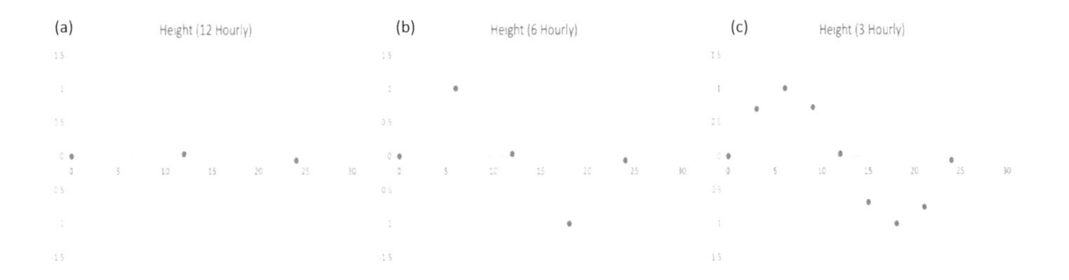

FIGURE 7.3 Shows an example of different amount of tidal data collected (blue dots) during a 24-hour period; (a) every 12 hours, three data points; (b) every 6 hours, five data points; (c) every 3 hours, nine data points.

As you can see in the diagrams, the 12-hour sampling doesn't really give us any insight into what's happening. The 6-hour sampling suggests that there is some form of periodic behaviour and the 3-hour sampling strongly indicates a sinusoidal form. Obviously, the more data we collect, the clearer the picture will become. If we had access to an expert, then we would know that tides rise and fall sinusoidally with an average period of 12 hours and 24 minutes in most locations in the UK.

With that knowledge, we could then construct a theoretical model of a sine wave for the tidal height that has three parameters: amplitude, period and phase (please see the vignette for more physics). We need three independent observations minimally for a perfect fit. In reality, due to random errors in observations, we may need to have more observations (say 30) to estimate the three parameters and the associated errors more accurately. **Without that item of knowledge,** *all we can do is collect more data* until we are confident that we understand what is going on.

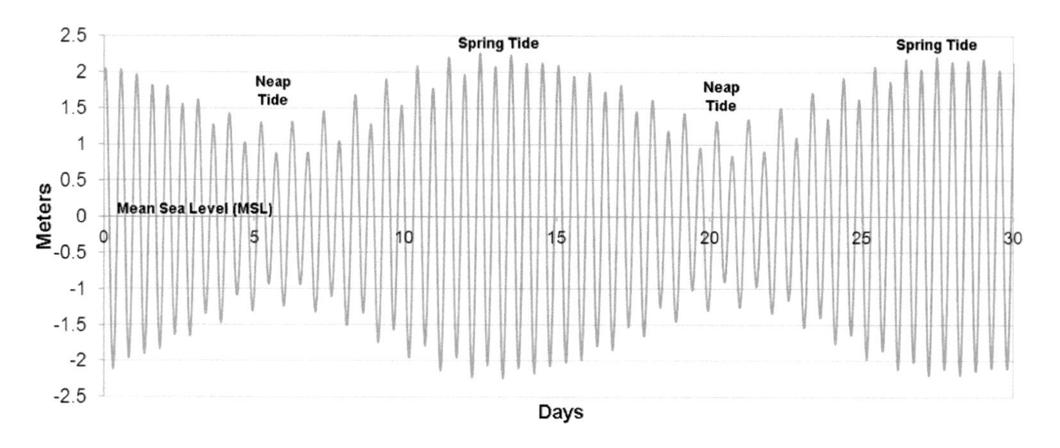

FIGURE 7.4 The tidal amplitude over a month from [10] reproduced with permission. This complex behaviour is due to the combination of the result of the principal semi-diurnal lunar harmonic (M2) and the principal semi-diurnal solar harmonic (S2). The envelope has a period of 29.5 days whereas the daily waves have a period of 12 hours and 24 minutes [8,9]. Please refer to the vignette, "Why tides rise and fall".

Unfortunately, if we only collect tidal data over a 24-hour period, we are missing some critical observations. The maximum and minimum height of tide varies over the course of a month … but remember we don't know that! If we monitored the height of tide every hour over a period of one lunar month, we would observe something similar to Figure 7.4.

Again, the more data we collect, the clearer the picture will become. If we had access to an expert, she would explain the theory behind this behaviour as due to the interaction between two different wave forms due to the influence of the moon and the sun, resulting in this 'amplitude modulation'. The underlying theoretical model will have a specific equation for the tidal height *with five parameters* related to the two waves (*see the vignette 'Why tides rise and fall' for more physics*). **With that knowledge**, we would only need to collect minimally five independent observations in order to construct a theoretical model, even though as before, we may need a few dozens more for estimating the five parameters with the corresponding errors. **Without that knowledge**, all we can do is collect more data until we are confident that we understand what is going on (*have we said that before?*).

Unfortunately, if we only collect data every 3 hours, we might be missing some critical observations! The challenge is that in some locations, other factors impact the height of tide. Geographic features may cause back eddies and all sorts of strange water flows. For example, there are no significant tides in the Mediterranean due to the narrow entrance at Gibraltar; the Straits act as a bottleneck and prevent the seawater flowing in and out of the Atlantic. Similarly, if there is a large river, the height of tide may be impacted by high rainfall. Figure 7.5 shows the tidal behaviour in the Solent, the stretch of water between the Isle of Wight and Southampton in the UK.

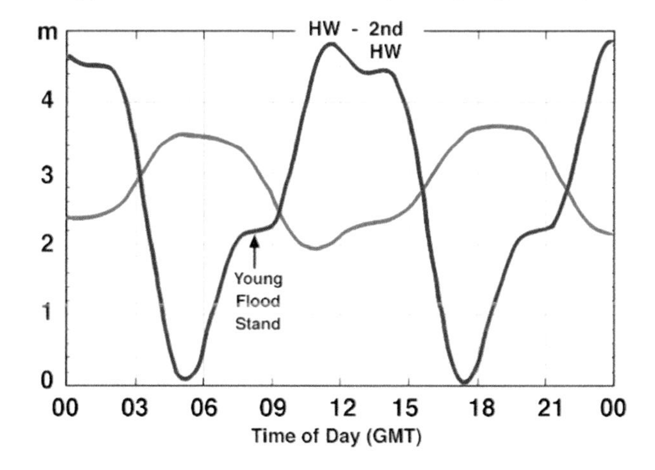

FIGURE 7.5 Tidal wave over a day at the Solent in the UK. The wave form is more complicated than a simple sine wave due to the local geographical features [11]. Reproduced with permission.

As you can see, this profile is very different to the sinusoid we have been talking about so far. The only way to discover that is to collect more data! The more data you collect, the more confident you will be that you understand what is going on (*have we said that before … twice?*).

This raises an interesting conundrum in the use of ML. The main reason for using ML is that we don't understand the underlying model and we need to create a model that can generalise using the data available. The more data we collect, the less we need to generalise and predict because we could, theoretically, just look up the data. ML will hopefully allow us to generalise and predict … but only if we have sufficient data. We're using ML because we don't understand the underlying model and we can't know if we have sufficient data unless we understand the underlying model.

So, how do we address this conundrum? First, ensure that whatever expertise does exist is tapped into! Whilst the experts may not fully understand what is going on, they will have knowledge that will help validate that sufficient data has been obtained. Second, ensure that an effective test strategy is adopted to ensure the behaviour of the AI in the production environment meets the business objectives.

For this simple tidal example, let's finish by considering the three factors we mentioned earlier.

Firstly, the complexity of the underlying model. If the tidal behaviour we are modelling follows a simple sinusoid, then we only need a handful of data to build an accurate model. If we are modelling the tides in Southampton, then we're going to need a lot more data.

Secondly, our theoretical knowledge of the situation. If we have a theoretical knowledge of tides, we can use this knowledge to inform our data collection strategy.

Thirdly, the accuracy required of our model. If we are sailing a yacht between the Solent and the Isle of Wight, then we can probably get away with a sinusoidal model. Yacht sailors generally don't need tidal data to be too accurate as they allow contingency and use their echo sounders. However, if we're manoeuvring an aircraft carrier, then we probably need a more accurate model and therefore we're going to need more data.

Example 2: Watching the Dice Roll (When Estimating How Many Examples to Train on, Consider All the Angles!)

For the next example, let us use the word 'features' to represent the key attributes of the model we are trying to build. Loosely speaking, you need to have enough examples in the data covering the various features representing the model. We will come back to a more quantitative example later in this section. For most AI systems, this is not an exact science. Let us take a seemingly simple AI challenge to teach a visual recognition system to read the value of two wooden dice being rolled in a box below the camera. How many training rolls would we need to make before the AI would have enough data to reliably interpret the dice spots into a number, reliable enough to go into production in a casino? *Spoiler alert: it's more than 11 (the number of possible totals) and it's more than 36 (the number of possible combinations) if you guessed either of those values.*

What you may have forgotten, in estimating your training data, is the number of angles the dice could be rotated by in the image the camera sees, a much larger multiplying factor. Yes, we humans know that four spots in a square mean 4 and it's the same value even when those four spots are rotated 45° to be a diamond shape. But, the AI might not, not that is until you have rolled enough training runs and given it the answer in training for it to start getting it consistently right. Camera angle is a feature in this example. If this was a real problem you were trying to solve, you might add a mechanical arm to sweep the dice into a corner (reducing the angle problem and therefore reducing the number of training runs you'd need) before engaging the AI to learn the more confined pattern of two perpendicular dice.

Example 3: Helping the Doctor (Not Enough Data to Meet Your Goal? Change Your Goal!)

Thought experiment. Let's say you were planning to build an AI to diagnose common illnesses. You might use the content of a medical encyclopedia to train your model. Your AI in tests could seemingly match with near perfect accuracy any list of symptoms with the corresponding disease/ailment. Is your new "Diagnozer" ready for real use? Does it know enough? Have you trained it on enough data?

It might seem so; however, a simple discussion with a medical doctor would inform you that whilst the encyclopedia in question did indeed hold great information on the illnesses that patients exhibit 95% of the time, it only held information on a small fraction of all the illnesses you need to know about to safely diagnose. This is because whilst the common illnesses represent 95% of all cases, these common illnesses represent only 20% of all possible medical complaints. The remaining 5% of cases represent the remaining 80% of possibilities, but more importantly many of the 5% patients have rare and deadly illnesses with symptoms easily confused with lesser ailments. Your Diagnozer could lead a critically ill patient to believe they had a benign condition. Your AI could kill!

So your "Diagnozer" is dangerous and worthless ... right? Wrong!

Business value is very much in the eye of the beholder, your AI tool is not suitable to be given to the general public for self-diagnosing that clearly would be a lawsuit waiting to happen. But consider your AI Diagnozer as a tool to augment a General Practice Medical Doctor and you have a massive timesaver/quality improver. Your tool would allow the doctor to instantly get to the most probable cause of a patient's illness, freeing plenty of time in the consultation to check for the less likely candidates. In the hands of an amateur, any tool can be dangerous; the same tool in the hands of a professional much less so. It is true of the cheapest medical scalpel, we should not be surprised it's also true for the most expensive medical AI.

Example 4: Predicting House Prices (Random Variables, Missing Information and Sparse Data...)

To make the data estimation problem more concrete, let us consider another real-world problem that most readers can recognise. Imagine we wanted to build an AI Realtor tool that was able to assess the achievable sale price of a house. For simplicity, we decide that the output of the AI Realtor for a house price will be either 'High' or 'Low' (instead of a number), relative to a boundary we do not specify. We all know that house prices are dependent on many different factors (we will call *Features*) ranging from the size and quality of the property to environmental factors such as proximity to main roads or noisy industrial partners. *Features* can comprise either raw data, such as the number of bedrooms in a property, or derived data, such as the average sale price of properties in the same neighbourhood or postal code. "Feature Space Volume" represents the combinatorial product of all values the features can take, since any one combination can occur in each observation. Let's imagine that the data set provided comprises the features defined below.

Feature	Possible Values	Number of Values
Number of bedrooms	1, 2, 3, 4, 5, >5	6
Number of bathrooms	1, 2, 3, >3	4
Proximity to school	<1 mile, 1–3 miles, >3 miles	3
Subsidence	Yes, no	2
Proximity to nightclub	Yes, no	2
Flooding	Yes, no	2
	Feature space volume	**576**

Because we have discrete values, it is possible to calculate the exact volume of the *Feature Space*. In this case, there are theoretically 576 possible permutations in the *Feature Space*. In the real world of course, numerical values are rarely discrete so the volume of the *Feature Space* is technically … possibly … potentially … infinite … but you get the idea!

The volume of data we require to understand this *Feature Space* depends on what happens at every possible point in the *Feature* Space. In particular, it depends on the probability of each possible output across the *Feature Space*.

Let's imagine for a moment that we decided to collect 576 examples, one record representing each of the cells in the *Feature Space* (*we will ignore the fact that in the real world, few if any, six bedroom houses exist with one bathroom*). For one particular cell, we find that the house price is LOW as shown below.

Number of Bedrooms	Number of Bathrooms	Proximity to School	Subsidence	Proximity to Nightclub	Flooding	Output
4	3	<1	No	No	Yes	LOW

However, if we were to collect ten records for each of the cells, then we might collect the data below.

Number of Bedrooms	Number of Bathrooms	Proximity to School	Subsidence	Proximity to Nightclub	Flooding	Output
4	3	<1	No	No	Yes	Low
						High
						High
						Low
						Low
						High
						Low
						High
						High
						Low

As you can see, when we collect more data, we find that for this particular set of input data there are five high-priced houses and five low priced. Knowing this particular set of data gives us a 50-50 chance of predicting the correct output value. If we had only used a single record, clearly, we would not understand the 50-50 distribution. A single record does not give us enough data on which to understand what really happens at that point in the *Feature Space*. In an ideal world, every time we collected data for an individual point in the *Feature Space*, we would observe the same output. If, as in the example above, the data reveals a 50-50 situation, then we can't really make a useful prediction. *Guess what? We need more data!*

If we're lucky, we could collect a thousand data points and discover that the distribution is actually 995 low to 5 high. That would give us a clear and useful correlation. If we're unlucky … and in this game, we usually are … collecting 1000 samples will still give us a 50-50 distribution. What does that mean? We need to collect more features as the features we are collecting are, as in the subsidence example above, insufficient to pull the data apart. Adding more features increases the volume of the *Feature Space* so … *wait for it … we're going to need to collect more data.*

So, how many records do we need to collect for each point in the *Feature Space*? It depends! It depends on the complexity of the underlying model and, generally, when using ML, we don't fully understand that model!

SO, WHAT FEATURES DO WE NEED?

Having considered the question of how much data we need, let's get back to the question of what data do we need. In Chapter 3, we talked about the domain of *Feature Definition and Extraction*. To understand this question, we return to the house pricing problem. In an operational solution, a *Feature Extraction* process will take raw data from multiple sources and construct the input data that is supplied to the AI. To keep it simple, let's start by taking into account two features: the number of bedrooms and the distance between the property and school. Generally, large properties that are close to schools will have higher valuations than small properties that are a long way from school. Figure 7.6 shows this "ideal world" situation.

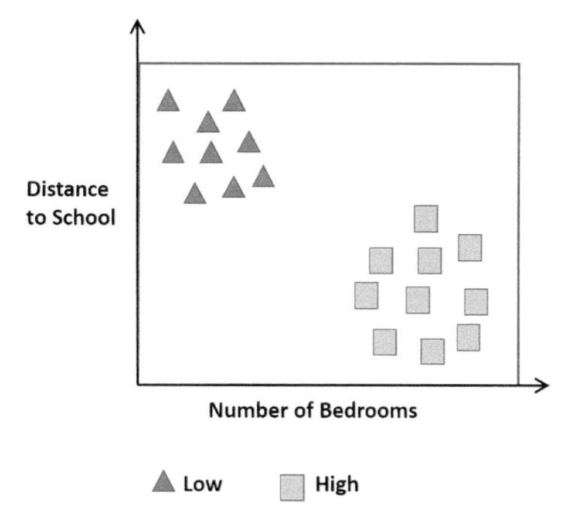

FIGURE 7.6 House price against two features, Number of Bedrooms and Distance to School, ideal case.

Unfortunately, the world isn't perfect. In reality, we will see a picture that looks more like that shown in Figure 7.7.

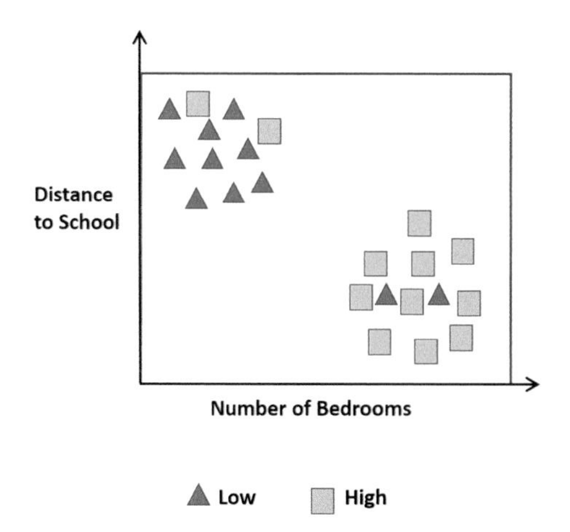

FIGURE 7.7 House prices against the same two features in Figure 7.6, reality.

It shouldn't be a surprise to anyone to see that we have a couple of cheap houses that are close to good schools and have a large number of bedrooms. Clearly, there are other factors which impact the price of a house so, at this point, we need to engage our Analysis Junkie mantra and collectively shout, "More data please!" (Figure 7.8).

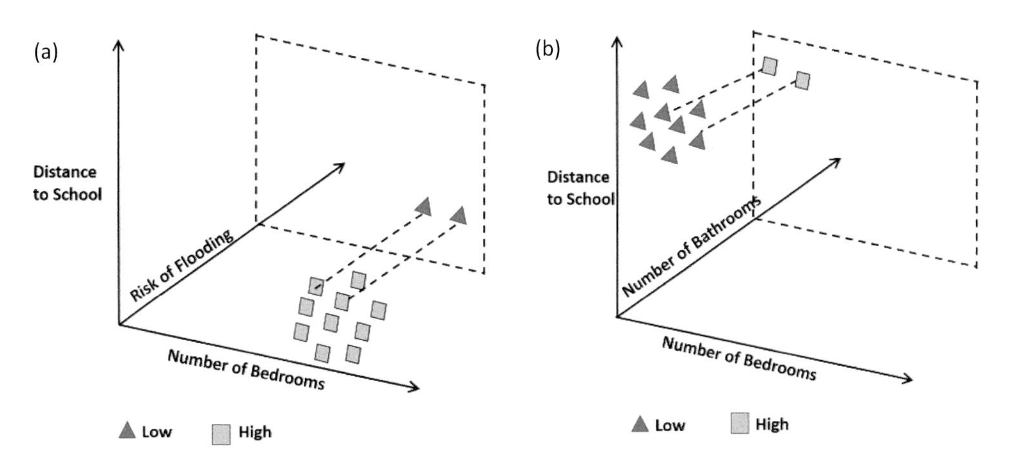

FIGURE 7.8 Explaining perceived anomalies with additional dimensions. (a) Low prices of the two houses are due to the risk of flooding. (b) High prices of two houses are due to the number of bathrooms.

The *Data Gods* hear our call for help and graciously provide us with more data and instantly we see that two of the unexpectedly cheap houses have a problem. They are the victims of flooding and are therefore much cheaper than other properties of the same size in similar locations. The two high-priced houses that are further away from the school happened to have a large number of bathrooms.

While the new features allow us to explain what is happening, it introduces a new challenge that we refer to as the *Curse of Dimensionality*. As we add more features, the size of the *Feature Space* increases. We therefore need more data in order to fully understand what is happening. In the case of a house price valuation, there are many different factors (See Figure 7.9) that need to be taken into account. Flooding is one, subsidence could be another possible factor that could justify a reduction in the value of the property. Alternatively, the house may have been constructed using dangerous materials such as asbestos or it may have been built in a location close to a busy road or a railway line or a noisy night club. The question is, how do you present this information to the AI in the most effective manner?

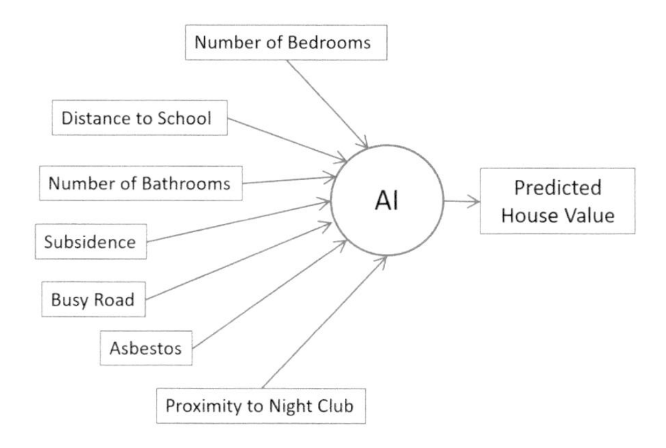

FIGURE 7.9 More data, but no data science or engineering.

Theoretically, you could just keep adding features. However, that isn't really an effective strategy for two reasons. Firstly, every time you add a new feature you are increasing the size of the feature space and you are going to need more and more data to ensure you cover the whole feature space. Secondly, no matter how hard you think about the problem in advance, once operational, you will always encounter new scenarios that you hadn't previously considered. It simply isn't practical to keep adding extra features.

One approach is to derive a new amalgamated feature that can be used for all inputs that can have a negative impact on the price of a property (Figure 7.10).

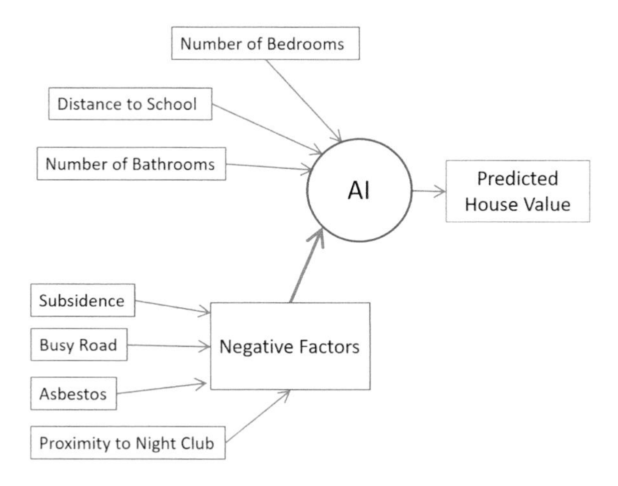

FIGURE 7.10 Less data, more data science/engineering.

The key here, this is a theme we will return to, is ask your experts what factors they take into consideration when they make a decision. Start by mimicking the information your experts use/need and build from there. *Hint* you may need to add more features than the expert initially identifies, often experts are good at their job because they have internalised their knowledge, e.g. they don't have to think about how they make decisions, they just do it based on experience. It is only after you circle back to the experts with a model that seems to do naive things that they remember, "oh yeah I forgot to say I always check to see if the house is on a registered flood plain before I give a valuation…"

The creation of this new amalgamated data feature requires data science and engineering.

By applying this engineering effort, we achieve two positive results. Firstly, we reduce the amount of data required to build an effective *AI Model* and, secondly, we ensure that the model is future proof. It is possible to add new features or remove old features without needing to re-train the *AI Model.*

This simple example highlights the importance of *Feature Definition* in determining the effectiveness of the solution. Like all computer systems, AI solutions are only as effective as the data they are provided with. In our house price solution, the ability to accurately predict the price of a property is directly constrained by the features available. If the solution has no access to flooding or subsidence data, then it simply cannot predict the lower price of a flood risked, or subsided house. This may seem obvious, but it's a point that is often missed in more complex case studies.

Within *Feature Definition*, we often talk about two specific techniques: *Feature Reduction* and *Feature Expansion. Feature Reduction* aims to take a large number of *Features* and reduce the number either by fusing them together, as in the above example, or removing them all together if they don't seem to add value. *Feature Expansion* aims to add new features in situations where we have insufficient data to perform the task. There are many different techniques used by data scientists to perform these tasks. Quite often, data scientists will use correlation techniques to understand the relationships between features. If, for example, you are measuring frequency and wavelength, those two features are directly related … they are effectively the same thing … so there is no point including both in the *Feature Space.* Data scientists also use a technique called Principal Component Analysis (PCA) to fuse multiple features into new features that are significant in making decisions. PCA is useful when you have a large sparsely populated feature space and you can't be certain which features to fuse or safely remove, PCA automatically starts to combine the features into new statistically correlated merged features, so your 100 sparsely populated actual features might come out as 10 PCA condensed pseudo features! Perfect for your ML training but not if "explainability" is part of your 'must haves'. Picture the scene…"Yes Mr Smith, we need to amputate your left leg, why? It's because the PCA3 variable which may or may not be based on a mathematical combination of seven symptoms that you may or may not have says so…" So PCA is great to prove a pattern exists and is great when there is no sensible way of manually reducing the feature set, but it is less useful if you really need to understand why your model makes a particular decision. Please refer to the vignette for conceptual details behind how it works.

The answer, once again, is to have a well-defined test strategy to evaluate the effectiveness of the AI. The good news is that, at this point, we can once again rely on the wonders of science and mathematics. *Whilst we can't predict how much data we will need, once we have collected data, we can determine if we have enough.* We refer to Chapter 6 for more details on some useful techniques.

The high-level messages are simple. AI is data greedy and, in everything you do, you should be ready to chant the Analysis Junkie's mantra and demand "more data". The amount and the distribution of the data you need will depend on the system you are trying to model. Always remember that just having a huge amount of data doesn't mean you have enough data! The amount of data you need depends on the complexity of the problem and the system you are trying to model.

As in all things to do with AI, it's important to apply the best scientific and engineering practices. AI is not engineering free so never throw data at algorithms without doing the science and engineering needed to understand exactly what is going on.

ENABLING EXPANDING FEATURE SPACES

Increasingly, AI Applications are going to need to respond to the provision of new data sources and the feature data that goes with them. In such situations, we need to consider whether to re-architect the AI Application. We also need to consider whether it is possible to build on the existing AI or whether we need to completely re-define and, if using ML, re-train the AI components. We refer to the vignette on how IBM's Watson Jeopardy project on Deep Question Answering dealt with expanding feature spaces.

WHAT HAPPENS IN THE REAL WORLD?

Bad things happen in the real world ... especially when it comes to data.

In textbooks when you are learning to use AI algorithms, the sample data provided is always complete and magically always has just enough records to answer the question you set out to tackle.....the real world is sadly not so kind. You will need to use your imagination, animal cunning and sometimes your cheque book to work around the inevitable data quality and sparsity issues.

Anecdotally, up to 80% of the resources assigned to AI projects end up focusing on accessing, cleansing, fusing, resolving and generally manipulating data. That amount of effort is required because data is never available in the clean and ready-to-go form that we would wish for.

Once again, the devil is in the detail so let's dig into the sorts of issues encountered in real projects.

Let's start with the challenges of **historical data**. Such data may have been collected over many years and, in some cases, decades. In medical analysis applications, for example, patient notes may be provided that go back seven or eight decades. The data recorded in those notes will vary in quality over time. An X-Ray from the 1950s will not be of the same quality as an X-Ray taken in 2020. Even basic test data such as blood pressure, weight and body temperature may vary in consistency due to the use of analogue technologies and unreliable sensors.

REDUCING DATA DIMENSIONS

This vignette addresses the scenario where you have too many attributes (i.e. features) and not enough data. Since each additional attribute requires additional data to cover the feature space, it is good to think of ways to reduce the dimension of the feature space for the available data. Principal Component Analysis (PCA) is an established technique for this purpose in statistics, and we will discuss the key concepts behind PCA briefly here.

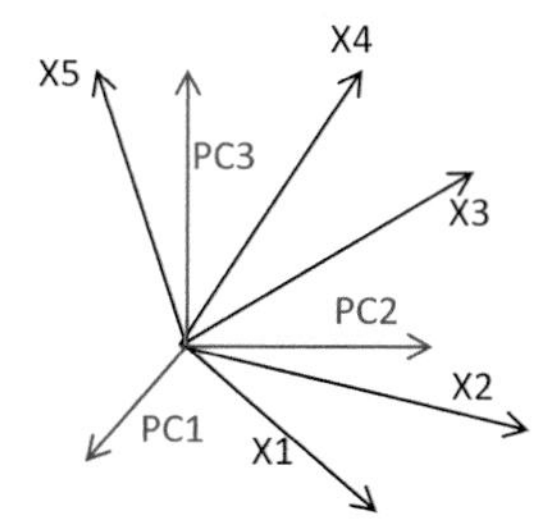

FIGURE A: An illustration of Principal Component Analysis. Three principal components (PC1, PC2, PC3, in red) represent the same space originally represented by the five axes (X1...X5).

Let us consider the housing price estimate problem again. We can imagine the various attributes of a house we discussed, viz., Distance to School, Number of Bedrooms, Number of Bathrooms, Subsidence, Busy Road, Proximity to Night Club, etc., as the dimensions $X1$, $X2$, $X3$, $X4$... in Figure A. Each house will be represented by one point in this plot. Since some of these attributes can be correlated with each other, such as Number of Bedrooms and Number of Bathrooms or Busy Road and Proximity to Night Club, it is not necessary to include them explicitly as independent attributes. We can combine them to make new composite attributes. The idea is to create smaller number of composite attributes i.e. Principal Components, which are less correlated with each other to describe the data than the original raw attributes. This is much like in describing the three-dimensional space by three Cartesian coordinates (x, y, z) which are perpendicular to each other, and any point or its motion can be described accurately in terms of the three coordinates. PC1, PC2 and PC3 in Figure 1 represents one such representation of the space spanned originally by the five dimensions $X1$, $X2$...$X5$. The principal components are independent of each other and they are derived from linear combinations of the original dimensions based on detailed statistical correlation analysis across the original dimensions. In general, the number of useful principal components will depend on the complexity of the data and the number of attributes available for analysis. Also, typically, a small number of principal components can account for most of the variation in the observations. Since creation of principal components is not trivial, most statistical tools [1] offer PCA as one of the key techniques to explain the analysis of variance in the statistical data. The tools give the explicit decomposition of the principal components in terms of the original attributes with weights as a part of the output and it is possible to recognise meaningful composite attributes on hindsight, such as Location and House Structures.

In the ML literature, there is an analog to PCA for reducing dimensionality called "Autoencoder" [2]. This is an unsupervised learning technique where the output of a neural network is asked to reproduce the input to force the network to learn the minimum representation to reproduce the original data. For example, this is a very efficient technique to reduce noise in images. An advantage of the Autoencoders over PCA is their ability to represent non-linear relationships among the attributes whereas PCA assumes linear dependency. However, they do have the problem of human understandability of their internal representations.

REFERENCES

1. F. Lima, "Principal component analysis in R," in R bloggers: https://www.r-bloggers.com/2017/01/principal-component-analysis-in-r/.
2. G. Dong et al., "A review of the autoencoder and its variants: a comparative perspective from target recognition in synthetic-aperture radar images," *IEEE Geoscience and Remote Sensing Magazine*, 6(3), pp. 44–68 (2018).

The interpretation of the data recorded in the notes will also change significantly with time as medical knowledge has developed. For example, the condition known as Systemic Lupus was unknown in the 1950s, and patients suffering from that condition would probably have been incorrectly diagnosed with Rheumatoid Arthritis.

The next problem to think about is **inaccessible data**. Quite often you will be told that there is a huge amount of data only to find that it is not in a form that is usable. It may be that the data is in the form of handwritten notes, or scanned image data, which cannot yet be transformed into an electronic or usable form. Alternatively, the data may exist in an electronic form but be missing the key index or reference number that enables it to be cross referenced with another key source.

The real world is continually changing and that is reflected in the continual creation of new data sources and new features. This raises an interesting issue relating to the scope of an AI Application and, in particular, whether to adopt a Fixed or an **Expanding *Feature Space***. In the most basic AI Applications, the features used are fixed. For example, an image classifier will be designed to process a specific size of image with no variance in the input data. However, in more sophisticated solutions, there may be a desire to add new data sources and features to the capability as they become available. In that situation, the *Feature Space* will be continually expanding. If **missing features** are an extreme form of **missing data**, then an **expanding feature space** is an extreme form of **missing features**.

Then we come to the problem of **missing data**. Data collection is rarely perfect for various reasons. People who are asked to fill in forms often fail to complete every field. Sensors and communication links fail so data that should have been collected is lost. Systems crash and hard drives are corrupted. When building data-marts, data-lakes and data-warehouses, data is often merged from many historical legacy data stores. However, when merging data, the data models are rarely an identical match so not all fields are populated.

An extreme form of **missing data** is **missing features**. In our house price example above, we realised during the analysis that we needed to collect subsidence data. Remember that if you subsequently add a feature like subsidence, because it helps the model, then you will need to answer that question for all the houses *already* in your set, if not, how will the model interpret the houses with missing values. If you are using historical data, data from a survey

WATSON JEOPARDY – EXPANDING FEATURE SPACE

In 2011, IBM demonstrated a massive break-through in Deep Question Answering when its Watson AI solution defeated two Jeopardy! champions, Ken Jennings and Brad Rutter, to win a $1 million prize.

For those not familiar with Jeopardy, it is a quiz show in the USA where contestants answer obscure questions. To give you a feel for the type of question, here's an example:

> Hard times, indeed! A giant quake struck New Madrid, Missouri, on Feb. 7, 1812, the day this author struck England.

The answer turns out to be 'Charles Dickens', since he was born on that day!

So, what was the big deal with Jeopardy and, in particular, what made the application different to conventional search or question answering? Firstly, the questions could not be answered just by keyword search. In most search engines, it is possible to simply extract keywords from the question and find a passage of text that is likely to include the answer.

Secondly, in conventional search applications, a massive proportion of the questions asked relate to a relatively small amount of the content. In Deep Question Answering, the application needs to answer questions that are very rarely asked and require the full breadth of content to answer. See Figure A for a comparison of traditional Search vs. Jeopardy.

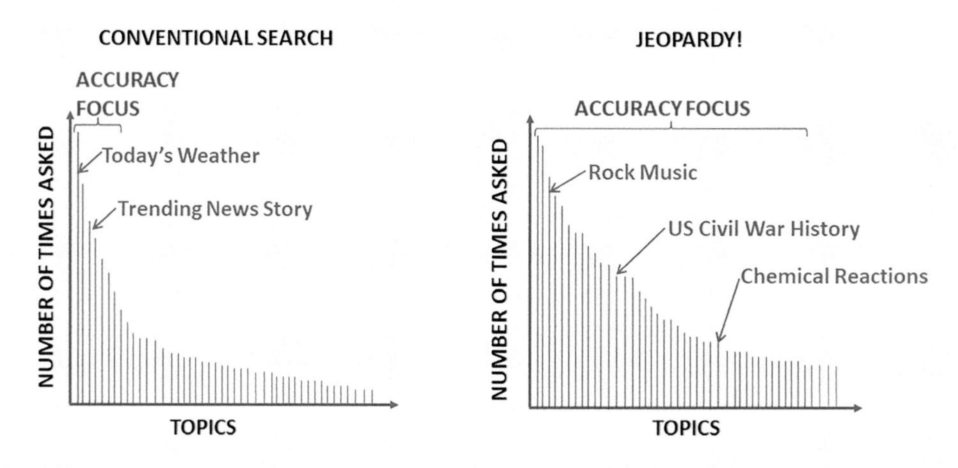

FIGURE A: Comparison of traditional Search vs. Jeopardy in terms of accuracy expected and topics covered.

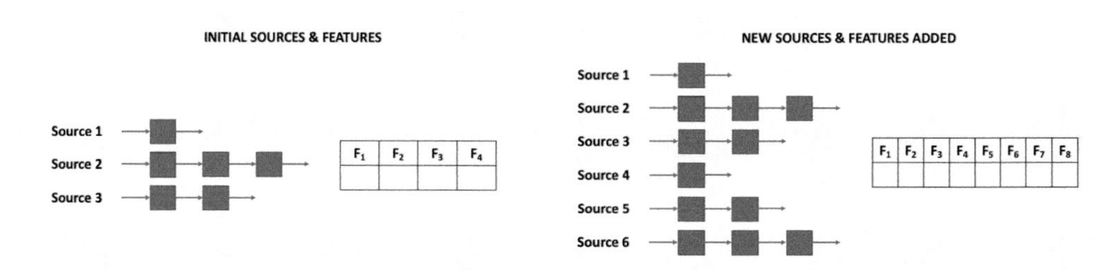

FIGURE B: An extensible approach used in the Jeopardy project to deal with the ever expanding feature space.

In developing the Watson solution, the IBM Research team had to identify and ingest content from a massive range of sources. Each of these sources required processing to extract features that were specific to each content source and the types of question that could be asked of each source.

With development taking several years, this meant that every time a new content source was identified and added, new features were also added. In effect, the feature space was continually expanding.

Given the scale of the Jeopardy challenge, it would have been completely impractical to consolidate and normalise the feature space every time a new content source was added. Instead, the team developed a new architecture whereby new content sources and features were added cumulatively to the application without the need to change any of the existing features.

ML was then applied as the very final stage of the process to learn which features were important for each type of question.

that didn't collect that information at the time, then you will get less value from subsidence as a model feature.

It may be possible to collect an entirely new feature going forward, but retrospectively finding such data is not always possible.

Data consistency is a huge challenge; especially with very large data sets and multiple annotators. If you ask a team of ten medical underwriters to assess the same 100 insurance applications, it's unlikely that they will all assess the same level of risk for each of the applications. When multiple people interpret data, they rarely produce identical results.

Finally, the **answer isn't always in the data**! That was an incredibly hard sentence for an Analysis Junkie to write. Unfortunately, it's true and it's important that we approach AI projects with a realistic expectation of what can be achieved. In some cases, the data simply does not exist to answer the question being asked. Criminal investigations are very significant in this respect. It's very common for large IT companies to volunteer their technology to law enforcement agencies in the hope of solving a major crime. Our intentions in doing this are genuinely good and our hope is that we can do something good for society. Quite often, however, you realise that the key information needed to solve the crime just does not exist in the data.

COPING WITH MISSING DATA

Sometimes you will be lucky, you will find a source of the missing data, merge that with what you have to complete all the records and crack on. Or you will have enough time in the plan to simply pause the project, whilst you start collecting the data you need from that point on, and reconvene the project a few months later with your freshly harvested complete data….but back in the real world, you will need to look for alternatives.

Data Imputation

If you have a percentage of records that lack information in certain fields, you can try to statistically impute the values. This of course depends on how many missing values you have versus complete, e.g. 10% missing is much better than 20%. You also need to care if

GOOGLE'S RETINAL SCAN CASE STUDY

This is a recent example of how AI that works well in the lab may not do so well in the real world. A team from Google Health built an AI-powered eye scanner that could help diagnose retinal diabetes [1]. Google Health's first opportunity to test the tool in a real setting came from Thailand. The country's ministry of health had set an annual goal to screen 60% of people with diabetes for diabetic retinopathy, which can cause blindness if not caught early. But with around 4.5 million patients to only 200 retinal specialists, clinics were struggling to meet the target.

The plan was to deploy the AI in rural areas, places where access to an eye specialist was limited. The AI had performed flawlessly in the lab, and all thought this was the perfect opportunity to utilise the power of AI for the benefit of mankind, especially as the lab results were so good. At this point, very few experts would have correctly predicted what was going to happen.

Unfortunately, the project team had forgotten to take into account the poorer quality of retinal cameras in use in the rural areas. The AI trained on a diet of clean high-resolution precise scans could not recognise the lower quality images, even though the human experts had no such translation problem. The trials predictably failed. Anyone who has built any real-world AI system will have similar tales. The AI engineers in this story were brave enough to publish lessons they learnt. We can only applaud their honesty and understand that they've now found a workaround.

This story of the retina scan illustrates another foible of AI, it doesn't think like you or I and doesn't always generalise in a linear way. The scans from the rural machines were probably not radically different from the high-tech ones. Certainly human ophthalmic doctors will have no problem interpreting them. The human would be able to extrapolate learnings from the clear image to the grainy one, not so the machine. The moral of this story is do everything possible to match the data diet of your training models to what you expect them to eat in the wild.

REFERENCE

1. MIT Technology Review, "Google's medical AI was super accurate in a lab. Real life was a different story," https://www.technologyreview.com/2020/04/27/1000658/google-medical-ai-accurate-lab-real-life-clinic-covid-diabetes-retina-disease/.

the missing values are not evenly distributed amongst your record types. If one type of record never has a value filled in, it can be problematic to impute it. It is also worth checking if some missing values actually mean zero or vice versa (in our house buying example an apartment record might leave the field for garden size blank rather than put a zero, whereas an auction property might have a blank for garden as the detailed information has yet to arrive).

In our house buying example if many of the historical records lack information on flooding, then we might consider it a candidate for Imputation, as it would be impractical to visit homes we had sold in the past to ask if they had this issue: Imputation is a statistical process that will use the complete data records to predict the missing value in incomplete records e.g. that house is too cheap given those features it probably has flooding or subsidence, etc. This may be a chicken and egg scenario for what you are trying to achieve, it obviously works best when the ratio of complete to incomplete data is high.

Proxy Data

If you cannot get the data you want, get creative, look around, is there some other data source you could get that might work as a substitute? Look for data values that might imply the feature you are after. For example, in our house price example could you source environmental data that provides a map reference data set of areas with flood risk, that's a good start but now you need to link house data to the map data to give each house a flood score. You will now need to enter the fun world of data wrangling!

In the real world for our housing example, we could use a geocoder service (software that calculates a map grid reference from an address) to establish the map references for each historical house sale record. We could then merge our historical house records with the environmental data to create a flood risk score per record. We could then run our models again using this data to see if the added score increased the accuracy of our model. Note from this point on, your housing price predictor would have to know or calculate the map reference for each new house and obtain the relevant flood risk score.

USE OF SYNTHETIC DATA

Given the challenges associated with collecting real-world data, it is not surprising that the generation of synthetic data can be very attractive. For many AI experts, this is a controversial issue ... in fact, it could be described as an incendiary issue ... raising the issue in some meetings is definitely a case of lighting blue touch paper and stand well clear! However, as with most things in AI, it's important to drop down a level and not apply a general argument to every situation. The general principle that many experts quote is that there really is no substitute for real data and they're right! Starting a project without adequate real (typically historical) data is at best foolhardy and at worst commercially fatal – don't do it! Synthetic data can never really be the same as the real world and using it in AI can lead to bizarre outcomes. If you use an algorithm to generate records, the AI will most likely reverse engineer the algorithm and not perform at all well on real data. You are not even safe if you try to use real data that you've manipulated to improve it, say for example, you have cleaned it to make your examples clearer, as you might for a human student. Guess what, you've damaged your model's chances of working in production, AI models will not rationalise situations as a human might. Maybe you are going to train on real address data, but you are rightfully worried about privacy laws, so you attempt to anonymise it so that the places and names are jumbled up, it's slightly better than pure fake data, but the AI will not learn the relationships between place names and postal/zip codes, or names and gender – either of which will mean you've built a sub-optimal model.

Even if you have managed to get real data, are you sure that it is truly representative of your intended deployment environment? Many AI projects have succeeded fantastically in the lab, only to fail operationally, when it transpires that the real-world data is much messier than the data the AI was trained on.

So, that's the general principle ... now let's drop down a level and consider the issues and challenges in more detail. Whilst using real data is currently the best practice in AI, that does not mean it always will be. Whilst previous attempts to use synthetic data may have failed, we need to understand why those methods failed and address those issues rather than simply dismiss the approach. This is important because, whilst using real data may be best practice, it is by no means perfect!

Why Do We Want Synthetic Data?

The decision to use Synthetic Data often starts with the realisation that the quality of the real-world data is poor. Missing values and inconsistent data can cause Engineers to decide that the real-world data is unusable. Unfortunately, *this is absolutely the worst reason to develop an AI using Synthetic Data.* If your AI cannot be trained on real-world data, then it certainly can't be applied to real-world data! As a rule, any data preparation and cleansing tasks that are applied to the training data must also be applied to the operational data. *The reasons we should consider using Synthetic Data are to address privacy issues and incompleteness.*

Synthetic Data to Address Privacy Concerns

Privacy is a huge issue for all IT Systems and the use of AI greatly complicates the situation. We have all become accustomed to the fact that IT systems store vast amounts of personal data. If we are required to use real data to configure and train AI applications, then that means we need to share the real data with the Developers of those applications. If we are to realise fully the benefits of AI, then that means we need to develop large numbers of applications and that means sharing data with large numbers of Developers.

The perfect example of this is the UK National Health Service (NHS). The NHS has existed for over 70 years and owns detailed medical records of an entire population. Putting all of this data into one centrally fused repository and making it available to AI Developers could potentially yield phenomenal value in terms of clinical best practice, drugs research, understanding lifestyle impacts and improving the efficiency of a national healthcare system. However, the cyber risk, privacy and ethical implications of doing so are massive. As discussed in Chapter 5, it's not enough just to identify an operational business value; any assessment of value needs to take into account ethical considerations, and the privacy concerns regarding the health records of an entire population should not be underestimated.

To protect us as citizens, legislation such as GDPR has been introduced to ensure our personal data is not misused. Included in this legislation is the right of an individual to ask for their personal data to be deleted. What does this mean for AI? Well, if we're using real-world data to train AI, then there's a good chance our training sets include personal information. What happens if the owner of that personal data demands that it is deleted? Naturally, the data must be deleted from the training set! However, what about the AI Model that was generated using the training set. Technically, an AI Model is a summary of the training data, and there are emerging techniques that enable the construction of training data from AI Models. As such, it may be the case that the AI Model must also be deleted/or rebuilt without those redacted records.

Now things really do get interesting … especially from a transparency and an explainability perspective. If we delete an AI Model that has been in operational use, we have no guaranteed way of understanding how operational decisions were made. If, after a model was deleted/rebuilt, a Customer demanded to know why they had (for example) been denied life insurance using the previous model, we could not explain why. In fact, we could not even be sure that the AI had made that decision as we could not reproduce the result!

Clearly, from an ethical and trusted AI perspective, there would be huge advantages in the use of Synthetic Data. Firstly, if done properly, Synthetic Data would not be subject to the same privacy concerns. Secondly, the explainability and transparency issue discussed above would be avoided.

Synthetic Data to Address Completeness

Now, let's consider completeness. Real-world data is not only messy, it's also incomplete! In fact, it's worse than incomplete because we don't know to what extent it is incomplete. Imagine that we wish to build a driverless vehicle and we need to collect training data. We equip a car with cameras, laser range finders, GPS and a whole raft of different sensors. We then drive the car around a town collecting data. How do we know when the data is complete? Over a period of years, there are many occasions when a dog runs out in front of the car. Do we have any data for cats running out in front of the car … and does it matter … will the car make a different decision for a cat instead of a dog? What about badgers? What about footballs bouncing across the road? Will the car only recognise certain colours of football because it's never seen a pink football? What about tennis balls? Will the car think that a tennis ball is a small football … and does it matter?

Completeness is a huge issue in AI and data analytics in general. It's a challenging subject for a lot of good, simple, practical reasons. Going back to healthcare, consider a situation where a child presents in hospital with a particularly unusual condition. The doctors immediately order a whole plethora of additional tests and gather data. However, there is no practical or ethical way for the doctors to gather complete data. They cannot immediately identify five additional children with the same demographic, social and lifestyle profiles who do NOT have the condition and put them through the same tests. As such, we only gather data for children with the condition and don't have a complete set of data to work with.

Using synthetic data in these examples, still risks the AI not working, but that risk is outweighed by the risk of privacy or completeness. If these issues are not inherent in your intended AI Project, in 2021, you will need to think carefully about choosing synthetic data because it can be a false economy.

The Future of Synthetic Data in AI

AI, at its current state of maturity, is analogous to aeronautical engineering in the 1920s and 1930s. Aircraft would be tested in real conditions by real pilots who would fly them into real weather. At that time, engineers and pilots would have argued that the only way to test an aircraft was to fly it … in real weather! However, those test flights were incomplete. The engineers would know that the aircraft operated safely in, say, 37 knots of wind but it hadn't been tested in 36 or 38 knot winds. Again, at that time, engineers and pilots would have said that wind tunnels and simulators were not effective as alternatives to real test flights. They were right! Just as experienced AI Developers are right to say that, today, Synthetic Data is no substitute for real-world data. However, aeronautical engineering did not stand still. Wind tunnels and simulators developed and improved to the point at which they became far more effective than conventional testing. The same thing needs to happen with Synthetic Data. Synthetic data applications and synthetic/training data providers are already on the market (2021) many show promising technical developments, especially for structured data, but synthetic unstructured data such as text narratives or images is still in its infancy.

We need to build the tools and methods to enable the generation and maintenance of Synthetic Data sets for use in configuring and training AI systems. We should not underestimate the significance of these tools as they will need to address the concerns about the maintenance of relationships. At present, in the case of unstructured content, our tooling is limited to text editors with search and replace functions. It should therefore not be a surprise that any Synthetic Data generated this way is not fit for purpose.

At this point, we need to acknowledge a significant argument against the feasibility of developing tools to produce Synthetic Data. The argument is that if you have the knowledge and automation required to generate Synthetic Data, then surely there is no need to develop an AI Model ... just use the data generation model instead! This is a very valid argument but, again, we must be careful about applying general arguments to specific cases.

Let's consider free text content. There is huge value in the provision of Synthetic Content for use in the development of entity and relationship extraction tools. The sensitivity of medical reports, police witness statements, customs documents and other free text content is such that sharing the content is difficult. Vast resources are available in academia and commercial organisations that could develop AI capabilities if provided with Synthetic Data that was perfectly representative of the real-world data.

In an extreme case, it would be possible for a human analyst to take a large corpus of content and manually obfuscate entity and relationship data in a way that was both consistent and representative of the original real-world content. This process is possible without developing a complete understanding of the underlying natural language model (because the human race has not yet developed a full understanding of how natural language really works). With improved tooling, this manual process could be made more efficient. It would still be labour intensive, but the rewards would be significant.

A further example exists in the world of simulation. Consider a very complex situation such as a motor race or traffic management or weather forecasting. In such situations, it is possible to build a simulator where the behaviour of individual entities is reasonably well understood, but the combined behaviour of all entities is too complex to understand. Having built a simulator to generate data, it is possible to use this Synthetic Data to develop AI decision makers. A key point to note here is that the AI decision makers are not being trained to learn the underlying model of the complex system. They are learning how to make the correct decisions to manage the underlying complex system.

In simulating complex systems, it is possible to run simulations with a much broader range of input parameters than would be experienced in the real world. For example, it would be possible to simulate a motor race where the fastest car is twice as fast as the slowest car. This ability to simulate extreme scenarios addresses the completeness issue mentioned above.

Obviously, the effectiveness of a simulator led approach is dependent on the quality of the simulation. Initially, a simulator may not be very effective in representing the real-world system. It is therefore important to invest the effort and resources required to converge the simulated environment with the real world.

The use of Synthetic Data in developing AI systems is a contentious issue. Many experts believe that Synthetic Data can never be sufficiently representative and, even if it could be, it would be impractical to generate the data. However, it's clear that there is a need for training data that is compliant with privacy regulations and complete from an engineering perspective. Developing the methods and tools to produce such data is going to be challenging. However, if we look at other fields of engineering, that's exactly what has happened. AI is not engineering free, so we should expect to invest considerably in tools, methods and data required.

MANAGING THE DATA WORKFLOW

In this chapter, we have discussed extensively various considerations regarding the acquisition and use of data in AI projects. The purpose of this section is to summarise the essential activities in terms of a Data Workflow. If you want to build successful AI applications, you are going to have to manage the Data Workflow.

Data Workflow is not just about supplying data to the AI model for development, it is about sustaining the AI throughout its operating life (potentially many years). Getting the data workflow right has more impact on the model performance and maintenance costs (business value), generally even more than the actual modelling step! Like modelling, it also contains parameters which should be tuned. Currently, managing the data workflow is a "black art" giving rise to many conceptual errors in practice. Data workflow must be understood as a core part of the AI process that must be optimised, cross-validated and deployed jointly with data modelling in order to ensure proper applicability. As the saying goes, "There is no AI without IA" [12]. IA refers to Information Architecture derived from data.

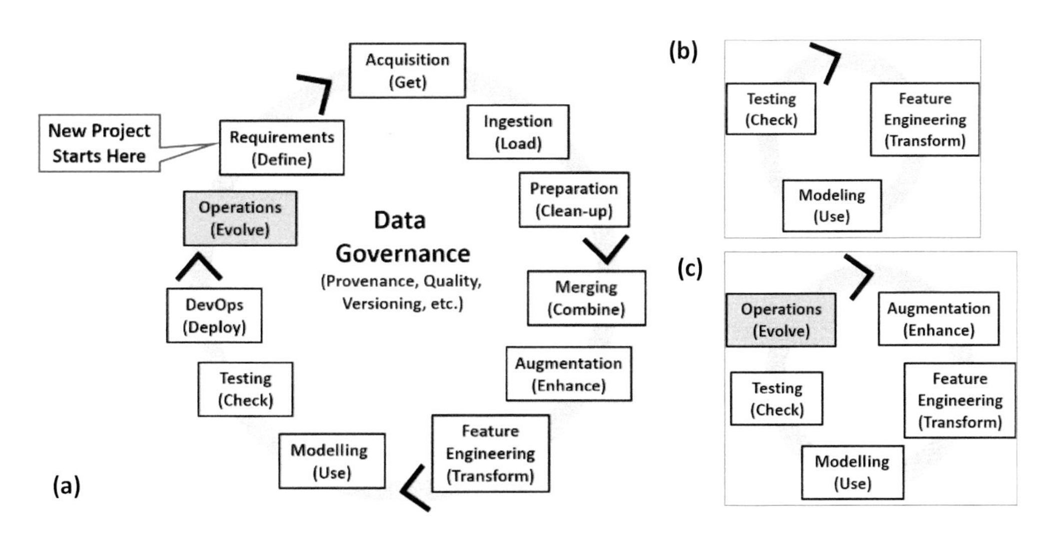

FIGURE 7.11 Data workflow for enterprise AI projects. (a) The cycle of all the activities in a logical sequence. In practice, there are many smaller cycles depending on the context. For example, (b) is a very common scenario of building and validating models. (c) The situation where a deployed model may be drifting due to previously unseen data or may need additional labelling. The details of these activities are described in the text.

Our recommendation for data workflow activities for an enterprise is shown in Figure 7.11a. Each rectangular box in the cycle represents an activity, with a verb in parentheses that captures the essence of that activity. The activities in the cycle are in a logical sequence. However, in practice, there are many smaller cycles depending on the context. For example, Figure 7.11b is a very common scenario of building and validating models. Figure 7.11c represents the loop where a deployed model may be drifting due to previously unseen data or may need additional labelling. Data governance is at the centre of the workflow, making sure that the project goals are met and data related risks are minimised. We will describe some details behind each activity to show its contribution and importance.

We start the discussion at the top left of Figure 7.11a, when a new project starts. As should be obvious, data workflow activities have a close relationship to the AI application development lifecycle.

Requirements (Define)

Data requirements stem from business requirements. The clearer the business objectives, the better the definition of data needed and the demands on the workflow shown above. Examples of business requirements and their data implications are:

Business Requirement	Data Implication
The application has to be deployed in the countries in the European Union.	Data governance has to support GDPR guidelines and associated IT responsibilities.
The medical AI application has to work in the international context.	The data corpus for training has to include international standards of practice, not just from one country.
Support Bot for users has to work in 10 major languages.	The training data (speech or text) has to contain the actual data from the users in the original languages, not a translated version from English. The original version will contain slangs, dialects, accents and other variations that have to be understood by the AI.
The mortgage assistant for the bank officer is to serve a diverse demographic population.	Need to check for systemic biases in the training data.
The image recognition software has to work with images from older devices with low picture resolutions.	Training and test data have to include low-resolution images.

Acquisition (Get)

Once you know what data you need, the next task is figuring out how to get it. This will depend critically on whether your company already owns the data or you have to go outside your company. Let us first consider the case, you own the data. If you own a retail department store chain (say, Walmart), you may already have the data from your past that can be used. It may need some data wrangling to make it usable, but that is within your control. If you do not have the data you need, but it is obtainable within your normal business processes, with some careful design, you can start collecting the right type of data over a period of time, before creating the AI application. If you are in a specialised domain (e.g. banking, insurance, manufacturing, legal, law enforcement, healthcare, etc.), you may not have a choice of an external data source for competitive or regulatory reasons. The only options may be the generic industry-specific sources (e.g. international banking standards such as BASEL III, HIPAA standards for healthcare, etc.) that can provide a basis for the domain knowledge *but not specific to your business.*

Now, if you do not own the data or you have to augment your data from other sources, the challenge gets a lot more complicated. If the sources are not for profit (e.g. Wikipedia, Data.gov), they will allow you to use their data for free, but for restricted purposes under a license, with proper governance. Most commercial data sources have strict license terms, usage policies and duration of use that need to be carefully considered, in addition to cost, before committing to them. Such sources are more likely to be in consumer areas such as retail, e-commerce and social media. A good understanding of the formats and the storage needs of the licensed data is also critical, before the decision to license the data is made. Getting a sample of the data ahead of time will help immensely.

Ingestion (Load)

Ok, you have the rights to the data, one way or the other! The next step is to actually get the data into your infrastructure in proper formats, volume, etc. One major option to consider is data virtualisation. This allows access to the data from multiple sources, without having to copy and replicate data, while making the most current data available for analysis at any time and reducing cost. The benefit of this approach can be significant if no subsequent data wrangling is needed. But, if snapshots of the data are needed to capture the data versions corresponding to the versions of AI models, some automation is needed to make the process more tolerable for day-to-day operations. If there is a real-time aspect to it (e.g. stock trades, IoT device outputs, etc.), a careful consideration of the ingestion rates, frequency of updates, etc., is also needed. In most cases, it is quite likely that you are going to need the data for the life of your application and so you need to make sure that the infrastructure is extensible for your future needs. If your business is successful, this can be a very long time!

Preparation (Clean-Up)

This is the step where you decide what part of the data you are actually going to use. This task requires a good understanding of the AI application you are building and the skills to do data exploration. Since more data invariably means more data wrangling, it is good to select sources and contents that fit the needs of the AI application the best. The clean-up requires various tricks, and the choices made in the preparation process will have implications when the model predictions are made later in the process. Let us say we want to predict the number of pandemic infections in a particular area for the next many months. The daily data may be too noisy due to statistical fluctuations or systemic issues (e.g. no reporting on Sundays!). The smoothing of the data can be done in many ways: moving averages over a time window (e.g. a week) and/or picking a larger granularity (e.g. grouping by county instead of town).

What if we have the test results for the individuals but do not have the demographic data (e.g. patient race or sex) from the same source? This is the next big challenge of *missing features*. In projects that use unstructured data, this may manifest as having not enough training examples of a particular type, such as lack of non-Caucasian pictures in developing AI applications for facial recognition [13]. Missing or inconsistent values in the data fields are the next challenge for clean-up. We have discussed some additional ideas on addressing the data preparation problems earlier in the "What Happens in the Real World" section.

Merging (Combine)

When the data from different sources are combined, it leads to various problems with data merging. The first task is to figure out the relationship between the data to get the basic story straight. Examples are:

- **Joining**: Linking rows across tables with unique identifiers (called 'key columns') that make it possible to expand the number of features in a rigorous fashion. Without the key columns, this is simply not possible. Example is a patient ID in hospital records across different processes such as billing, physician's records and appointment scheduling.
- **Normalisation**: The same entity may be called by different names in different data tables. Example is under the column 'Company', it may contain three names: International Business Machines, IBM and IBM Corporation for the same company. The merging process will typically create a new column called "Normalized Company" to represent using the normalised name, may be just 'IBM'. We will have a similar problem when we have textual descriptions from different sources involving the same entity using different names. This is called "Entity Resolution" in NLU.

Augmentation (Enhance)

Sometimes the available data needs to be enhanced to meet the needs of the AI algorithm. In supervised ML (see Chapter 3), in the training data we need the inputs as well as the corresponding output labels. An example is identifying objects in images. So, we need to label the object that goes with each image for the algorithm to use. This is typically done by humans to start with, but once the algorithm learns how to do this with enough accuracy, labelling can be fully automated or semi-automated with partial human validation. Similar need exists in NLU where we need to identify the various entities and their relationships from reading unstructured textual documents. Initially these had to be manually annotated by humans for the machines to learn from. Recent advances in Natural Language Processing [14] have made it possible to preprocess the text to do multi-level automatic feature representation learning. In business domains where the language models are different from the popular language corpuses, domain-specific manual annotations may still be needed. If the annotations are done by a group of individuals (e.g. crowd-sourcing via Amazon Mechanical Turk), there needs to be enough checks to make sure that the labels by different annotators are consistent and correct. It may be also necessary to add synthetic data where it makes sense (see earlier discussion in this chapter).

Feature Engineering (Transform)

In our discussion of the housing price model earlier in this chapter, we discussed some aspects of feature engineering, i.e. when we have to add features and when we have to reduce features (curse of dimensionality). Techniques such as PCA help with this task. In the end, the data scientist has to decide what features are critical to create the best model with the available data. In DNNs, the latest algorithms are able to identify internally what features of the inputs are needed to perform the task and hence one less thing to do for the data scientists.

Modeling (Use)

The goal of the modelling task is to create the best model with the available data. You have to be careful to make sure that training data represents the operational data in feature distributions. There are two aspects to modelling:

i. Try different models with the available data and pick the one that gives the best performance in terms of metrics of relevance.
ii. Identify if the model performance can be improved by either additional quantity or improved quality of data. This needs a careful debugging of the model performance and a judgement to ask for additional data help.

Cross validation is a technique that uses different subsets of the training data to build and validate models during the modelling process to give a better view of the quality of the model created.

Testing (Check)

Primary goal of testing the AI system is to get the confidence that the model generalises the outputs to inputs previously not seen in the model building task. This is best accomplished by using a 'hold-out' test data set that represents the operational data well and evaluating the model performance. To be deployed in practice, when tested with feature values beyond what is in the training set, the application should recover gracefully, even if the model fails to generalise.

DevOps (Deploy)

From the data governance perspective, DevOps activity needs to make sure that the versions of the model and training data are tracked and documented, in case the model performance in the operations requires a diagnosis of the modelling process. If the model performance degrades during operations, it may be necessary to regress to an earlier model version with confidence. Lack of a clear definition of a 'defect' in AI systems due to their statistical nature challenges traditional notions of success and failure in integration testing and deployment. Running automated tests prior to deployment has to be complemented with some debugging activity.

Operations (Evolve)

In AI systems, monitoring is not an option, but a required activity. This is because the AI model behaviour can drift with time due to changes in the input data and the data distributions in the deployment. The specific details on what to monitor and how often will depend on the specific use cases and business goals. One common challenge with automated model evolution during deployment is that the data collected during deployment is not 'labelled' (i.e. the correct output is unknown) and hence not directly usable for model retraining. Techniques such as active learning [15] can be used to get users to provide the labels directly; semi-automated approaches combine automated labelling and human validation. As the Microsoft Tay bot example demonstrated [16], continuous model learning has its own challenges in validation.

Data Governance

As it should be obvious from the descriptions of the data workflow activities above, data governance is a critical activity in an AI project. It has to start even before a project starts and last as long as the AI application is deployed. It covers business aspects (data regulations, licensing constraints, etc.) as well as the technical aspects (the data infrastructure, tools, testing and development practices, etc.). Here are some critical areas of data governance:

- **Data provenance**: the quality of the AI applications critically depends on the veracity and quality of the data used in the modelling. Data scientists typically use data from wherever they can get it and do not question its trustworthiness. In the heady pursuit of a performant model, they can often make ad hoc assumptions and data transformations in the process, which are not recorded for reproducibility.
- **Trust**: it is now well known that business data contains implicit and explicit biases. AI applications bring them to the foreground. There needs to be a clear recognition upfront on the business priorities with respect to the importance of bias and fairness in decisions and the requirements on the explainability of the AI outputs. The ability to check for the data quality across the entire data workflow is essential to meet these objectives.

Data tools for AI are in their early infancy, and they are critical to help with the data workflow tasks shown in Figure 7.11. Gradual progress is being made in this area.

IMPROVING DATA QUALITY

In improving data quality, there are really three areas that need to be considered.

First, if you can't measure it, you can't fix it. As discussed in Chapter 5, any AI Application should be deployed as part of a broader business strategy underpinned by Measurability. By measuring volumes and quality of data, we should be able to identify areas for quality improvement with appropriate prioritisation.

Second, once we know what is needed, we can use a whole plethora of tools [17–19] to cleanse existing data and improve its quality. Data cleansing as a massive subject in its own right.

Third, it's important to fix the problem at the source as well! Often data cleansing is seen as a one-off exercise when, in reality, it never ends. It's always frightening to see requirements for new data repositories because they invariably include some sort of vision statement that the new repository will bring together all the historical and legacy data into one single new, all singing, all dancing repository. Unfortunately, if the user input at the source isn't simultaneously fixed, when the data in the new repository goes live, it too becomes a legacy repository. When cleansing existing data, it is imperative that the sources of that data are understood and actions put in place to improve the ongoing quality of data received from those sources. For example, if source data is often incomplete or inaccurate then perhaps the user experience of the source data system needs to be reviewed to improve the quality of data capture.

IN SUMMARY – IT REALLY IS ALL ABOUT THE DATA!

Data is at the heart of all AI systems, and if there is one message that you take away from this chapter, it should be that you need an effective data strategy. Here are some key points:

- The current Data Tsunami offers both challenges and opportunities.
- Depending on the context, you may need structured or unstructured data or both, to develop your AI applications.
- In addition to the data owned by the enterprise, there are many public and proprietary sources of data for the enterprise use.
- Decision about data source should consider the data availability for the life of the AI application.
- Data wrangling is the task of making the raw data useful for AI model building. This can take considerable effort in a project.
- The quality and quantity of the data needed depend on the complexity of the AI task.
- Feature engineering is an important task to address "the curse of dimensionality".
- Synthetic data can be useful when privacy is a concern or the data is incomplete.
- There are many emerging tools to help with various data science tasks.
- Managing the data workflow and having the proper data governance throughout the project are vital for its success.

REFERENCES

1. "What happens in an internet minute?" https://lorilewismedia.com/.
2. IDC Global DataSphere, Forecast: 2021–2025 The World Keeps Creating More Data - Now, What Do We Do With It All? (IDC #US46410201, March 2021).
3. List of datasets for machine-learning research: https://en.wikipedia.org/wiki/List_of_datasets_for_machine-learning_research.

4. The EU General Data Protection Regulation (GDPR), https://ec.europa.eu/commission/priorities/justice-and-fundamental-rights/data-protection/2018-reform-eu-data-protection-rules_en.
5. California Consumer Privacy Act (CCPA): https://oag.ca.gov/privacy/ccpa.
6. Neuro-symbolic AI: https://mitibmwatsonailab.mit.edu/category/neuro-symbolic-ai/.
7. T. R. Besold, et al. "Neural-symbolic learning and reasoning: a survey and interpretation," https://arxiv.org/abs/1711.03902.
8. UK National Oceanographic Center, "An introduction to tidal numerical modelling," https://www.noc.ac.uk/files/documents/business/an-introduction-to-tidal-modelling.pdf.
9. Z. Kowalik and J.L. Luick, "Modern theory and practice of tide analysis and tidal power," https://uaf.edu/cfos/files/research-projects/people/kowalik/Book2019_tides.pdf.
10. Oceanography 101, Chapter 11 – Tides. https://gotbooks.miracosta.edu/oceans/images/tides_neap_spring.jpg.
11. The National Coastwatch Institution, "Solent tides and currents," https://www.nci.org.uk/stations/solent-tides-and-currents.
12. S. Earley, "There is no AI without IA," *IEEE IT Professional*, pp. 58–64 (May/June, 2016).
13. J. Buolamwini and T. Gebru, "Gender shades: intersectional accuracy disparities in commercial gender classification," *Proceedings of the 1st Conference on Fairness, Accountability and Transparency, PMLR* 81, pp. 77–91 (2018).
14. T. Young, et al., "Recent trends in deep learning based natural language processing," *IEEE Computational Intelligence Magazine*, 13(3), pp. 55–75 (2018).
15. M. Elahi, F. Ricci and N. Rubens, "A survey of active learning in collaborative filtering recommender systems," Elsevier, *Computer Science Review*, 20, pp. 29–50 (2016).
16. G. Neff and P. Nagy, "Talking to bots: symbiotic agency and the case of Tay," *International Journal of Communication*, 10, pp. 4915–4931(2016).
17. S. Schelter, et al., "Automating large-scale data quality verification," *Proceedings of the VLDB Endowment*, 11–12, pp. 1781–1794 (2018).
18. E. Breck, et al., "Data validation for machine learning," In: *Second SysML Conference* (2019).
19. S. Shrivastava, et al., "DQA: scalable, automated and interactive data quality advisor," *IEEE International Conference on Big Data (Big Data),* pp. 2913–2922 (2019).

CHAPTER 8

How Hard Can It Be?

It's amazing … we put a small team of grads on this and in just a few days they achieved so much. They took some data, used a Cloud based AI and built an image classifier that could really transform our business.

The Client was right to be impressed because the graduates had done a great job and the solution that they had demonstrated was impressive.

Three years later, the capability that the grads built still isn't being used operationally.

DEMONSTRATIONS VERSUS BUSINESS APPLICATIONS

No matter how excited we are about artificial intelligence (AI), people buy solutions and that means we need to think about all the things that make AI useful, including the boring stuff! The graduate students in the story above had shown what could be achieved from an AI perspective in a very short timeframe. What the team had not done was take live data from a live system. Nor had they built the integration to write the results back to a target system for consumption by other users. Nor had they made the system secure as there was no need to provide enterprise security in a demonstrator. Oh, there wasn't any audit because that wasn't needed either. Just like backup and recovery which weren't important in a demonstrator. The user experience was great but had only been tested amongst the students themselves with no comprehensive user testing. User training and business change were also still on the "to do" list.

It's great to test ideas and build demonstrators as long as everyone remembers that business needs complete, end-to-end solutions! So, let's talk about the challenges of building complete, end-to-end solutions and what you need to do to deal with those challenges.

DOI: 10.1201/9781003108498-9

THOUGHT EXPERIMENT: HOW SMART CAN IT BE?

I was on my way to speak at a conference in London when a very simple experience really showed me the power of human intelligence. Whilst leaning against a handrail on a busy tube train (headphones in and ignoring everyone else), I glanced down at the floor and saw a leaflet lying by my right foot. The leaflet was upside down and at a skewed angle. There was a black and white photo of a child. It looked like the sort of photo you see in famine appeals with what appeared to be an African landscape in the background. There was a balloon over the child's arm showing what appeared to be some form of a worm coming out of the arm. There was an icon on the arm that looked a bit like a pair of glasses.

In the couple of seconds that I looked down at the floor, my brain did some incredible things. It rotated the leaflet so that I could understand the detail. It classified the leaflet as a charity leaflet. It classified the photo to determine that it showed a child in Africa. It identified the concept of a magnifying bubble and identified what appeared to be a worm embedded in the child's arm. It classified the icon as looking like glasses.

My brain then went a step further! It recalled some knowledge from a trip to Africa many years earlier; I remember someone telling me about a type of worm that causes blindness. My brain then concluded that this was a leaflet for a charity that worked in Africa to prevent and cure blindness caused by this parasitic worm.

On arriving at my destination, I quickly searched the internet and there was the charity ... Sight Savers!

What's remarkable about this experience is how unremarkable it is! Every day, we mere humans experience this type of thinking over and over again. We glance at images and can rotate them ... we drag knowledge out of our past experience ... we hypothesise about what something might mean ... we fuse data, information and knowledge from a massive array of sources ... and in the process we improve our knowledge and experience.

AI is becoming exceptionally good at the low-level tasks and, increasingly, we're starting to think about how to string these tasks together into overall thought processes. However, even the simplest task can be a lot more complex that you at first think ... especially when considering the edge cases (i.e. outside the common experience).

Therefore, how about trying a thought experiment? Over the next few days, pick a couple of every tasks from both your professional and personal life. It could be dealing with a situation whilst driving to work such as a road blocked by the bin lorry. It could be cooking a meal ... we recommend potato salad (see the vignette in Chapter 3 on 'Food Analogy to explain AI Terms'). It could be handling a customer request at work. Take two or three ordinary, boring tasks and then think about how you would break each one down into a series of sub tasks that could be performed by an AI. In doing so, you will gain an interesting perspective on what it really takes to deliver an AI application!

James Luke

SETTING EXPECTATIONS ... YOURS AND OTHERS!

There can be a terrifying moment when discussing AI in any organisation ... it's the moment when someone in the room decides that the AI needs to be used to solve a really hard problem. After all, if AI is as good as everyone thinks, then surely it should be able to handle the toughest challenge. This desire stems from the perception that AI must be more intelligent than human beings. That's certainly the case in the SciFi movies, and it is certainly the image that some companies are encouraging in the media. This in turn leads our Algorithm Addicts to believe that AI can make sense of huge and massively complex data sets and discover insight where mere humans fail. Whilst tackling the most difficult problems may offer the biggest potential return in terms of business impact, it is also a very high-risk strategy. Given that AI is currently a long way from emulating human levels of intelligence, it is also a flawed strategy. The first step in successfully delivering AI is to *change our expectations to match what the current AI technology can actually do and focus our projects on those most likely to succeed* [1]. In Chapter 4, we proposed a first-pass filter aimed at assessing whether a particular task was suitable to be replaced by AI. In this chapter, we give more details on how to assess the reality of delivering AI projects.

DO WE NEED AN INVENTION?

For many years, engineers have joked about boxes in architectures labelled "clever stuff happens here". By definition, an AI project is going to need a box in which clever stuff happens. However, there is a massive difference between "clever stuff" and "impossible stuff". When evaluating potential AI projects, it is important to understand whether the "clever stuff" required already exists or whether it needs to be invented.

We are most definitely not telling you never to attempt a solution that requires an invention. Quite the opposite! The Apollo mission, Bletchley Park and the Manhattan Project were all massive programmes that required inventions in order to achieve their objectives. In today's context, the development of driverless vehicles also fits into this category. Work on driverless vehicle technology has been underway for at least two decades and has resulted in many inventions. We think projects on this scale are incredible and should definitely be attempted ... as long as you can meet two critical criteria.

First, you need to appreciate the scale of your undertaking and be willing to resource accordingly. You need to understand that the project may take 30 years and may cost hundreds of millions. Second, you need a top-notch leadership team to work with you and you should consider the authors of this book!

Seriously though, the scale of the invention required to deliver a solution will naturally determine its likelihood of success. This is often a judgement call so it is important to bring in experienced technical experts to assess the level of invention required and to ensure you have the right people, resources and organisation in place to maximise your chances of success. Given the seven decades of research and development, let us explore the current state of AI in the next section.

CURRENT STATE OF AI

Neural versus Symbolic

Much of the excitement in the last decade in AI is due to the advances in statistical learning from data using Artificial Neural Networks (ANNs), particularly Deep Neural Networks (DNNs). The idea that machines can 'learn' on their own from data without human intervention has been proved beyond doubt. Due to their statistical nature, these systems learn from large amounts of data and perform narrow tasks very well. The knowledge in these 'neural' systems is captured in internal representations of the statistical models and not easily understandable by humans. This is in contrast with the expert systems of the last century (discussed in Chapters 2 and 3) that were 'taught' explicit human-created, machine-readable rules and knowledge in any specific domain. These are called 'symbolic' systems. Since these rules and knowledge representation were created by humans, it was easy for humans to define and understand the behaviour of such systems. Logical reasoning using human-friendly concepts came naturally with symbolic systems.

The reality we face today in AI is that we have centuries of knowledge captured in human-friendly 'symbolic' form (i.e. languages, documents, etc.) and we do not know how to integrate this knowledge with the 'neural' models that learn statistically from data. Key point on the 'neural' learning is that it is based on statistical properties of data and not based on human-level abstractions. For example, if you type 'bathroom' in a search engine, it will suggest various options, 'bathroom vanities', 'bathroom ideas', 'bathroom remodel', etc., based on probabilities derived from the data from millions of people who have used those combinations in their searches. *Believe it or not, the AI behind the search engine actually does not know anything about a bathroom.* The fact that it is one of the rooms in a house where people take baths. It has a sink, tub, shower, toilet, cabinet, etc. These are the attributes captured in a symbolic knowledge representation of a bathroom. So, the challenge today is how to combine 'neural' aspects of statistical learning with the 'symbolic' aspects of human knowledge and reasoning. Such a 'neuro-symbolic' AI [2] will be able to do a lot more than what is possible now.

What Can AI Do Well?

Given the dichotomy between 'neural' and 'symbolic' AI discussed above, there are things we can do in either of the approaches, but at least for now, natural integration of the two is an active research area [3,4]. On the symbolic side, there are many established techniques for specific purposes e.g. Rule-Based Systems, AI Planning, Optimisation Techniques, etc. [5]. The advantage of all these techniques is that you can easily understand how they work and come up with natural explanations for their outputs. While having sufficient data can help to create and validate algorithms, they are not critically dependent on large volumes of data. On the neural side, since AI needs large quantities of data for learning (see Chapter 7), depending on the type of data (structured tables, text, images, etc.), there are many narrow tasks that AI can do well. Tasks range across search, question answering, interactive bots with speech or text modes, recommender systems, image and text classifications, etc. Chapter 2 has many examples of AI applications both in the consumer and enterprise domains that represent the tasks that are possible today.

Examples of AI Challenges

However, AI has a long way to go before it can come close to emulating true human intelligence. As in many other fields, there are still some fundamental problems that have not yet been solved! It is fair to say that AI has fallen between two different approaches: 'Symbolic AI' of the prior decades that incorporates compositionality and construction of cognitive models, but inadequate at learning. 'Neural AI' of the past decade that is very good in learning, but poor at compositionality, abstractions and building cognitive models of the world.

Our current AI technology works well in carefully scoped environments on constrained data and feature spaces. Here are some examples where the current AI techniques have trouble. For a more detailed discussion, we recommend the book *Rebooting AI* by Gary Marcus and Ernest Davis [1].

- **Data Needed**: As we have emphasised in various places in this book, due to their statistical basis, Deep Learning (DL) techniques need data in sufficient quantity and quality to perform well. Most often, you can find the data inadequacy only after sufficient experimentation in a project.

- **Long-Tail Problem**: Algorithms do well with common items for which there is a lot of data but have difficulty with large number of rarer items that have very little data. This is the typical scenario with enterprise AI.

- **Proclivity to Errors**: DL systems are based on the concept of statistical learning to create the underlying models. Inherently, even the best optimised models will inevitably make mistakes at some fraction of outputs. The enterprise has to decide on what percentage of errors is tolerable to the business and *how to manage the risk, not if, but when the errors occur.*

- **Objects in Unusual Poses or Surroundings**: The training data for images of objects typically come from their natural poses and surroundings. In testing or actual use, algorithms have trouble identifying the same objects, if they are placed in unusual surroundings or in uncommon orientations (e.g. a school bus lying on its side or a diagonal view of a cube).

- **Synthesis of Concepts**: Even the most advanced natural language applications are looking to match exact set of words or synonyms and their 'statistical closeness' based on correlations. They cannot synthesise information from many documents that do not use the exact same combination of words or synonyms. The interesting example is the difficulty with listing the seven Horcruxes from just reading the Harry Potter books, where they do not appear together in the same passage or items explicitly identified as a Horcrux in sufficient proximity.

- **Common Sense**: Machine learning (ML) techniques do not have the common sense understanding of relationships between real-world objects and so they cannot follow a chain of inferences that are implicit. The example from [1] is *Elsie tried to reach her aunt on the phone, but she didn't answer.* There are many challenges for statistical algorithms here. (i) 'Reach' has many different usages in practice beyond communication,

COMMON APPLICATIONS OF NATURAL LANGUAGE PROCESSING (NLP)

Language is critical for communication among humans. Given the ubiquity and the importance of institutional and archival knowledge captured in natural language documents, it is critical that we leverage AI to "understand" their contents. This is not an easy goal to achieve due to the variety and nuances of languages. The purpose of this vignette is to give common examples of application of Natural Language Processing (NLP) in the business context. We refer to [1] for a deeper technical review.

Information Extraction

This is an essential task that is required before implementing many of the other applications. Given a corpus of unstructured text (e.g. pdf documents, wikipedia, etc.), how do we extract explicit or implicit information from it? Commonly extracted information includes named entities and their relationships (e.g. companies & their CEOs), events and their participants (e.g. Wimbledon and Roger Federer) and temporal information (e.g. World War II events, dates and relative order). The outputs are typically stored in relational databases for efficient retrieval and analysis.

Search

The purpose is to help the user find the information most relevant to the input words as quickly as possible and rank the retrieved items by relevance. You can think of your favorite search engines.

One-Shot Question Answering

User asks a question and the system provides the best answer it can find from the relevant body of knowledge. Examples are IBM/Watson Jeopardy!, Amazon/Alexa, Apple/Siri, Google/Home, etc.

Interactive Bots

User inquiry needs an extended dialogue and context for the system to provide the right response. Since questions in open domain (i.e. any arbitrary topic) are difficult to answer, commercial bots are trained in specific domains/skills (e.g. banking) and/or specific tasks (e.g. money transfer).

Text Classification

This is a popular application of NLP in business where the AI classifies the input free-text documents into predefined categories. Imagine a news agency classifying the various streams of incoming news items into Politics, Sports, Culture, Weather, etc.

Text Generation

Here are four examples of NLP tasks that require the generation of human-like language.

- Given non-textual data (e.g. images), the goal is to create text (e.g. captions) that represents the contents of the data.
- *Extractive Summarisation* that finds key elements in textual documents and produces a summary of the most important content by selecting and rearranging text taken directly from the documents.

- *Abstractive Summarisation* produces a conceptual summary of textual documents, possibly using words never seen in the original documents.
- *Machine Translation,* which involves the translation of documents in one language into another. Even difficult for humans, this task requires proficiency in both languages and in areas such as structure, syntax, and semantics and cultural sensitivities.

IBM's Project Debater [2] is a recent example of advances in NLU for a live debate with a human.

REFERENCES

1. D. W. Otter, J. R. Medina, and J. K. Kalita, "A survey of the usages of deep learning for natural language processing," *IEEE Transactions on Neural Networks and Learning Systems,* 32(2), pp. 604–624 (2021).
2. N. Slonim, et al. "An autonomous debating system," *Nature* 591, pp. 379–384 (2021).

(ii) 'aunt' is not physically 'on' the phone which is the common usage of 'on' to specify relative positioning of objects and (iii) 'she' refers to the aunt because Elsie cannot call herself. There is really no model of the relevant world in DL.

- **Dealing with Negation**: ML techniques have trouble with understanding the general usage of '*not*'. A search of 'a restaurant that is not McDonalds', brings back a bunch of references to local McDonalds!
- **Composition**: ML techniques do not have the concept of composition, i.e. bringing various parts together to make a whole; you compose the image of a cat with head, feet, body, tail, etc. Head is composed of eyes, ears, mouth, etc. The current DL systems get rattled by changes to a few pixels of an image and mislabel an object, even though to a human it looks just fine [6].
- **Causal Reasoning**: ML cannot do causal reasoning [7]. As a human, you can look at a situation and decide that it is dangerous. You know that certain objects are heavy or hard and that they will cause harm if they come into contact with humans or animals with a high level of force. You know that certain situations cause objects to fall, or to be catapulted, at high speed. You know that other circumstances make floors and walkways slippery. You are able to look at a structure, such as a ladder leaning against a wall, and decide that the setup does not look very secure. You can fuse together all these different scenarios and decide that what is happening is dangerous and that it would be safer to move away from the scene. The current breed of AI technology is not able to perform that type of prediction.
- **Ambiguity in Language**: One of the reasons it is so hard for machines to understand human language is that it is packed with ambiguity. Even something as simple as a word can have multiple different meanings depending on the context. The word 'bat' could refer to either a flying mammal or a wooden instrument used in playing baseball or cricket. As human beings, we have an incredible ability to resolve

these ambiguities by taking into account the context and using our extensive human knowledge (and perhaps some common sense) to figure out what was really meant. Even then, a phrase such as "Mary was watching the Yankees game when she was unexpectedly hit by a flying bat" could be interpreted either way. The recent advances with pretrained models from massive data corpuses in the public domain [8.9] can often give people the impression that AI is very close to 'understanding' natural language. Unfortunately, it is a long way to go before that happens, if at all.

THE IMPORTANCE OF DOMAIN SPECIALISTS

Understanding whether a human can perform the task is an extremely useful exercise because it will help you identify the Domain Specialists that are critical to your project. No matter how clever the AI and no matter how sophisticated the ML, the inclusion of subject matter experts in your project is essential to success. Whilst Domain Specialists are obviously essential in the preparation of training data for ML systems, their usefulness extends much further than just labelling data. Domain Specialists often know more than they realise and their value often increases as projects progress. Their contribution will be needed in the definition of features, understanding and managing data issues, articulating exception cases and ensuring the solution is focussed on the correct issues. As discussed in Chapter 5, Domain Specialists will almost certainly be essential Stakeholders. To ensure their support, it is important to consider the impact of the Application on the business processes and experts in your organisation.

BUSINESS CHANGE AND AI

The delivery of any solution using advanced technology is always going to include some element of business change. In the case of AI applications, the first consideration is whether the application is being delivered as part of an existing business process or if a completely new business process is required. The main advantage of delivering AI into an existing business application is that there is an existing framework within which to evaluate and prove the AI. If there is a process involving human analysis and decision-making, then it is possible to capture the input data and resulting decisions for use in defining the AI replacement. In such a situation, the primary business change consideration is the level of AI augmentation proposed. Is the intention to automate the business process and replace the existing analysts, or is it to augment the existing analysts such that they can focus on high-value tasks?

As discussed in the Stakeholder section of Chapter 5, it is important to ensure the role of the existing analysts in any AI development is understood. If the analysts are essential to the delivery of the AI, then asking them to make themselves redundant is going to be difficult. Even in situations where the aim is to fully automate an existing business process, we recommend embedding the AI in an infrastructure that is monitored to ensure the behaviour of the AI remains within expected limits. Consider an application that evaluates social security claims and automatically determines whether a claim is valid or not. During the initial development and operation of the application, it is important to establish its normal behaviour. If an application that normally rejects 25% of claims suddenly starts rejecting 40% of claims, then perhaps something has changed. It may be that there has been a change in one of the data sources or something in the environment is causing a higher number of claims. A monitoring framework is essential in identifying unexpected changes in behaviour so that the behaviour of the application can be assessed and, if necessary, corrected.

If a completely new business process is being defined, then the standard business change considerations apply. However, in the case of an AI application, it is critical to ensure the AI is accurate enough for the business process to be effective. This requires careful definition of the AI evaluation and, in particular, it's important to ensure that the data used for the evaluation is representative of the data that will be encountered in the new business process. AI applications are massively data sensitive and even small changes in data can have a dramatic impact on accuracy.

In Chapter 5, we also talked about the business motivations and the basic requirements behind creating Trustworthy AI. Implementation of these requirements needs adoption of appropriate ethical principles and careful evaluation and selection of the relevant technology approaches. Here is a table summarising the popular trust requirements and useful references for help with implementations.

Trust Requirement	**References (as Examples)**
Fairness/Bias assessment and Bias mitigation across the model lifecycle	R. K. E. Bellamy et al. [10] https://aif360.mybluemix.net
Explainable AI for various stakeholders	https://aix360.mybluemix.net/
Model robustness evaluation and mitigation	https://github.com/Trusted-AI/adversarial-robustness-toolbox
Uncertainty quantification	https://uq360.mybluemix.net/
Transparency and AI Governance	https://aifs360.mybluemix.net

A key, but often overlooked, aspect once you have put an AI into production *and* remembered to put monitoring infrastructure in place is now you have to establish what you will do if your AI starts to underperform. Knowing your AI is failing is only half the story, you will need a plan "B". You could keep an army of experts, just in case, to take over with manual routines in reality that's not practical. The best strategy is to keep a small team of data scientists employed to not just monitor your production AI but to keep alternate models in training, evolving with the data, ready to deploy if the main algorithm starts to wane. A doubly good tip is to keep at least one of those models in a form that is explainable, even if it is sub-optimal. If your production AI is pulled due to a potential ethical bias, you may not be able to replace it quickly with another that cannot prove it isn't biased.

Similarly, in considering transparency and explainability, any application needs to include appropriate tools to understand, and demonstrate to Stakeholders, why decisions were made. These tools should include the DevOps capability to identify which model was in use when a specific decision was made. For ML models, we should also be able to trace the training data used to generate the model and the test data was used to validate it. In cases where a decision has been challenged, we should be able to identify all cases in the original test and training data that are similar to the data on which the contentious decision was based.

AI IS SOFTWARE

In the enterprise context, the purpose of the AI is to do a specific task to support a business application. Therefore, it is natural to expect that the invocation of the AI is through some traditional application program interface (e.g. REST-based microservices). In short, it is just another software component, albeit with some special attributes. Thus, from the system or software management point of view, it has all the same expectations as any other software component. Figure 8.1 shows the recommended system and software quality models and attributes from the ISO/IEC 25010 process standard [11]. Even though the specific interpretation may have to be refined for the purpose of AI components, the utility of the basic structure is immediately evident. The quality attributes in use (in the left column) i.e. effectiveness, efficiency, satisfaction, risk and context coverage do represent the relevant dimensions for consideration. Inclusion of 'Trust' under the 'Satisfaction' category is fortuitous in hindsight, since it has taken a more profound meaning for AI components. The product quality attributes on the right are essential for product owners. Notably, the common metric used by AI algorithm owners is accuracy which relates to 'Functional Correctness' in Figure 8.1, and it is only one of the more than two dozen attributes in the ISO standard. It is important to evaluate an AI component against these attributes to understand the requirements they place on the use of an AI component in a software system.

THE GREAT REUSE CHALLENGE

Reuse in Software

One of the biggest challenges in conventional software engineering is understanding how to reuse existing software when requirements for a new development are similar but not identical. In cases where the same developers are consulted, there is a high likelihood they will adapt the existing software and leverage the previous investment. If the development is

ISO/IEC 25010: 2011 Recommended Software Quality Attributes

Quality In Use	Product Quality		
Effectiveness Efficiency Satisfaction Usefulness Trust Pleasure Comfort Freedom from risk Economic risk mitigation Health and safety risk mitigation Environmental risk mitigation Context coverage Context completeness Flexibility	Functional suitability Functional completeness Functional correctness Functional appropriateness Performance efficiency Time behavior Resource utilization Capacity Compatibility Co-existence Interoperability	Usability Appropriateness recognizability Learnability Operability User error protection User interface aesthetics Accessibility Reliability Maturity Availability Fault tolerance Recoverability	Security Confidentiality Integrity Non-repudiation Accountability Authenticity Maintainability Modularity Reusability Analyzability Modifiability Testability Portability Adaptability Installability Replaceability

FIGURE 8.1 ISO/IEC 25010 system and software quality models [11].

undertaken by a new set of developers, there is a good chance that they will feel it is easier to start from scratch. Let us face it, no one wants to debug someone else's code!

Despite the challenges of reuse, the software industry has progressed a great deal in recent years. Whereas systems developed in the late 1990s were largely bespoke, more modern applications are increasingly built using Consumer-Off-The-Shelf (COTS) products such as databases, application servers and message brokers. The advent of Cloud-based services encompasses a huge level of reuse based on publicly available application program interfaces (APIs). The growth of Open Source Software (OSS) has also introduced a new twist. Since the development of OSS is visible to the community, the utility and quality of the resulting software can be judged directly. There are numerous examples of successful OSS projects in the last two decades, Linux operating system, Mozilla Firefox Browser, Apache Tomcat Web server, to list a few. OSS licenses allow full-scale commercial use of software at low costs, and hence, they are very popular in the industry. The reuse of open-source components and libraries for specific purposes has also become very common.

Reuse in AI Applications

Due to significant differences in the processes for developing AI applications versus traditional software applications (to be discussed in more detail in Chapter 10), reuse in AI deserves some discussion. Whilst AI software components may be generic, AI applications are highly specific. It is important to remember that enterprises want to solve business problems and not just invest in AI technology. When buying or building an AI application, an enterprise may ask for a fraud detection component … or an image classifier … or a speech-to-text transcription tool … or a language translation service … or any of the other thousands of potential AI components. When procuring such components, it is only natural for Stakeholders to expect reuse of existing capabilities. Someone working for a pharmaceutical company may see a new report about a tool being used to translate chemistry papers for a university. It is only logical, in such circumstances, to expect the tool that works on chemistry papers to work on the pharmaceutical content. Unfortunately, AI technologies are very, very sensitive to data, and even small changes in the data can have a big impact on accuracy.

Figure 8.2 shows a high-level decomposition of AI application development from a reuse point of view. The boxes in grey represent traditional software components that may also be a part of an AI application and their reuse opportunities are guided by the general software discussion we had above. The orange boxes represent AI-specific topics we discuss below in more detail.

AI Applications

Application needs are generally unique within an industry (e.g. banking, insurance, manufacturing, etc.), and even companies within the same industry typically strive for differentiation. There are still some AI applications that can be reused within a given industry or even across industries with some minimum customisation of data and workflow. The obvious example for these is a Cognitive Assistant or Service Bot. AI agents to help clients with questions on information about specific product offerings, Frequently Asked Questions (FAQs), ordering, active online support, etc., have become very common [12].

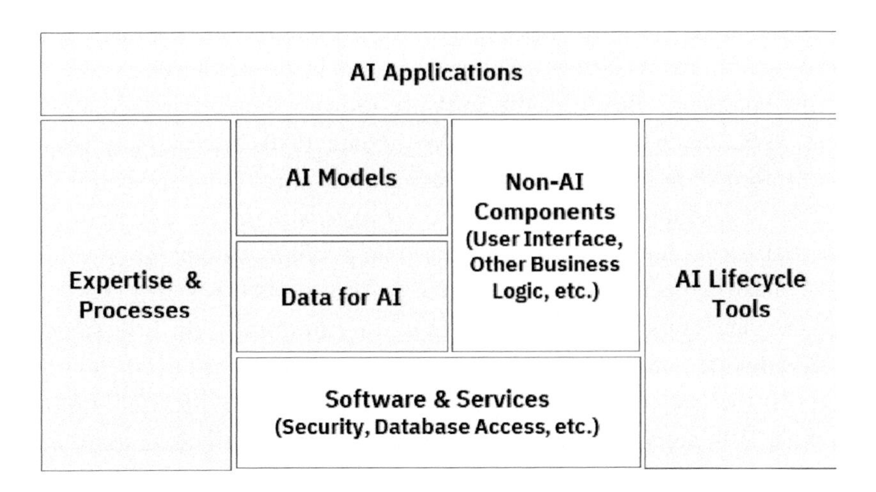

FIGURE 8.2 Reuse components in AI Application development.

AI Models

There are two scenarios for reusing AI models. (i) Reusing model within the enterprise and (ii) reusing a model from an external source such as open source or a commercial vendor. Within the enterprise, ideally, it should be possible to use an AI Model from one application directly in another similar application. In reality, even if the applications are similar, it is highly unlikely that an AI Model will be directly reusable unless the data being processed share at least some features and their distributions. One possible solution to this is the use of Transfer Learning which we described in Chapter 3. Transfer Learning is a powerful capability; however, the key point still applies. It is incredibly rare for an AI Model to be re-usable in different, albeit similar, applications without some form of revision.

For the scenario of using an external source, many AI capabilities are generally available as generic components, typically as cloud services. Many commercial vendors such as Google, IBM, Microsoft and Amazon provide access to commercial services (e.g. speech-to-text, text-to-speech, sentiment analysis, image recognition, etc.) supporting AI application development.

Data for AI

Given the importance of data in an AI project, it is reasonable to expect to reuse data as much as possible. Consider an insurance company that offers life, home and car insurance. An AI model developed for some aspect of life insurance underwriting will almost certainly not be re-usable in a car insurance process. However, the customer data may be directly transferable. This core data will, of course, need to be complemented with other data such as health data for life insurance or vehicle data for car insurance.

As we had mentioned in Chapter 7, public data sources available in unstructured and structured formats [13] include images for various purposes, videos for action recognition, text data (news articles, messages, tweets, etc.) and structured multivariate data in a tabular form in various domains (financial, weather, census, etc.). However, some caution is needed to avoid potential poisoning of the publicly available data by adversaries.

REUSE OF ENTITY EXTRACTION SOFTWARE

Every time I received an Invitation to Tender (ITT), a chill would run down my spine.

I was the Chief Architect for IBM's Text Analytics Group and, every week, I saw multiple ITTs from Clients wanting to procure entity extraction tools. Invariably, the ITTs included a question about availability of re-usable libraries and models for entity extraction.

Many companies responded to this by actively marketing their libraries and models. The marketing material would refer to the "Law Enforcement" or "Healthcare" entity extraction package. The message to Clients was simple; if you buy our product, you are getting a massive head start because we already have an off-the-shelf package.

I found it massively difficult to explain to Clients that re-using such packages was very difficult. On almost every project I worked on, I found some unique set of requirements that made reuse of generic packages very difficult.

For example, with one police force, we had a package that was highly effective at labelling dates. However, when we started working with the police force, they didn't want just dates. They wanted dates of death, date of arrest, immigration dates, dates of birth, dates of marriage, dates of conviction, dates of bail and so on.

If that wasn't enough, in what set of documents we found dates written in an eight-digit format that was unique to that particular corpus. The 22 November 1997 would be written as 22111997. Stakeholders naturally assumed that the off-the-shelf package would still give us a head start. Unfortunately, that wasn't the case because entity extraction models are interlinked in strange ways. Modifying the package to correctly identify the eight-digit dates impacted the phone number identification and correcting that issue impacted the visa entry entity identification.

It was incredibly difficult for Stakeholders within the Client to understand that reuse was not always the head start it should have been. It was also hard to convince them that, with the correct tooling, starting from scratch was not a huge investment. In the end, I found it easier to not argue the point and simply say that we were committed to reuse wherever practical … then in reality to do what was most effective.

James Luke

AI Lifecycle Tools

The lifecycle of an AI application is strongly affected by the demands of the ML model(s) being used. Due to the significant differences from traditional software components, the creation and sustenance of ML components require a wide range of tools across the lifecycle. We refer to [14,15,16] for current approaches to manage AI model lifecycle. The toolset covers the various phases in an AI project i.e. Data preparation, Model Building & Validation, Testing & Verification, Imbuing Trust, Deployment and Monitoring. We will discuss more details of the engineering lifecycle in Chapter 10.

Expertise and Processes

The practical challenges in creating and supporting AI applications that use ML cannot be underestimated. From their experiences at Google, Skully et al. [17] concluded that ML applications carry significant technical debt. Due to lack of clear functional boundaries, the behaviour of a ML component can be summarised as "Change Anything – Changes Everything". Data dependencies are more costly than code dependencies. Systems can also become overly complex due to glue code needed to support various models, data processing and ubiquitous experimentation. Maintenance is expensive due to inevitable changes in data and models with time. Amershi et al. [18] reported on a study of software teams at Microsoft developing AI-based applications. They had three key observations:

i. Managing the AI data lifecycle is harder than other types of software engineering,

ii. Model customisation and reuse require very different skills that are not typically found in software development teams

iii. AI components are more difficult to handle than traditional software components due to the difficulty of isolating error behaviour.

Another study at Microsoft by Kim et al. [19] reported on the technical and cultural challenges of integrating data scientists into software development teams.

From the description above, it should be obvious that while bringing some exciting new capabilities, using ML in AI applications also brings significant risks. An enterprise needs to have sufficient knowledge and experience to develop AI applications. Sharing lessons learnt and the practical aspects of deploying tools and practices across AI project teams can add considerable value to the enterprise.

THE AI FACTORY

One approach to maximise reuse opportunities is to see AI as a cross department issue and take a "factory" approach. The AI Factory is a combination of skills (otherwise known as people), assets, tools and processes aimed at improving the efficiency of AI delivery. An AI factory will help ensure you have a consistent approach and rigour to all your AI projects across your business, this will dramatically reduce your maintenance and risk. An AI factory needs a blended team of data scientists, traditional IT folk and most importantly business domain experts.

Getting started on an AI journey can be challenging for many enterprises simply because of the resources required to get started and the time taken to demonstrate value. In order to evaluate the feasibility of an application, some form of evaluation environment is required. This may involve provisioning hardware or virtual cloud infrastructure. Evaluations are complex and require both skilled evaluators and tooling if they are to be conducted properly.

The AI Factory aims to build a core team of skilled resource supported by ready provisioned infrastructure and tools to allow rapid evaluations. A team of people that can both build new AIs and maintain your organisations growing stack of operationally deployed AI. The overall aim of this is to reduce the cost and time taken to evaluate ideas and

AI FACTORY – A CASE STUDY

The problem was scale!

Every Client we talked to wanted to build an AI application (or ten). With a relatively small team of specialists, it simply wasn't possible to resource all the projects. Not only were we short of bodies, but each individual project required Cloud infrastructure and a whole set of tools configured before any evaluation could be started. Then the evaluation itself was a cumbersome with very little tooling to run evaluations and compare output files with expected outputs.

The answer was to build our AI Factory.

We assembled a small team comprising a lead architect and three developers. They became the human element of our Factory; now all we needed the infrastructure, tools and processes.

From an infrastructure perspective, we built what became known as the AI Factory Floor. The Factory Floor was the place anyone could go to engage with the Factory, and it was designed from the outset to make it easy to get started.

The first thing we added to the Factory Floor was a repository for artefacts. The repository brought together AI tools, services, models and data so that we had a start point for any new applications.

We also configured a set of Cloud servers using Red Hat's multi-cloud technology. The aim was to significantly reduce the startup effort for any new application. If, for example, a Client asked us to evaluate the application of a particular AI tool to a particular data set, we had virtual servers ready to go and a team of people who knew how to access them.

In addition to infrastructure and artefacts, we developed a series of tools. Again, the aim was to reduce the cost and time taken to evaluate AI capabilities. A simple example would be a tool to compare the output of an AI service with a set of 'truth data'. Other tools did simple stuff such as ingest files into a common format or break files up into smaller chunks.

Finally, we defined standards so that we could bring multiple technologies and partners together.

The overall aim is simple. If a Client asks for an AI Application, we can go to the Factory Floor and find existing capabilities and data. We can stand everything up very quickly because we've already provisioned the servers and we have the magic multi-cloud technology. Then we have tools to run the evaluations so we can get an answer quickly. All of this is run by a small team of specialists. The whole point being to transform the initial phases of an engagement from several months into just a few days.

One question we are often asked is how does the AI Factory compare with the online environments provided by the Cloud companies? The answer is simple … it's not a proprietary environment! The AI Factory uses services and products from the Cloud companies and from other suppliers. It's an ecosystem that aims to make the best capabilities available in the simplest and easiest way. It's also important to understand that it's not a huge investment. With just two or three people, we were able to create a very powerful environment very quickly and immediately generate savings in the reduced time and effort taken to develop AI applications.

James Luke

capabilities such that an enterprise can fail fast and deliver the benefits expected of an AI programme.

An AI Factory requires several elements:

- A team of specialists tasked with growing their experience in evaluating, building and operating AI services.
- Provisioned infrastructure comprising servers, software and services that can be rapidly configured to build and evaluate an AI asset.
- A library of assets that can be reused as the basis of a new application.
- Tools and processes to enable rapid evaluation, identification of reuse opportunities and development of new capabilities.
- DevOps processes and tools to enable the deployment of capability in the relevant operational environments.

When an enterprise identifies a new potential AI application, an AI Factory should allow a significant improvement in the effectiveness of the evaluation and delivery. Without a Factory, the enterprise would need to start from scratch in defining the evaluation process, setting up a team, building an evaluation environment and all the other steps required to take a potential application from initial idea to working capability. With a Factory, a new idea can be presented to an experienced team with detailed knowledge of previously deployed applications. The team can identify which existing capabilities are most appropriate for potential reuse and use existing tooling to evaluate. All evaluation, development and deployment activities can be undertaken on already provisioned servers and using best of breed tools. Where gaps in capability are identified, development tools are already available to adapt, or create new, capability. Finally, the DevOps tools are in place to enable the deployment of any capability into an operational, production environment.

Developing an AI Factory may not be justifiable for a small enterprise with only a limited number of potential applications. In such cases, a Factory may be operated by a consulting organisation. However, in cases where an enterprise has an extensive analytics requirement, with many dozens of potential AI applications, then the establishment of an internal AI Factory should be considered.

IN SUMMARY – IT CAN BE AS HARD AS YOU MAKE IT

The attraction of ML is simple … just throw some data at the algorithm and magic happens. We hope we have dissuaded you of that opinion and shown that AI can be delivered successfully if you approach it in the right way. In particular, always remember:

- Demonstrations are much easier than actual deployments of AI applications in production.
- It is important to set the right expectations for all the Stakeholders.

- Choosing the hardest problem for AI may not be best strategy for success.
- Understanding the current state of AI technology is important to achieve higher probability of success.
- There are some unsolved (hard) problems in AI … understand when your solution requires you to solve one of them with a breakthrough invention.
- Domain specialists are critical for successful delivery.
- Understanding the impact of AI on the business process is important.
- Instrumenting trustworthiness requires careful implementation.
- Since AI is software, you also need to manage the traditional expectations of software (e.g. maintainability, reliability, etc.)
- AI Solution needs a monitoring framework to make sure that it is operating in a trustworthy fashion.
- Extensive tooling support is needed to manage the AI model lifecycle.
- Reuse of assets is possible, but it comes with a set of caveats.
- Transfer learning is a realistic alternative, but it depends on the closeness of application domains.
- Any significant AI programme should be supported by an AI Factory.

REFERENCES

1. G. Marcus and E. Davis, *Rebooting AI*, Pantheon Books, New York (2019).
2. H. Kautz, "The third AI summer," Robert S. Engelmore Memorial Lecture, AAAI 2020 https://www.youtube.com/watch?v=_cQITY0SPiw.
3. T. R. Besold et al., "Neural-symbolic learning and reasoning: a survey and interpretation," arXiv preprint, arXiv:1711.03902 (2017).
4. A. d. Garcez and L. C. Lamb, "Neurosymbolic AI: the 3rd wave," arXiv preprint, arXiv:2012.05876 (2020).
5. S. Russell and P. Norvig, *Artificial Intelligence-A Modern Approach*, Pearson (2020).
6. H. Xu, et al. "Adversarial attacks and defenses in images, graphs and text: a review," *The International Journal of Automation and Computing*, 17, pp. 151–178 (2020).
7. J. Pearl, "The seven tools of causal inference, with reflections on machine learning," *Communications of the ACM*, 62(3), pp. 54–60 (2019).
8. T. Young et al., "Recent trends in deep learning based natural language processing," *IEEE Computational Intelligence Magazine*, 13(3), pp. 55–75 (August, 2018).
9. X. Qiu, et al., "Pre-trained models for natural language processing: a survey," *Science China Technological Sciences*, 63, pp. 1872–1897 (2020).
10. R. K. E. Bellamy et al., "Think your artificial intelligence software is fair? Think again," *IEEE Software*, 36(4), pp. 76–80 (2019).
11. ISO/IEC 25010, "Systems and software engineering - systems and software quality requirements and evaluation (SQuaRE) - System and software quality models," (2011) https://www.iso.org/obp/ui/#iso:std:iso-iec:25010:ed-1:v1:en.

12. E. Adamopoulou and L. Moussiades, "Chatbots: history, technology, and applications," *Machine Learning with Applications*, 2, Elsevier (2020).
13. List of Datasets for Machine-learning Research: https://en.wikipedia.org/wiki/List_of_datasets_for_machine-learning_research.
14. J. Thomas, "Operationalizing AI — managing the end-to-end lifecycle of AI," https://medium.com/inside-machine-learning/ai-ops-managing-the-end-to-end-lifecycle-of-ai-3606a59591b0.
15. K. Ishizaki, "Does AI model lifecycle management matter?" https://www.ibm.com/cloud/blog/ai-model-lifecycle-management-overview.
16. MLOps: Continuous delivery and automation pipelines in machine learning https://cloud.google.com/solutions/machine-learning/mlops-continuous-delivery-and-automation-pipelines-in-machine-learning.
17. D. Scully, et al., "Machine learning: the high-interest credit card of technical debt," *Software Engineering for Machine Learning Workshop*, NIPS (2014).
18. S. Amershi, et al., "Software engineering for machine learning: a case study," *ICSE-SEIP '10 Proceedings of the 41st International Conference on Software Engineering: Software Engineering in Practice*, pp. 291–300 (2019).
19. M. Kim, et al., "Data scientists in software teams: state of the art and challenges," *IEEE Transactions on Software Engineering*, 44(11), pp. 1024–1038 (2018).

CHAPTER 9

Getting Your Priorities Right

Many years ago, I was running a workshop for a major airline. These sessions can be very dull, so I like to lighten the mood by telling the odd joke or teasing the audience. On this occasion, I chose the latter! To get the audience fired up, I suggested that no matter how funny or exciting or glamorous pilots were, deep down they were really quite boring people. In fact, I said that the average airline pilot was the sort of person who would, even if they had driven over a hundred miles from their home, immediately rush back home if they suddenly thought they'd left the oven on.

At that point, a pilot in the audience stole the show! He didn't get upset nor did he send any abuse back in my direction. He just calmly said, "James, I would know that the oven wasn't left on because, before leaving the house, I complete my checklist".

James Luke

This chapter represents the culmination of the contents presented in the previous chapters. We expect you to have already gone through **Doability Method Step 1** in Chapter 4 and passed the Artificial Intelligence "(AI)/Not AI" decision diagram (Figure 4.1). Here, we introduce "**Doability Method Step 2**" also called "**Doability Matrix**". The goal is to help your AI project get off the ground on a successful path and keep it there. In case you chose to come here directly from Chapter 4, we believe you will still benefit from the methodology described below. You can always go back to the other chapters for clarifications and more details on business value and doability. *We recommend that you read this chapter carefully since this can stop your project from crashing and help you build AI solutions that people will want to use.*

DOI: 10.1201/9781003108498-10

Key Ideas in This Chapter Are:

i. An AI Project Assessment Checklist comprising Value & Doability parts.

ii. A Doability Matrix for evaluating your specific project idea or a portfolio of ideas.

Whilst the evaluation of projects in terms of *Value* and *Doability* may be normal for complex engineering projects (apart from the use of the word *Doability*), there are many nuances introduced by the use of AI that need to be understood. An effective approach is to conduct a *Doability* workshop with introductory AI education for all participants, especially the business stakeholders, before you settle on one particular project. Use this chapter and Chapter 4 to guide your workshop content.

AI PROJECT ASSESSMENT CHECKLIST

The use of checklists has been proven very effective in preventing mistakes in other domains such as healthcare and airplane pilot procedures [1]. Along those lines, we have collated key questions from our combined experience in AI projects and a careful analysis of what typically goes wrong in publicly reported AI projects. Typical causes of AI project failures and mitigating strategies are to be found in the other chapters in this book. This chapter is about spotting the icebergs in the path of your planned project and help you navigate around them.

The AI Project Assessment Checklist consists of 21 questions overall. There are ten Value questions addressing the possibility that your project stakeholders may not see your AI quite as you imagine; so it may not become as valuable as you hope. Then, there are 11 Doability questions that try to rout out the hidden costs of building your AI including the feasibility/complexity of your idea and your team's capability to build it. The goal is to provoke serious consideration of aspects of your project related to Value and Doability that will be critical to your project success.

The checklist targets the two key phases of your AI project: the building phase (is it cost effective/feasible to build?) and the operation phase (will it really work as you intend?). *You need to answer all the questions, even if you have not written a single line of code yet.* We know these may seem like a lot of questions but answer them honestly. For some questions, you will need to imagine your AI project is already built and deployed into the real world with real users.

AI projects are, in most cases, complex systems engineering projects. As discussed in detail in the previous chapters, beware of algorithm addiction … it's dangerous to assume that all you need is an algorithm and some data. AI algorithms need to be trained or configured and integrated into either existing or new business processes. Data needs to be accessed and the user experience needs to be designed. In addition, for many AI automated systems, a separate quality monitoring system and parallel manual failsafe process may also be needed. There may be a huge element of business change and stakeholders need to be managed.

Value Questions

Theme	Questions	Yes/No/May Be
Business	1. **Is the business problem clearly defined?** Early on in an AI project, it is good to start with a general sense of direction rather than a fixed goal. The truth is that no AI goal can be properly defined until you have experimented with the data. BUT… you must move as swiftly as you can to a solution that targets a specific business problem (and get stakeholder buy-in) otherwise you could end up spending project time and money on something nobody needs or wants.	
	2. **Once the system is operational, will you be able to measure the impact on the business?** You've done it! Your AI customer help system is up and running in production… but sales are dropping! Is your AI system to blame, or is it actually doing a good job retaining customers that would have left with the herd? If you can't answer that question, because you haven't designed a way to measure the value of your AI system demonstrably, your system is likely to be next for the chop.	
Stakeholders	3. **Are the Stakeholders clearly identified?** Who is going to pay for the system, who is going to own and operate the live system and be responsible for fixing the system when it breaks, who is liable if the system is sued, who owns the data? Who are the users? Do not forget the external stakeholders, the ones you can't choose, the press, the regulators, your competitors, etc. If you go ahead and build a system with uncertainty around stakeholders, you may well encounter large hidden costs. For example, if you ignore the fact that your great AI idea is going to require 24-hour monitoring and a team of six experts to maintain it at short notice, you will undoubtably miscalculate the business value and fail to address the requirements of your stakeholders.	
	4. **Is the AI application morally acceptable to all your Stakeholders (i.e. Users, Society, Employees, etc.)** Do you care if people object to your system on moral grounds? You might, if it starts to affect the uptake of your new product/service, or your employees refuse to work on it. Spend time as early in the project as you can, identifying and mitigating any potential threats, real or perceived that could damage your business plan.	
	5. **Can you be sure that the system will not completely replace the people you need to create it (e.g. customer support)?** Tempting, huh? Thinking of savings in wages? Who is going to train your AI? Who is going to know when it has generated an answer that is subtly wrong? Who is going to step in and run the business process manually when your production AI needs to be fixed? It is one thing to look for efficiencies, but if your business plan is predicated on experienced knowledge workers leaving the business then it will be hard to secure their buy-in and, once they've left, re-employing them will be expensive. A safer ambition is to see your AI as a force multiplier, something which can make the employees more productive.	
	6. **Have you confirmed that there are no legal concerns (e.g. privacy, data ownership, etc.)?** This one should be obvious, but you would be surprised how many projects plough on regardless, thinking it will all be ok in the end. It never is. Address this issue immediately or your AI project will not succeed and/or you may become famous for all the wrong reasons.	

(*Continued*)

Theme	Questions	Yes/No/May Be
Trust	7. **Have you made sure that the AI output is not mission safety/life critical?** This is not necessarily a showstopper, but it has a massive impact on cost. It is not just the extra testing and safety subsystems you will need to design and build. Can your project afford or avoid the lawsuit that may follow an erroneous AI decision?	
	8. **Do you know the AI performance metrics (e.g. accuracy, latency, bias, etc.) to achieve the business objective?** This one is easy, but fundamental. You must diligently check and understand your AI model performance and aim for a business solution that is within its limits. For example, will the owners of your new driverless car accept it crashing 10% of the time? Probably not. Will the users of your pop music recommender system be happy with only 9 out of 10 tunes being fantastic? Probably yes.	
	9. **Do you have plans to understand and remove any unwanted systemic bias in the training data?** Not all bias is bad and the trick is to know it exists and act accordingly. If you don't know if your training data has some form of selection bias or not, then it probably does. Unknown bias in your training data will mean your AI decisions will also be biased and capable of harming your systems credibility. Save money on lawyer fees later and test your training data now.	
	10. **Do you understand the form of explainability required for the system stakeholders?** Supporting evidence, equivalent training data, rule output, legal justification … etc. This is not always a showstopper. Sometimes you do not need to provide an end user explanation. Do you know, or care, how Alexa picks tunes you may like? But if your system needs to be able to justify its decisions, it can get expensive if you leave this till the end of your AI project. This might not seem important when you are in the heady days of AI development, with your new AI wonder tool yet to amaze the public. But if you do not consider this while you have the training data and your data scientists in hand, that amazement could manifest itself as fear or mistrust. A public that mistrusts AI decisions is unlikely to turn into customers of that AI.	

Doability Questions

	Questions	Yes/No/May Be
Trust	1. **Once the system is operational, will you be able to prove that the AI is working correctly?** Will it be technically possible to measure the accuracy (without the ground truth of training data)? This is a real gotcha problem. It is so easy to ignore this issue until too late in a project's lifecycle, often a month or two after "go live" and after all the developers have left the project. Sometimes this problem is impossible to fix, even if you consider it early enough! For example, if your AI recommends a patient a drug, and the patient dies, does that mean the AI failed or would the patient have died anyway?	

(*Continued*)

	Questions	Yes/No/May Be
Data	2. **Can you guarantee the supply of the data into the future (e.g. do you already own or license it)?** Everybody now knows that data is the new precious resource. If your AI runs using somebody else's data, the data owner may want some compensation. Just because you found your data on the internet doesn't mean it is free. If someone owns the data that makes/made your profitable AI work, they may feel they have some legal right to your income. If the owner of the data stops supplying it, or changes its content, or regulators stop you using it, your AI is unlikely to survive. *Secure your data sources before you go live.*	
	3. **Is your data labelled?** Having loads of data is great, but you still have work to do – this work can eat into your projects budget, 80% of your development time could easily go to data wrangling – make sure it's in your plan. If your data lacks a clear target variable (the thing you'd like your AI to provide), or if you want your AI to interpret meaning from unstructured data (text, pictures, audio, etc.), then most likely you are going to have to engage human experts to start labelling the data so it can be used for training the AI. *Spoiler alert: this can be time-consuming and expensive.*	
	4. **Is the training data representative of the operational data?** Be worried, for your future in AI, if you find yourself reading this question and asking why? Do not train your speech recogniser on PBS newscasts and then be surprised when it cannot decipher high-school playground chatter.	
	5. **Is your internal development/testing environment representative of your target operational environment?** Whatever data conditions you created to train your model (cleaning the data, aggregating it, cross referencing it to list data etc.), you will need to recreate that for each and every record at *run time*. This can be a problem if you have developed your lab prototype on a supercomputer and expect it to operate on a cell phone.	
	6. **Does the training data for the AI have all the information that a human would use to do this task?** This bear trap is not always set, there will be exceptions (typically when you have millions of rows of data with limited variation), but those are very rare. For all your other AI projects, if the decision task needs information that is not in the data (e.g. the side-effects of a medicine not just its name), you are going to have to bake enough of that data into your training and operational data to help the AI differentiate in the same way. This will mean merging the reference data with the individual training data records. Remember (as per question 5) you are going to do that same data merge in the production environment as well.	
AI Expectations	7. **Is the task for the AI simple enough that it does not need the skills of a very experienced human?** It is not that you shouldn't go ahead with your project. Just be aware it is unlikely to be cheap or quick. Using AI to replace a boring, repetitive task is a great idea, using AI to replace a complex task, one that takes a human many years to master is rarely easy. This is especially the case if the human has to deal with rare anomalies (sparse training data examples) and/or needs common sense/external context (no training data).	

(*Continued*)

Questions	Yes/No/May Be
8. **Are you confident that your idea does not require an invention?** Inventions take time to create and test and they may not work. They rarely conform to business application delivery schedule. You want to build an AI to find a cure for cancer, fix that thing your best engineers couldn't do, or decode alien radio signals, or maybe select perfect soulmates ... How are you going to know the AI is fully trained and ready for deployment, when you do not know what or if the answer exists? The only fallback you have is to exhaustively test the results, which will be expensive, assuming there is a test you can devise. Even then how will you know it's complete and will work for all cases... Expecting AI to do magic is not good business; try to stick to fixing problems that can be measured.	
9. **Can the system be implemented without it needing any complex situational awareness?** This is less of a bear trap and more of a money pit for your project. If your AI needs to understand context or common sense for the given task or requires situational awareness involving multiple data streams, it will need a more complex architecture possibly including multiple AI components. Managing one AI component is already difficult, managing multiple AI components and the interactions between them is even more difficult and therefore expensive.	
10. **Have you confirmed that the deployed AI will not feedback and corrupt the data source you are using for training?** If you are using today's share prices to predict tomorrow's share prices, you need to know your fun probably won't last (unless you put a lot of effort into re-training new models every day). The simple fact is that by financially investing in those AI recommended shares, you will affect tomorrow's market price in a novel way, which your model will not have been trained to recognise, and it will probably start to fail. You could always build a model and not use it to invest, but what would be the point?	
11. **Can the AI be tested without major changes to the existing business processes and systems (e.g. run in parallel with existing systems)?** You have just created an AI to recommend online adverts to your website customers. The problem is you have never sold these products before nor do you have a way to see if the adverts are influencing the customer purchases. Are they buying because of the adverts, or are they buying less because the adverts are poorly targeted? Are they using the adverts info to investigate cheaper deals elsewhere? The bottom line is, "how are you going to justify your investment in the AI, if you can't test its performance"?	

USING THE DOABILITY MATRIX

Doability Matrix (see Figure 9.1) is a convenient way to visualise the responses to the Value-Doability Checklist. The horizontal axis represents the 11 questions on Doability and vertical axis represents the ten questions on Value. At the basic level, once you have the responses to the Value-Doability Checklist for a project idea, you just count the number of 'YESs' for each of the two dimensions (i.e. Value and Doability) and locate the project on the matrix.

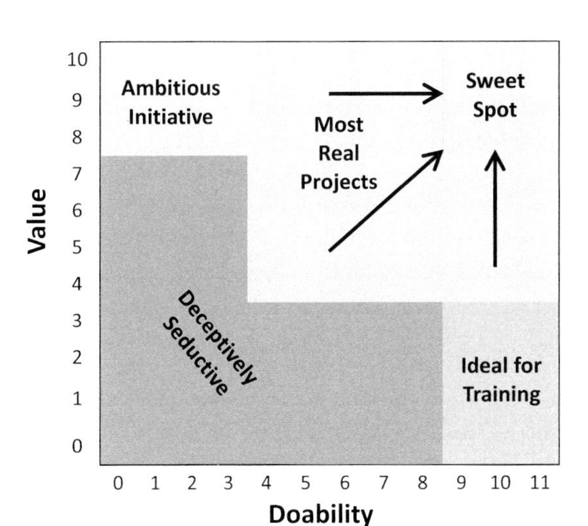

FIGURE 9.1 Doability matrix to place project ideas for comparison and assessment.

If you answer YES to *all* the questions in the checklist, your project is on solid ground and has a high chance of success – Congratulations!

If your project elicits even a single unavoidable "NO", then the score for that dimension (i.e. Value or Doability) is reduced to ZERO (sorry, but it's for your own good, no amount of YESs can fix a NO). If you are familiar with the children's game "Snakes & Ladders", you can think about the NO answers as snakes that always take you back to zero.

You can add some depth to your project evaluation by including "Maybe" as a third response. This typically means you have more work (and thinking) to do before your project can be operationalised. More May Be's mean more effort, the project with the least May Be's is better. We will return to specific nuances of "No" and "Maybe" responses later in this chapter.

To give you a perspective of the broad possibilities of AI projects, we have identified specific areas in Figure 9.1 in different colours with special names to indicate their essential attributes. You can use Figure 9.1 to evaluate an individual project or compare competing project ideas. The box you end up will help identify the winners and losers; do not be surprised when the Cinderella project wins!

Sweet Spot

Projects that are evaluated to be high *Value* and high *Doability* are in the top right-hand box in green. These are rare. Enabling the scaling of a human task is a great place to start an AI project. The best opportunities for AI are in scaling tasks that humans can already perform. In this respect, consider two different types of scale.

The first type of scale is when there is a human-led task such as classifying images or textual contents. If we can leverage the human expertise to train an AI, then we can do the task on a massive scale with an AI. An example of this would be in the field of radiography. A human being can evaluate an X-Ray and identify potentially malignant growths. However, evaluating each X-Ray takes several minutes and so it's only possible for a human being to evaluate a relatively small number of X-Rays in a day, not to mention the potential

SWEET SPOT

A great example of a "Sweet Spot" project was the classifier built for a military Client and mentioned briefly in Chapter 5.

The Client organisation had a team of four Analysts who each spent their days classifying rows of data received via a live feed. Even though the dimensionality was high, the rows of data were easily presentable in a spreadsheet, and there was a vast history of labelled data produced by the Analysts. As a result, it was a relatively straightforward task to extract training and test sets from the mass of available data. After we had a simple Machine Learning model using a set of training data, we were able to apply the model to a set of unlabelled test data. It was a trivial task to evaluate the accuracy of the classification, and the Analysts were able to confirm the effectiveness of the model.

We conducted this exercise just as the business was preparing to implement a major change programme. The change programme would massively increase the volumes of data received via the live feed. Our ability to prove the effectiveness of the Artificial Intelligence convinced the leadership team that Artificial Intelligence was the key enabler for the delivery of the change programme.

One of the biggest reasons that this project was so successful was that the Analysts bought into the vision from day one. They saw immediately that the AI we were building would NOT replace the Analyst. In fact, the Analysts would become the Data Scientists responsible for building and maintaining the Machine Learning models. Rather than replacing the Analyst, the Artificial Intelligence enabled them to scale and made their role more interesting in the process.

Other factors also contributed to the success of the project. There was an existing business process in place; we didn't need to collect any additional data and/or face any integration challenges. We weren't asking the Client to change any existing processes or systems and the business case was easy to demonstrate. We were able to test the AI very easily using a simple extract of the data and with little additional effort from the Client. The fact we could test and prove the new capability with so little effort and no immediate operational impact placed this project well and truly in the "Sweet Spot".

James Luke

fatigue that could lead to errors. In short, the human can complete the task in a reasonable time but can't complete the required numbers of tasks.

The second type of scale is where a human can describe how to perform a function, but actually completing a real-world task requires the chaining of that process many thousands or millions of times. The most obvious example of this can be found in the earliest applications of computing to code breaking. A human being can take an encoded message, test a specific key and evaluate whether that key is the correct key to decode the message. What a human being cannot do is repeat this process for all permutations of keys. Early forms of AI exploited the ability of computers to scale simple functions to deliver AI applications in fields ranging from code breaking to scheduling algorithms. However, even with brute force computation many of these applications were only made possible by detailed algorithmic work to reduce the size of the search space and therefore the level of scaling required. It's easy to dismiss these applications as "not really AI"; however, in creating AI, we need to leverage every advantage at our disposal. Whilst we should strive to avoid brute force algorithms that cannot scale, it would also be unwise for engineers to ignore the fact that computers are really good at doing simple tasks repeatedly. Murray Campbell's vignette on 'Chess Programs, Then & Now' in Chapter 4 on the future of chess playing presents an interesting perspective on this point.

A classic feature of a "Sweet Spot" project is when you are evolving and enhancing an existing business process. This is important because it means you will have easy access to labelled data. Then you can extract training data from the operations, train and evaluate machine learning models very easily. This will enable you to demonstrate the operational value of the AI and have that value confirmed by a team of supportive Analysts.

Unfortunately, it's very rare that a project is blessed with a clearly defined business problem complemented with a large set of labelled historical data that can easily be used for model development and evaluation. *Sweet Spot* scenarios are rare... especially when AI is proposed to enable a completely new business process.

Deceptively Seductive

The opposite of the "Sweet Spot" is in the bottom left of Figure 9.1, ... this part of the matrix (in red) is referred to as "Deceptively Seductive", representing those difficult projects of questionable value. The ideas for these come for two main reasons. (i) From the perception that AI needs to be smarter than your average human and perform some brilliant task way beyond human capability. Quite often, the proponents will want to select a problem to test the sophistication of the AI, rather than support a relevant business task. (ii) The influencers in an organisation want to 'do' AI as soon as possible because everyone else on the planet is 'doing' it. This forces the creation of AI projects without due consideration of Business Value or Doability. These factors naturally pull you into the "Deceptively Seductive" region. You have been warned; at all costs, stay out of the red region.

Ambitious Initiative

The top left box (in gold) covers projects that are extremely hard to do, but of high business value. This could be a research project that is hard to do because we don't yet have the technology or a project that requires a whole new infrastructure. We tend to think about this box as the *Manhattan Project* or the *Apollo Programme* box. In the domain of AI, the quest for driverless vehicles falls into this region.

These are not necessarily projects to be avoided as long as you have the vision, commitment and resources to go for it. They are the types of projects that are undertaken by a visionary with the conviction that the business return or societal impact is worth the investment. To be clear, these projects are still worth consideration if you do two things ... first, invest properly to ensure success and, second, call us to join the project because big, ambitious projects can be very exciting!

DECEPTIVELY SEDUCTIVE

Example 1: No Existing Business Process to Support AI

When a major bank decided to give online investment advice using a virtual assistant, they were developing a completely new business process. Their existing business process relied on human advisors visiting customers in their homes and having one-on-one conversations. These conversations were not recorded and the advisors were threatened by the prospect of their jobs being taken over by an AI. Without an existing business process, there was a lack of data with which to evaluate the AI. The business case required proof that the AI capability would work before the new process could be implemented; however, without the business process there was no data with which to prove that the AI could do the job. A lack of data and unsupportive Stakeholders meant that this was not the right place to start an AI programme!

Example 2: Picking the Hardest Problem for AI

I remember one particular Client meeting where we were brainstorming potential ideas with which to evaluate the AI. Lots of ideas were presented and, in each case, there was a reason why that particular idea should not be taken forwards. There were issues over regulation … or data availability … or resource constraints … and so on and so on. All of these 'anti-bodies' were actually indicators that the stakeholders were not committed to the project and were actually working under duress; they had been instructed to use AI but weren't personally bought in to the project. Some way into a long and painful brainstorm, one of the participants suggested, "why don't we try that really hard problem we stopped working on three years ago?"

That statement alone sums up the *Value* and the *Doability* of the proposed project! It's "really hard" and the fact they "stopped working" on it suggests its low Value. Be careful! If you are desperate to convince a sceptical audience that AI can add value, it's difficult to say no when someone offers you a "really hard" problem. "Deceptively Seductive" indeed!

James Luke

Ideal for Training

The bottom right box (in purple) covers projects that are very easy to deliver but have very low business value. There is no reason to work in this area as the return is so low. However, this is a useful area for skills development and training the technical teams in AI.

Often when presented with a *Deceptively* Seductive project, the answer lies in the bottom right. Rather than battle to secure the investment required to attempt a challenging project, recommend starting with a more doable project even if it is of low value. If your *Deceptively Seductive* project proposal is a fully interactive customer assistant requiring extensive knowledge, then suggest starting with a simple question answering system for customers using a defined set of frequently asked questions. Creating a simple question answer Virtual Assistant was a great way of getting started, developing skills and educating the Stakeholders in the art of the possible.

Developing a project in this "Ideal for Training" region of Figure 9.1 may not revolutionise your business, but it will give you the confidence to attempt more challenging projects with greater expertise.

Most Real Projects

That leaves us the middle region of Figure 9.1 in yellow! This is the area where most candidate projects lie, and the trick is to identify which of these projects you can move, with appropriate actions, into the top right. So, how do we evaluate projects and how do you get them to the ***Sweet Spot***?

Applying AI in the enterprise is to understand the nuances of AI that will either deliver success or completely derail the project. Projects need to be evaluated, both at initiation and continually throughout development, understanding where they map onto the matrix, will give you a good idea of their potential. Make a note of your responses and assumptions, it is wise to revisit the checklist as your project evolves.

Back in the real world, you are probably going to answer NO or MAYBE to a few of the questions; this could be because you haven't gotten around to that part of your planning yet or *it's a topic that you don't think applies to your project.* This second response is quite common till the teams learn it the hard way. Do not ignore the questions that you answer NO to or they will come back and bite you.

The ratio of YES's to MAYBE's will give you a sense of how much extra project work you need to do before your AI can go live. Don't panic, assuming you do not have any immoveable "NOs" (i.e. showstoppers), most things are fixable, and whilst it's not cost free to add work to a project, these extra tasks are much cheaper to fix if you identify them early.

Remember, **do not proceed, with an individual AI idea, until you have resolved all the NOs,** do not carry on with the other parts of the project in the hope the "NOs" will resolve themselves – tackle them first. It is not uncommon to have to radically change the direction of an AI project when one of these shows up. Do not despair it is better to identify these early and avoid expensive failure later.

Sometimes all it takes is for your project to reset its goals, maybe scale down its ambition. Would your plans for a "fully autonomous robot surgical doctor" be more successful with the public if it dropped the 100% autonomous feature and was developed as a surgical training simulator, or a surgery augmentation tool to help a human surgeon speed up operations?

For every MAYBE answer (and most NOs), there are things you can do to improve the answer. The short explanation below each question in the Value & Doability Checklists above, and the rest of this book is all about that. Circle back to the checklist when you've addressed those issues and try it again. You will need to factor that new effort into your project plan … do not be a victim of "Algorithm Addiction" and assume that some magic AI will figure it out!

This is not a static evaluation. The status of projects will change as old problems are resolved and new problems emerge. The devil is in the detail, and your view of the detail will evolve as the project develops.

Be prepared to modify your business goals and project ambitions. It is our experience that the projects that change early, to reflect the reality of the data, are the *only* ones that ever succeed.

IN SUMMARY – NEVER TAKE OFF WITHOUT COMPLETING YOUR CHECKLIST

Selecting the right project shouldn't be a lottery. It takes a bit of discipline and some honest question answering to get it right. The key points are:

- The AI Project Assessment Checklist has 10 Value questions and 11 Doability questions covering five themes: Business Problem, Stakeholders, Trust, Data and AI Expectations.
- The Doability Matrix can be used for assessing and managing risk in one AI project or for prioritisation across a portfolio of ideas based on business value and underlying technical risk.
- Chapters 5 and 6 addressed the topic of Business Value and Chapters 7 and 8 addressed the questions on Doability.
- Any organisation dealing with a portfolio of business ideas for exploiting AI should use Chapters 4 and 9 to prioritise and manage the risk and outcome.

REFERENCE

1. A. Gawande, *The Checklist Manifesto: How to Get Things Right* Picador (2010).

CHAPTER 10

Some (Not So) Boring Stuff

Everything not invented by God is invented by an engineer

Prince Philip, Duke of Edinburgh

Science is fascinating! Engineers love applying science to solve real problems and change the world for the better. The science that is being done in research labs around the world is essential in furthering our knowledge of AI. However, to deliver real AI that has a real impact on the scale that society expects, we need to put a lot more effort into the engineering.

We have already discussed how AI projects are different in many other aspects, ranging from ethics to stakeholder management in the previous chapters. In this chapter, we consider the fundamental differences in the engineering of AI applications compared to typical enterprise software projects.

TRADITIONAL ENGINEERING

For many decades, engineers have been building complex systems such as automobiles, trains, ships, planes and even spacecraft with amazing success. Each of these systems consists of many subsystems i.e. electrical, mechanical, engine, communications, sensors, etc. Over the recent decades, the software content in these systems has been growing steadily with increasing levels of functionality. Table 10.1 gives some examples of well-known systems/applications with the size of software in them [1].

This table gives a sense of the increasing importance and complexity of software systems over a few decades. These increasingly complex software systems have been delivered using traditional systems and software engineering practices [3-6] that have also evolved over decades.

Figure 10.1 shows an example of a typical traditional software application lifecycle. High-level business or system requirements (top left) are decomposed into functional (subsystem) components and their constituent code modules that individual programmers

DOI: 10.1201/9781003108498-11

TABLE 10.1 Some Well-Known Systems or Software Applications with Estimates of Total Software Lines of Code

System/Software	Lines of Software code (in Millions)
Boeing 787 Dreamliner (avionics & on-board support)	6.5
Android (Mobile Device Operating System)	12–15
F-35 Fighter Jet (2013)	24
Microsoft Office 2013	45
Facebook (2017) not including backend code.	62
Volvo automobile (2020) [2]	100

write. These program units must have clear boundaries and known expected behaviours (i.e. specification of input vs. output) at the design time. Special attention is paid to the appropriate user interfaces. Testing at various levels (unit, function, and application) of integration is meant to expose any unexpected behaviour. The enterprise IT infrastructure may include on-premise systems, private or public cloud options, which can also support external calls to services from other sources or vendors. Suitable DevOps (Development/Operations) processes facilitate a smooth deployment of the application

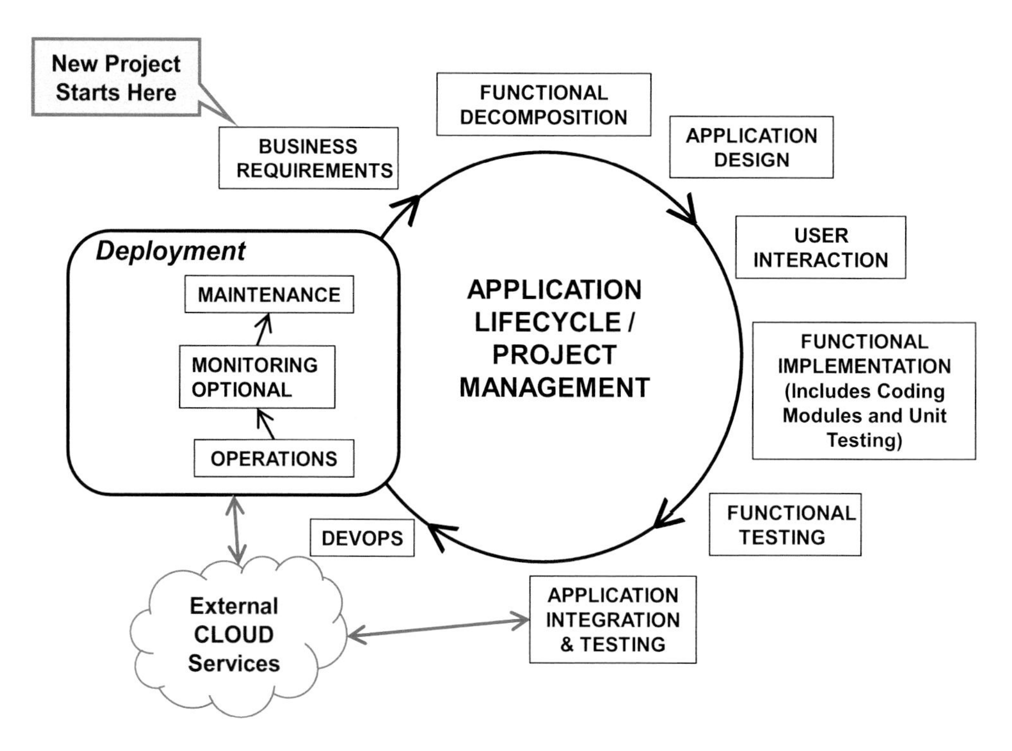

FIGURE 10.1 Traditional Application lifecycle. A new project release cycle starts with Business Requirements and ends with Maintenance, only to start the next cycle for the next release of the application. Use of agile or iterative processes within the application lifecycle is not shown for simplicity. The underlying infrastructure must support the needs of the in-house development activities and application deployment. Details are described in the text.

WHAT CAN MICKEY MOUSE TEACH US ABOUT AI?

A good metaphor for understanding Artificial Intelligence's (AI's) current capabilities and weaknesses is to look at Mickey Mouse's broom and bucket spell in the Disney movie *Fantasia*, based on 'The Sorcerer's Apprentice' story [1] by Johann Wolfgang von Goethe. Mickey introduces us to all that is cool and not so cool about AI.

As the apprentice, Mickey has the tiring job of carrying the water from a well to a cauldron using a pair of buckets. In Sorcerer's absence, he puts on the magic hat and trains the 'now magically alive' broom to fill up the buckets of water from the well and deliver them to the cauldron, but that's all he teaches it. Within a few minutes, the broom can do the work all by itself and Mickey can sit back and relax. Unfortunately, Mickey has not really designed his solution too well. For one, he does not know how to stop the broom; he also does not recognise that the broom has no common sense and will keep filling the cauldron with water even though it is obviously overflowing, thus causing a flood. He uses an axe to break the broom to disable it, only to find out that the many broken pieces become clones of the broom and continue overfilling the cauldron even faster. The story ends when the Sorcerer returns and admonishes Mickey for his untrained use of magic and sends him back to his original task of manually filling the cauldron.

The similarity of magic in this story to AI in the society today is truly uncanny. If Mickey had really wanted a quiet life and/or wanted to launch that broom as a consumer product, taking advantage of that scalability with broom clones (surely every home would want one?), he would have needed to spend time identifying and teaching the broom all about the anomalies and edge cases that it might encounter; an empty well, a bucket with a hole and so on. He would also probably have to create another spell to monitor the cauldron's water level and an alerting spell to override the broom's spell – in short, all small things but probably more effort than filling the cauldron himself. It could be said Mickey's business case for using his spell to do a one-off piece of work was lacking, with a bit more planning and design, who knows, he could have beaten Scrooge McDuck to being the first cartoon billionaire!

At present, AI in business feels a lot like Mickey's bucket spell, media stories of chaotic failures and project overspend abound. The overspend is often caused by projects recognising too late that expensive 'guard rail' subsystems need to be in place before you launch any AI that is expected to perform a non-trivial function into production. Hopefully, with this book, your wizard AI idea can avoid these common calamities.

REFERENCE

1. The Sorcerer's Apprentice https://en.wikipedia.org/wiki/The_Sorcerer's_Apprentice.

and its components from development to the operational environment in an automated and efficient manner. Application monitoring has two main purposes: (i) Understand user behaviour to improve system design. For example, using two different user interfaces (i.e. A/B testing) to evaluate the conversion of items in shopping carts to completed purchases. (ii) Collect logs at specific events in the application execution (e.g. user login, writing to disk, request for resources, etc.) to provide debugging information for troubleshooting when something goes out of the ordinary or an instrument alerts and metrics to help with diagnosing performance issues in the running system. However, depending on the maturity, many systems can function quite adequately without such active monitoring. *In short, monitoring is optional.* The primary need for application maintenance is to address the tickets raised by customers on the deployed application. Depending on their severity, tickets are resolved as soon as possible or scheduled for fixes in the next planned release. The frequency of software releases is managed by a process (i.e. agile, waterfall, etc.) to meet the expectations captured by business value, pending customer requirements, planned new functions, bug fixes, etc.

Any deviation from the expected software behaviour is the definition of a defect (bug), at the heart of any software quality management program. Detection, diagnosis and resolution of software bugs are key activities across the software lifecycle. *Data only serves as inputs to programs. Programs and data are managed separately.* The software behaviour is deterministic (i.e. we know the expected output for the given input) and the development team follows good design practices such as modularity, encapsulations and separation of concerns. There are tools to support various activities (e.g. code analysis, debugging, data flow, change management, bug tracking, test harnesses, DevOps, etc.). Even though there are still many challenges to execute complex software projects successfully [7,8], it is fair to say that there is enough actual evidence to show that *we know how to build complex software systems with adequate quality for real use.* You will see below why that may look 'boring' when compared to what we face in projects that include AI components.

WHY IS ENGINEERING AI DIFFERENT?

If you are reading this book for background information (perhaps you are going to only be responsible for the project management or business application of the AI and not the technical bits), you may feel you can skip this chapter and leave it to the geeks. If you were to skip it, you would be missing an insight into why so many AI projects fail. It's probably not what you think, often their core AI is sound, but in the hubris to develop the shiny "clever" AI, the project teams forget to plan (and cost) in all the extra pieces that an AI needs to be successful once it leaves the lab.

Building operational AI introduces a number of extra responsibilities that you are going to need to factor into your projects. You can get a good idea from this chapter. If you want to go deeper into this subject, there are a few publications [9–17] that address this topic in greater depth, ranging from the basic technology to project management aspects.

To make this discussion more concrete, let us say you want to build an application where you think an AI component can perform a specific task (e.g. object identification, text classification, machine translation, etc.). Here is a list of topics that you will need to consider in addition to usual software development topics. We will delve into these topics deeper in this chapter.

- **Uncertainty in Project Estimation**: Due to the experimental nature of the AI model development, and its complete dependence on the data content, quantity and quality, estimating the effort and schedule is more difficult. If we had problem with project estimation before [18], it just got worse! The safest approach is not to commit to a grand project until you have really proved the AI will work.

- **AI Task Selection**: The selection of the task to be performed by AI has to be done carefully, based on business requirements such as risk, required accuracy and availability of pertinent data (*See Chapter 4*). Data can be fickle, it doesn't always hold the answers that you want, but it always holds something, be prepared to alter your business plans based on what the data can actually do, not on what you wish it could do.
- **No Specifications**: Since machine learning (ML) algorithms learn from training data containing inputs and corresponding outputs, there is no need for a specification document i.e. "Data is the new specification". At first glance, this may sound like a good thing! The problem this creates is one of practicality. Since we have not written down the specifications of the expected functional behaviour in terms of inputs and the expected outputs, we are completely at the mercy of 'statistical learning' on how the AI is going to behave. While it may not be a good engineering feeling, it can be empowering, if you are flexible in your business planning.
- **No Simple Way to Define a Bug!** Since ML models are statistically learnt, there is implicit uncertainty in the model outputs, and hence, they are not deterministic. That means, you can get different outputs for the same input. Without a predefined behaviour captured in a specification, there is no clear definition of a software bug. There's no magic way out of this one, you just need to make sure your application has monitoring and contingency built in from the start.
- **Debugging Is Complicated!** When there is an unexpected behaviour in the output (e.g. wrong classification), figuring out the cause for the behaviour involves understanding the model performance and the training data. To make matters worse, if you are re-using a prebuilt AI model e.g. "transfer learning", you may not have access to the original training data, which makes understanding why a model is failing practically impossible. *It is no longer finding the relevant lines of program code resulting in the observed behaviour, since such a code does not exist! If you can keep your original training data somewhere safe, you may need to go back to it to fathom a bug.*
- **Traditional Testing Will Not Work**: Since there is no specification, there is no expected behaviour. Consequently, traditional testing approaches that rely on verifying expected outputs for specific inputs do not work. *Making sure your training data is an accurate reflection (noisy, dirty, etc.) of the data your AI is going to meet in production is the best defense.*
- **Fault-Tolerant User Interaction**: In applications such as Support Bots (aka AI Assistants) that need more natural user interaction that involve AI (e.g. text, speech, gesture interfaces), the interface needs to be tolerant of potential user errors and input variations (spelling mistakes, unknown accents, etc.). For any serious AI/human interaction applications, you are going to have to factor in a parallel process with the

ability to hand over to a human assistant seamlessly when the AI cannot cope. By analyzing the failed interactions, you can improve the AI performance quickly and the need for a human to take over will diminish.

- **AI Models Can Drift over Time**: Once deployed, AI model behaviour can drift over time due to previously unseen data. All AI applications need to be monitored closely after deployment to understand whether they are behaving appropriately, and there needs to be a plan for what to do, if they're not! In addition, increasingly they will need to stand up to scrutiny of audit and be able to explain their actions. In the event of continuous learning during deployment, new patterns of relationships in data can emerge unknown to the model owners. You can think of an AI deployed in fast-changing business environment as a soccer player in a tough match with you as the manager; you are going to need a bench full of substitute players ready to take over if he/she becomes injured or start to underperform.
- **Dealing with Bias**: If the AI model learns from existing historical data, there is a distinct possibility that any inherent bias in the training data or in the algorithm will become visible in the model. Systematic approaches to detect and mitigate any unwanted bias are necessary for the application to be trusted (see Chapter 5).
- **Need for Explanations**: Since the ML models are complex functions that are not available to the users, they are just 'black boxes'. Consequently, there is a critical need for explaining the output to various stakeholders (see Chapter 5).
- **Robustness of AI Models**: There is overwhelming evidence that the outputs of AI models are susceptible for various types of adversarial attacks. A whole new discipline of how to attack and protect AI models is emerging. We discussed AI robustness to adversarial attacks in Chapter 5.
- **DevOps is Complicated**: Deploying model changes from development to production requires versioning of both models and associated data. Instead of the traditional automated regression testing to validate the application behaviour before deployment, validation of the AI model is statistical and more complex, and therefore more involved.
- **Frequent Application Maintenance**: In addition to the model drift mentioned above, dependency on cloud-based AI services and continually improving AI capabilities may warrant more frequent refresh of the underlying technology components for AI applications. This can be difficult to manage in regulated industries (e.g. banking).

So, there you have it ... as you can see, so many aspects of traditional application development activities are affected by the use of ML components. In this chapter, we will describe in some detail the various AI lifecycle activities and best practices.

PROOF OF CONCEPT TO PROTOTYPE TO IMPLEMENTATION

Delivering my first operational AI application was a massive undertaking! It took nearly 3 years in total and required three phases.

The PoC phase started with a disk of data. We took the data and preprocessed it for ingestion into our research environment. Our research environment was built on a mathematical modelling tool and was perfect for data science and experimentation. We constructed training and test sets, built an ML model and then applied the model to the test data. When the experts sat down and evaluated the output, the conclusion was just amazing … the AI was achieving an accuracy of 96% and, if we could achieve that in production, it would have a massive impact on the Client's operations.

After several weeks of negotiation, we agreed to build a prototype. The prototype would be built on top of the mathematical modelling environment. That decision would minimise cost and enable rapid prototyping but meant that there would be no actual integration with operational systems. It would also limit our ability to understand two key factors. Firstly, the performance requirements in terms of throughput of data. Secondly, the scalability in terms of number of classes processed. This latter point was important as our system was a classifier. Our PoC had evaluated the ability of the AI to classify records into 25 different classes. In the operational environment, we would have to consider ten times that number!

The prototyping was a fascinating experience as key stakeholders, the operational users, demanded more control than we expected. They wanted to visualise the data in ways that they were familiar with but that we felt were unnecessary. They wanted the AI to ask their permission before making certain types of classification. Whilst the demands of the users could be frustrating, listening and responding to those demands was critical in ensuring the support of these key stakeholders.

In the end, we were very successful and the Client signed a contract for a full-blown operational application … and then the fun really started!

The Implementation phase was probably the hardest thing I have ever done. We ran into all the usual challenges of building a software application and then hit the AI challenges. When we started developing the AI models, it became obvious that what worked with 25 classes did not work with 250 classes. Theoretically, it shouldn't have been a problem but, for reasons, we couldn't understand, the learning algorithm wasn't learning.

As the issues became apparent, the same question came up time and time again! Why wasn't this discovered during the PoC or Prototyping phases? The answer was simple. Because we couldn't access all the operational data until we'd built the operational solution.

Whilst stress levels went through the roof, we managed to stay calm and work the problem! By introducing a pre-classifier, we transformed the problem from one of classifying records into 250 classes to ten problems of classifying records into 25 classes. We delivered a working solution to budget and timescale. I know that it was still operational 10 years later … and could potentially still be operational today.

James Luke

FOUR PHASES OF AN AI PROJECT

For an enterprise starting to use AI technology in business applications, it is useful to think about the following four phases. If you are considering multiple AI components in the application, each one of them deserves this level of consideration.

Proof of Concept (PoC)

This goal of this phase is simply to demonstrate that AI can be successfully applied to the selected task. The work includes alignment of AI task with business goals, assessment and evaluation of quantity and quality of available data and experimentation with various models and their evaluations. The duration of this phase will depend on the complexity of the problem, skills and resources available. If the proof of concept (PoC) does not succeed, you can be happy that you failed fast and move on to a different AI task.

Prototyping

If successful with the PoC, the next question is if you can build an end-to-end application. This phase will enable evaluation of all the broader issues that need to be considered. These range from integration of the data pipelines for repeated use to the definition of the user interface (if there is one). A prototype will allow you to test the feasibility of the solution in an environment that is as close to the operational environment as possible. Often the prototyping phase is timeboxed and designed to identify potential issues ahead of implementation, rather than actually address them. Whilst this helps to limit the duration and effort of the prototyping phase, this can be a false economy. Deferring the resolution of issues to the implementation phase is, in reality, kicking the can (and the risk) down the road.

Implementation

The goal of this phase is to build the actual application and deploy it in an environment akin to the actual production environment. Uncertainty at this stage is often introduced by scale and performance against realistic workloads. Whilst the PoC and Prototyping phases were meant to reduce this risk, practical limitations (e.g. lack of resources, aggressive schedule, etc.) may mean that it is not always possible. Typically, the risk manifests in this phase as the realisation that the real scope is substantially different from that assumed in the earlier phases, such as the number of entities for extraction or the number of classes for classification is orders of magnitude larger! If the 'trustworthy' aspects of AI (e.g. bias, explanation, robustness, etc.) were not considered in the earlier phases, you will have a serious rude awakening here.

Monitoring

This is the phase when the operational application is monitored to ensure it is working within its expected parameters and behaving as expected. Changes in the environment and, most importantly, in the data being processed can have a massive impact on performance. In the event that these changes break the AI Model, it will be necessary to spot what's happening, intervene and address the issue. This topic is discussed in more detail later in this chapter.

In the following section, we will consider in some detail the tasks that are undertaken in each of these phases.

DEVELOPING AN ENTERPRISE AI APPLICATION

To help us with the discussion of engineering AI applications, we will use the activities captured in Figure 10.2. For simplicity, we assume that the business application we have in mind will use one ML component (e.g. object identification).

At the outset, in contrast with the traditional application lifecycle in Figure 10.1, there are now two overlapping sets of activities: one lifecycle for creating the business application on the right and another lifecycle for the ML component on the left. The grey box in the middle represents the integration of the two and the deployment of the business application. If there is more than one ML component, each one of them will need a parallel set of activities like the one shown. *The diagram does not imply a waterfall approach; in fact, waterfall project plans are typically suboptimal for AI projects.* The diagram merely indicates all the different activities you will need to define your development process in a logical order.

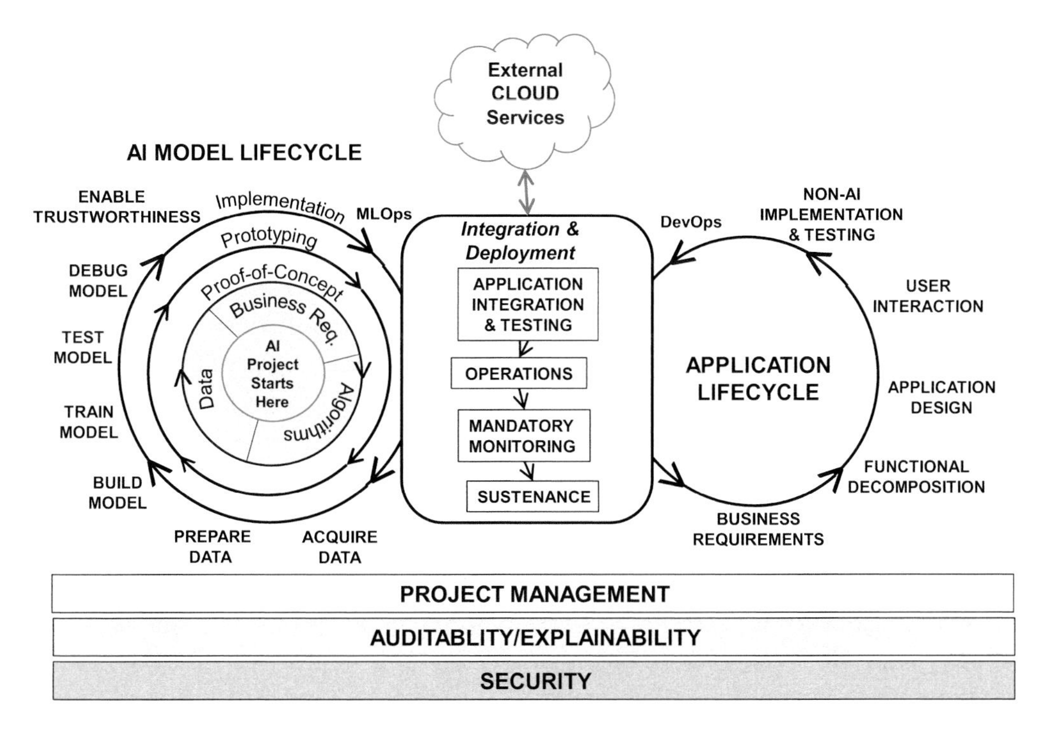

FIGURE 10.2 AI Application Development with one AI component. Left side of the figure represents the AI model lifecycle and right side represents the AI application lifecycle. Inclusion of business requirements in both lifecycles is key to the success of the AI project. Grey box in the middle represents the integration of the two processes and deployment of the application for production use. Three horizontal bars at the bottom, Project Management, Auditability & Explainability and Security, touch all activities. Use of agile or iterative processes within the lifecycle is not shown for simplicity. Details are described in the text.

The ML model lifecycle may go through many iterations before delivering a usable model to support the specific AI task for the business application. Project planning must include these false starts when deciding on resources and schedules. Also, in reality, any ML components may be relatively a small piece in the overall application. There are many other non-AI functions needed to create the business application (e.g. user interface, access management, customer relationship management, network connections, data retrieval, traditional analytics, visualisation, etc.).

We will discuss the activities in Figure 10.2 in the following sub-subsections:

- AI Model Lifecycle
- Application Lifecycle
- Integration & Deployment
- Project Management
- Auditability & Explainability
- Security

AI MODEL LIFECYCLE

Acquire/Prepare Data

We refer to Chapter 7 for a detailed discussion of data related topics. Raw data acquisition may involve licensing, security, privacy issues as well as proper data governance after the acquisition. ML modelling currently requires large amounts of labelled data that may have to be acquired from live user inputs or domain experts or crowd sourcing. For AI applications, proper preparation is needed to avoid bias and ensure fairness & trust. Feature extraction is a critical task in this process, which can help remove redundant data dimensions, unwanted noise and other properties that degrade model performance.

Build/Train/Test/Debug Model

This step aims to produce the best model that meets the business requirements with the available data. We refer to Chapter 3 for a broader discussion on the choice of algorithms. In practice, various AI frameworks (e.g. TensorFlow, PyTorch, Scikit-Learn, etc.) are used to create the model code. These frameworks typically provide some tool support for the coding process. However, as should be evident, even if the model code does not have any errors in it, that does not mean that the model is good for the business purpose. Another important step in building ML models is the separation of training data and validation data so that the model's ability to generalise can be evaluated accurately [19-20]. k-fold cross-validation is a standard practice, which partitions the available data randomly into k subgroups, and each subgroup is validated against the model trained with the other remaining data. This helps to tune model parameters, select data features and tweak the learning algorithm. The data needs to be drawn from the same distribution for the training and validation sets. Unfortunately, debugging of the ML models is complex [21,22,23] since ML behaviour can be the result of the model-inferred code and the underlying training data. Breck et al. [24] discussed 28 specific tests and monitoring needs, based on experience with a wide range of production ML systems at Google.

Enable Trustworthiness

As we discussed in Chapter 5, the black box nature of the ML models invokes questions about the trustworthiness of ML model outputs. These concerns must be addressed during the model building activity since the preparation of the data or the specific algorithm chosen can be significantly affected as a consequence. There is often a reluctance to tackle trustworthiness early in the project since it may appear extraneous to the main function. But this is AI. Doing this whilst the training data is fresh is much easier/cheaper than finding out too late that you have lost the opportunity to bake in trustworthiness – which is the difference between people using your application or not. We emphasise four specific aspects of trust below.

- **Explainability**: In many business-critical applications, the outputs of the black box ML models also require explanations to meet the business objectives. There are many motivations for explanations [25], and it is important to know the need so that the appropriate approach can be used. There are examples of open-source packages for implementing explainability [26,27] in business applications.
- **Bias/Fairness**: Due to the potential sensitivity of the outputs of the ML models to biases inherent in the modelling data, there is a critical question of the fairness of the algorithms [28] in extracting the model from the data. There are examples of open-source packages for understanding and mitigating biases [29] in business applications.
- **Robustness**: The owner of an ML application needs a strategy for defending against adversarial attacks. Xu et al. [30] provide a comprehensive summary of the adversarial attacks against ML models built using images, graphs and text and the countermeasures available. Reference [31] describes an open-source software library, designed to help researchers and developers in creating novel defense techniques and in deploying practical defenses of real-world AI systems.
- **Transparency**: Given the abundance of AI components (i.e. algorithms, services, libraries, frameworks) available from open source and commercial offerings, it makes sense for a company to reuse the available software component in its application. There are two different aspects to consider:
  - i. If using an AI component from another source, making sure that a proper review of the detailed information about the component is done to manage the potential risk.
  - ii. If creating a new AI component for the enterprise use, documenting the necessary technical and process details behind the component is needed for potential reuse or an audit for internal or external reasons.

There are techniques (e.g. FactSheet [32]) to help with defining what information to collect.

MLOps

Like the DevOps process for software code and its configurations, MLOps represents the step of moving the validated AI model from development to operational use and managing it in a production environment. We refer to [33,34] for detailed descriptions and underlying practices. MLOps is a complicated extension to DevOps in that the ML models are easily changed, yet have a significant impact on business decisions and must therefore be carefully managed.

To minimise the coupling between the model lifecycle and application lifecycle, models are typically made available as microservices in the enterprise operational IT environment. Application integration process can invoke the model services with hold-out test data sets and make sure that the performance of the model is adequate for the business purpose before integrating with the rest of the application components. If the model performance is found lacking, the AI model lifecycle starts again. MLOps activity also needs to keep the model and training data versions in sync so that any changes to the model in the future can be adequately tracked and audited.

Consider a situation where an ML model is trained using a set of training data and deployed onto a production environment. The AI then makes business decisions such as offering discounts to particular customers or estimating insurance premiums. At some stage, new data is used to train a new model that is then deployed into production. The new model will almost certainly behave differently, so some customers may no longer be offered a discount or insurance premiums may be higher for one group of customers. From an operational perspective, the enterprise needs to understand which model was in use at any point in time and what training data was used to develop that model.

In some ways, this could be considered to be just standard configuration management and best practice in configuration management so DevOps is certainly the best place to start when considering ML Ops. However, it's also important to think about some of the subtleties that ML brings to this problem. For example, what if a customer makes a request under GDPR for their personal data to be deleted? If that personal data exists in a training set, it needs to be deleted in accordance with the regulatory requirement. At that point, there is an immediate question about the ML model. The model is technically a summary of the training data. Hence, is there a requirement for the ML model also to be deleted and rebuilt? If the model does need to be deleted, then the behaviour of the system with that model is no longer re-producible. Consequently, we have an operational system that made operational decisions and we cannot re-create the conditions under which those operational decisions were made. If there are any historical concerns about those decisions, we can't understand exactly how and why they were made! These are important questions that the application owners have to consider and make business decisions that make sense legally and ethically.

APPLICATION LIFECYCLE

Now you have mastered the AI component, it is time to integrate it into your enterprise application. Since we have already discussed traditional application development at the beginning of this chapter, in this section we only focus on how the software development activities are influenced by the inclusion of the ML component in the application. We use the right side of Figure 10.2 and start at the bottom right with Business Requirements and go counterclockwise.

Business Requirements

This is a critical activity in the application lifecycle since it matches the application need to the capabilities of the AI component. The requirements should reflect the business goals (e.g. support multiple languages or countries), use cases (e.g. end user vs. domain expert) and system performance expectations (e.g. accuracy, runtime performance, etc.). It should also include any specific concerns about robustness, security, bias, ethics, human-level explanation and system transparency explicitly. These requirements must be considered during the PoC and Prototyping stages of the AI Model Lifecycle before embarking on the AI application.

To add to the complexity, there are situations when the requirements are uncertain. A practical example is when the system needs to work in the 'open world' (e.g. self-driving cars, again). It is simply not possible to anticipate all the different variables and their impact on the system performance. One way out of this is to think about defining properties (e.g. safety, security, etc.) of the system important for the business/mission and make sure we can characterise them and address them. Chechik [35] proposes a framework for managing uncertain requirements in three phases:

i. Identify phase that defines uncertainties.

ii. Assess phase that evaluates the degree of uncertainty and its relevance to the property of interest, such as safety.

iii. Address phase which comes up with an option to tolerate the uncertainty or reduce it by explicit action such as introducing redundancy in the architecture.

Functional Decomposition

This activity requires careful consideration of which task in the application can be reliably executed by an ML component. If the output of the ML task has a high consequence (e.g. human life) and the confidence in the ML output is low, then the task is not suitable for AI. Conversely, if the output has low consequence and high confidence, then it is an ideal task for AI. A critical requirement is the availability of data of adequate quality and quantity before an ML model building process is attempted. Any assessments and/or decisions about reusing existing AI components either from internal projects or from external sources are also needed. These considerations have significant impact on the PoC and Prototyping stages of the AI Model lifecycle.

Application Design

This activity must address all the components needed to build the application, not just the ML component. As discussed in Chapter 2, key contributors to the application complexity such as the number of AI components, interdependency of AI components and event-driven or context aware behaviour need to be assessed and understood. The application architecture must support the use of other mechanisms such as rule-based checkers as guardrails to make sure that the ML components are meeting the business/mission-critical objectives. The demand for computing resources at run time (i.e. when the users are actually using the application) has to be sensitive to form factors of edge devices (i.e. cell phones, tablets, etc.) and other networking limitations.

User Interaction

With the proliferation of smart phones and their role as popular application delivery channel, applications need to support productive human-machine collaboration and a better user experience. Beyond the traditional graphic user interface, the opportunity to leverage other user interaction modes that use AI (i.e. unstructured text or speech) is very attractive. As an example, many companies allow speech as the interface to interact with an automobile for specific tasks. There is no reason for today's business applications not to exploit this technology. This may even help with age old accessibility challenges with technology. The decision on which interaction paradigm to use depends on the quality of the AI component available as well as on the ability for a graceful recovery in the event of its potential failures during the interaction such as spelling mistakes and unfamiliar accent. If a decision is made to include an AI component (e.g. speech-to-text or text-to-speech) to support user interaction, it should be treated an additional ML component in the AI application lifecycle and put through proper AI development and evaluation process using realistic input data.

Non-AI Implementation and Testing

The role of non-AI components in building AI applications cannot be underestimated. As we noted already, significant IT infrastructure is needed to support the data pipelines and modelling activities of the ML component. In addition, depending on the sophistication of the business process supported by the AI application, there will be many critical functions (e.g. customer relationship management, user interfaces, privacy and security, visualisation, traditional analytics, etc.) to be developed. It is important to make sure that all the complementary pieces of the application are ready for integration with the AI components.

Application DevOps

Even though DevOps for traditional software development is generally well understood [36,37], the integration with the MLOps from the AI lifecycle needs some care. In the iterative or agile model of development, the full capability of the application may not be realised in the early stages of development, but only after many iterations/sprints. It will be necessary to keep track of versions of relevant non-AI software components and their versions to complement the versions of AI models and data to provide a view of the evolving application capability over many iterations.

APPLICATION INTEGRATION AND DEPLOYMENT

This represents the activities inside the grey box in the middle of Figure 10.2 where the '**Implementation**' phase of AI modelling (from the left) and the other application components (from the right) are integrated to create the full AI application and deployed for the enterprise. Such an integration may also need cloud services from external sources as shown at the top of Figure 10.2.

AI Application Testing

Testing AI applications is still an open research area, and it requires many innovations to be practically useful. In this section, we point out some common practical approaches and specific challenges.

Black Box Functional Testing

The model created during the AI model lifecycle using the training data needs to be evaluated to see if it can generalise to realistic data obtained during normal business operations. Consequently, black box testing requires people with a good understanding of the business data requirements. Typically, this step uses separate data sets (i.e. hold-out sets) *that are not used* in the model building process [19,20] and that the data distribution in the hold-out set must represent the business requirements. These 'hold-out data sets' may need to go through a separate process for verification, e.g. in a system that reads job application resumes to sift which candidates had the right qualifications, you would deliberately send a small percentage of the resumes to a human to validate the decision recommended by the AI model. If the model does not meet the business requirements, the development process must go back to the ML model lifecycle activities for improvements.

Errors Are Inevitable

Since ML models use the technique of 'statistical learning', there are some basic trade-offs [38] in the model building process that cannot be avoided, giving rise to some percentage of errors. If the input data contains some scatter in the output for the same input value, there is no way to avoid this in the inferred model. *This is called* 'irreducible error'. If the model used to represent the training data is consciously chosen to be simple (e.g. a linear regression) for easier explanation, it may not explain the detailed behaviour of the training data well. *This results in errors.* If the model selected is very complex with a lot of parameters (hundreds to millions depending on the problem), it fits the training data extremely well but cannot generalise to the test data easily, consequently *introducing errors.* So, you see, *errors are inevitable.*

Let us explain this with an example of recognising cats in pictures. You have trained a model to recognise cats during the ML lifecycle. During the application testing activity, you can give it more previously unseen pictures, some of which are cats, to test if it recognises them successfully. You can decide that it is production ready when it gets this blind test 100% correct. But (and it's a big but) even then you will never know if (a) it will recognise *all* future cat pictures when you put it into production (e.g. maybe your training photos did not include any ginger cats, there is no guarantee that the AI's definition of "catness" is not dependent on fur colour) or (b) if presented with a random picture say of a school bus it won't yell "it's a cat, it's a cat". Again, this is caused by the inscrutable nature of what pattern the AI has learnt from the training pictures you supplied. Added complexity comes from statistical nature of the ML algorithms which select outputs for given inputs typically based on confidence levels and hence not in a deterministic way.

Testing without Specifications

As we had mentioned earlier, the ML functions use the training data to map inputs to the outputs and so do not need explicit specifications. Testing such systems using the traditional verification techniques is simply not possible since there is no description of what the system is supposed to do. If errors are encountered, the first challenge is to decide if it really is a defect; this changes our whole thinking about quality management practices, as they cannot be defect centric.

Absolute testing is hard for any type of cognitive system, including humans! Can an examiner absolutely know that a rookie driver can handle a car in "every" situation they might encounter after passing the standard driving test? If we expect AI to deal with complex situations that require human-level judgement with many possible outcomes, it is simply not possible to define every possible behaviour. For example, we could define some form of statistical system that used weighted voting on recommendations from many subsystems (e.g. many sensor subsystems in a self- driving car) to make decisions (e.g. to stop or not). Such systems may not be using ML; however, it is still not possible to validate every possible output for every possible permutation of input. To be honest, if it was, then it would not be an AI application.

Knowing the Right Answer

A further consideration is that in many AI applications, it may not be possible to know what the correct outcome should be. In current web AI applications, it is relatively easy to know if the image classifier is correct or not. Did it correctly identify the cat in the photo? However, in more sophisticated applications performing tasks such as medical diagnosis, it is not that straightforward. If an AI decides a particular treatment plan, how do we know that an alternative treatment plan would not have had a better outcome?

The secret to AI in the enterprise is to understand early how you are going to measure performance when it goes live. Sometimes it is possible to build in a secondary test for the answer. The codebreakers in the Bletchley Park story had a simple secondary test: Was the decrypted message recognisable? That is, was it readable as a German language message? If no, then you hadn't broken the code. If yes, then you probably had. The probability of getting a correct decrypt was 1/150,738,274,937,250 [39], the chance you had got it wrong and still produced a readable but incorrect German sentence was even greater odds.

Consequence of Personalisation

Some AI applications (e.g. product recommenders) require customisation of the AI output to match the user profile. The correctness of such an output cannot be validated by a generic 'user acceptance test', but only by explicit or implicit user feedback. This can be captured by the acceptance of the recommendation by the users or by their disregard. Therefore, the validation of the AI model can only be done during the application monitoring activity during deployment. This is an interesting use of monitoring to be discussed in more detail below.

Testing for Trustworthiness

If the business requirements include concerns about trustworthiness of the AI application (i.e. fairness, explanations, robustness, transparency, etc.), it is important to have an independent team of testers (with a good understanding of the business and the application users) to test for them and document the results and any shortcomings. These are the new manifestations of potential defects in an AI application.

Operations

There are three areas in which the inclusion of an ML model in the AI application affects the traditional operations activity.

- The need for parallel management of the evolutions of ML model versus the rest of the application. The model needs close monitoring and more frequent updates compared to the rest of the application.
- Depending on the nature of the application, it may be difficult to judge if the observed application behaviour is correct or not. This is because we may not know the right answer at that time.
- Debugging needed to resolve problems with the behaviour of ML models is intrinsically more complicated (involving the details behind the training data and the ML algorithms) compared to looking for specific lines of offending code or configuration parameters.

Mandatory Monitoring

In AI systems monitoring is not an option, but a required activity. We refer to [40,41] for more details on the need for monitoring and the various practical approaches. Here we just summarise some key ideas.

Need for Monitoring AI Systems

Well, the bottom line is that an ML model can misbehave during deployment. The primary reason for this is 'Data Skew'. Simply put, the training data for the model is not representative of the live data seen in the deployed system, this problem can creep in over time. Here are some examples of how this can happen.

- *During model creation, the training data was not chosen carefully.* Distribution of the features in the training data does not match the distribution seen in the production.
- *A feature that was used as input to build the model is not available in production.* The only option here is to remove the feature in the training data or use an alternate feature that is already available in the production data or can be derived from the production data, with additional feature engineering.

- *Data from different sources do not match.* Data used to train the model came from a different source from the one used during production. Due to differences in feature engineering in the data pipelines, even the same features in the testing and live data may have different values.
- *Dependencies on other systems.* It is not uncommon for data scientists to use the data from other systems not directly under their supervision for model training. If these systems change the way they produce their data, unknown to the data scientists, there will be a negative ripple effect in the data use. For example, if there are changes to the government guidelines on benefits eligibility (e.g. age for social security changed), the data before and after the change will have inconsistencies that have to be reconciled.

Beyond these practical data and modelling issues, there are other reasons models may get stale or behave badly.

- **Changes in Business Context**: The historical data used to train the models does not represent the current behaviour of the population. For example, the business drivers have changed (e.g. increasing use of the internet for schools, work, etc.) since the eruption of the COVID-19 pandemic.
- **Changes in Social and Cultural Behaviour**: Over a period of time, behavioural patterns change in various domains (e.g. politics, fashion, music, entertainment, etc.). These are particularly significant in the behaviour of consumer-facing recommender systems.
- **Adversarial Attacks**: As described in Chapter 5, various types of attacks by adversarial actors can make your model behave badly. The impact of these attacks can be severe depending on how much familiarity they have to the modelling and data environment.
- **Continuous Learning During Production without Guardrails**: This goes back to Microsoft's Social Bot Tay that had to be shut down due to unacceptable behaviour due to learning directly from hateful user input data. We refer to the vignette on "Continuous Learning" for more discussion on this topic.

Hopefully, we have convinced you that *monitoring AI applications during deployment is absolutely critical for business success.* Since every change in the deployed model can be risky and expensive, it is important to monitor appropriate factors, detect any significant changes in model behaviour and put a process in place to correct them in a timely fashion. Now let us discuss some ideas on what to monitor more specifically.

CONTINUOUS LEARNING

For many people, AI is all about ML. Most non-AI specialists assume that all AI systems must include ML and quite often believe that learning in AI is a continuous process just as it is in human intelligence. There is a natural assumption that any AI system must be continually learning and continually improving its decision-making.

In reality, the need for predictable and assured behaviours in AI applications makes continuous learning impractical for most applications. There are three fundamental reasons for this.

Firstly, we need to think about the model itself. Being able to verify, analyse and explain historical decisions is important in delivering trustworthy AI, and to do this, we need to have access to the model used at the time a decision was made. That means either maintaining a continuous set of backups or implementing a mechanism to be able to re-create the model for any point of time. Both approaches require considerable resources and sophisticated engineering.

Secondly, there is the issue of testing and assurance. Normally, any model would be tested before being deployed to ensure that it operates as expected. If we update a model continuously, how in practice do we undertake the required level of testing in the time available?

Thirdly, AI applications can (and do) learn bad behaviour. In previous chapters, we have talked about bias and the risk of ML applications learning a bias that exists in historical data. If an AI is learning continuously, then there is a risk of learning biases that exist in society. Sometimes, this can be the result of a deliberately malicious act as experienced in 2016 when Microsoft's Tay Bot acquired the prejudices of Users.

Whilst continuous learning is attractive from the visionary AI perspective of truly emulating human intelligence, its use should be focussed in the right areas. Those AI applications that require fixed, version-controlled models that have undertaken some form of testing and assurance are clearly unsuitable for continuous learning.

Other applications, in particular recommender systems such as those used by companies such as Netflix and Amazon, are more suitable. Applications that are suitable are generally those where the decisions made are not critical enough to warrant repeatable and explainable decisions. In addition, it is helpful if the User is either constrained in their ability to influence the ML or invested in its success. For example, in shopping recommender systems, the ML is leveraging the human's buying decisions so there are limited opportunities for the User to mislead the system. In other applications, the recommendations are made to professionals such as doctors who are strongly invested in the proper use of the system.

What to Monitor

Ideally, we want to know the accuracy of an AI model running in production for every instance of its use. Only in some simple scenarios (such as in recommender systems), do we know the validity of the AI output, if there is live user feedback. In most other cases, we do not know the accuracy of a model immediately. Examples are fraud detection, healthcare treatment recommendation, predicting future housing prices, etc. Given these constraints,

MONITORING ENTITY EXTRACTION

The importance of monitoring AI applications was brought home to me in the early days of entity and relationship extraction. The business requirement was to develop an entity extraction tool for use in analysing a database of English language documents. The system was performance critical, and there was a very tight specification detailing the content that existed in the document repository.

The system extracted over 30 different entity types from the content and the most important entity requirement was to extract names. Recall was more important than precision and the Client stated that they did not wish to miss any names.

One of the most basic rules for identifying potential names is to test whether a word appears in the dictionary. If a word doesn't appear in the dictionary, there is a high probability it is a name. This simple rule is highly effective in finding names and is often used as a default in applications where recall is more important than precision.

The delivered solution included a monitoring component that generated statistics about the length of each processed document, the number of names extracted, the number of verbs and a whole host of parameters. Suddenly, one morning, the monitoring system flagged an alert. The documents being processed included a very high proportion of names for their size. On investigation, we discovered that a whole set of foreign language documents had been loaded into the repository.

Fortunately, an effective monitoring component ensured that we were able to intervene and correct the issue quickly.

This happened many years ago, and a modern-day entity extraction tool would almost certainly perform language identification as part of its processing. When our solution was deployed such an approach was not feasible.

Similarly, it could be argued that other controls, such as checking language at the point of ingestion, would have avoided the other issue. The key point, however, is that things change and it's not uncommon for an AI application to suddenly be applied in a context that it was not designed for. It is not always possible to anticipate every possible change of circumstances so it's important to ensure a monitoring capability is in place to identify when an AI may be applied incorrectly.

James Luke

we can only exploit what we can observe in the live system to form an indirect assessment of the quality of the AI models being served. Here are some examples of things to track:

- **Model Versions**: In a realistic scenario of frequent updates to models, there should be a clear mechanism to track the current version of the model being deployed with the corresponding version of the training data and the specific data pipeline used. This will allow a quick diagnosis of any configuration errors or auditing of the model outputs for specific instances.
- **Model Inputs**: If the input to the model is structured data consisting of attributes (features) and values, many standard data quality checks [42,43,44] such as missing or inconsistent values will apply. Simple analysis techniques such as statistical

distribution of numerical features or frequency tabulation of categorical features can help to get some understanding of the range and variance of the model input data during deployment. These can be compared to the similar analysis of the input data during training. Direct comparison of input data distributions in the training and operational data is hard due to high dimensionality of the data (i.e. large number of features) and potentially non-standard distributions. If the input to the model is unstructured data, say images, it will still be useful to capture some metadata (e.g. image size, contrast, resolution, etc.) and make sure that they match the attributes of the training data.

- **Model Outputs**: Since the correctness ('ground truth') of the output labels is not typically available during the operations, it is difficult to sense any model drift directly. However, in cases where the application output expects a user action (e.g. purchase the recommended product), it is easier to understand if the model is having the desired effect. In applications where the AI is recommending a decision (e.g. mortgage approval), it is possible to check for model fairness; analysing input data and output predictions against known features (e.g. race, gender, etc.) for bias detection and potential mitigation.

Sustenance

As mentioned in Chapter 2, we call the upkeep of the AI application 'Sustenance' instead of the traditional 'Maintenance' for software applications. There is an 'optional' nature to maintenance of traditional applications; application owners decide what changes go in during operations (e.g. hot fixes), what goes in the next release (e.g. bug fixes, enhancements, new functions, etc.) and how often to put out bug fixes or new releases. When it comes to AI applications, the 'Sustenance' starts immediately after the deployment of the AI application and continues till the application is withdrawn from operations. The constant monitoring, ongoing assessments and immediate actions that are needed to support AI applications should remind you of a 2-year-old mischievous child who never grows up.

There are further complicating factors to AI sustenance. Since the AI technology is changing rapidly, it is necessary to evaluate the available technology on a regular basis and update the components to reflect the state of the art. As discussed in Chapter 2, applications must be architected in such a way that switching of components is painless. If the enterprise is using external technology providers to support the AI application, it may be necessary to change technology providers more frequently based on their current capabilities. All these changes must be reckoned with the testing challenges mentioned above.

PROJECT MANAGEMENT

In Figure 10.2, Project Management covers all activities in the AI application development. Typical project management involves definition, assignment and management of resources; tasks; schedules and stakeholders to meet the business goals. Tracking progress, managing the risk & outcome through project execution are key elements. The introduction of an AI component creates some new twists to this otherwise 'established' process.

- **Two Lifecycles to Manage**: Application lifecycle versus AI Model lifecycle. While the AI Model lifecycle lifecycle is critical to the project, it also bears the most uncertainty on the quality and timeliness of the delivered AI model(s).
- **Data Strategy & Governance**: Availability of data in adequate quality and quantity to build models, data provenance throughout the life of the application (which can be many years), data storage & retrieval, relevant data privacy and security are all new aspects introduced by the AI component.
- **Business Goals versus AI Performance**: Typical data scientists claim success based on the accuracy of the AI models they create. Business goals (see Chapter 5) revolve around value of the technology in a tangible way such as increased revenue and reduced cost.
- **Managing Trust Expectations**: Understanding the trust requirements (fairness, explanations, etc.) of an AI application at the beginning of the project is critical for its success. This is something completely new for the AI applications.
- **Persistent IT Infrastructure**: Due to the demands for reproducibility and regulatory auditing, the data pipelines, model versions and the supporting infrastructure have to be maintained for the life of the application.

AUDITABILITY AND EXPLAINABILITY

Throughout this book, we have talked at great length about the differences between conventional software applications and AI applications. One of the most fundamental differences is the fact that AI systems make decisions that are more akin to the judgements made by human decision makers. The fact that the decisions made by AI applications may be contentious, together with the importance of the decisions, means that Auditability and Explainability become critical considerations.

If we are to trust AI Applications, we must be able to review decisions retrospectively. For example, if the system is shown to have been biased, it may be necessary to identify and correct historical decisions. This is not as easy as it sounds when the system is constantly evolving in behaviour as in a continuous learning environment, or where you have failed to track model changes, or have not kept the training data. The dynamic and statistical nature of AI systems creates massive audit and traceability challenges.

To enable retrospective reviews, we are going to have to log far more information than in conventional software applications. We are going to need to keep thorough logs of all data used in designing, developing, testing and operating the AI application. Any changes to the environment such as hardware configurations can alter model outputs such that it may not be identical to the original model; but it should always be close.

Ultimately, when leading an AI project, the key question you need to consider is whether you can recreate the circumstances in which a contentious decision was made. Are you able to produce the training data that was used to construct a ML Model? Do you know how the data was labelled and can you validate that the labelling was undertaken by a suitably qualified person? In cases where the ML algorithms are sensitive to the order in which training data is presented, different hardware configurations can change the resulting model. Are you able to re-run the training process and recreate the model that was responsible for the contentious decision? Are you able to recreate the operational environment in which the contentious decision was made?

In setting up your AI project, it's critical that you consider these questions and ensure you have the correct auditing capabilities to be able to explain why your application behaved the way it did.

SECURITY

Security is another of those (not so) boring considerations that must be considered throughout the delivery of any project. In terms of security, AI is just like many other emerging technologies; it creates a whole new set of risks.

For anyone enthralled by AI, understanding emerging security risks adds a whole new dimension to this fascinating subject. A detailed study of the sub-field is beyond the scope of this book. For those who are interested in a deep dive, we recommend [45].

To whet your appetite, here are some simple examples of things you may wish to consider when running your AI projects:

- AI Models are, in many cases, a summary of the training data used to create them. Techniques are emerging that allow an attacker to re-construct training data using only the model. This means that you must be careful about who is able to access the Model and the data used to construct it. If you develop a Model using highly sensitive data and then deploy the Model in a public environment, you may be giving away sensitive data.

- Data is everything in AI (we may have mentioned that before) and data actually determines the behaviour of many AI applications. If your source data is not secure, an adversary may be able to tamper with the data and, as a result, change the functionality of your application.

- Whilst an adversary may not have direct access to your development environment, they may actually be the supplier of your data. This effectively gives an adversary indirect access to your development environment. Consider a military system that operates with an ML-based classifier. An adversary may deliberately behave in a certain way in peace time, for example, transmitting on certain radio frequencies, to ensure the classifier behaves in a certain way. In war time, the adversary may then change their behaviour in war time in order to fool the classifier.

- People are often the weak link in IT security. AI projects are heavily dependent on people to label data. This creates an opportunity for an adversary to have an unwanted influence on the effectiveness and behaviour of your application.

As in conventional application development, security is critical yet may feel like an expensive investment with little return. The only return is that nothing happens. However, it is absolutely critical and don't underestimate the importance of ensuring your data and models are protected.

IN SUMMARY – THE BORING STUFF ISN'T REALLY BORING

The use of AI brings a new perspective to the routine tasks required in any software project.

- Applications with ML components are very different from traditional software applications.
- ML components do not have traditional functional specifications that map program inputs to corresponding outputs. The functional behaviour is completely determined by training data.
- Statistical learning implemented with ML models is bound to produce errors allowed by the limitations of data and algorithms – you just have to accept this fact and build mitigation plans.
- Understanding, as early as possible, the level of trustworthiness needed by your target application could make the difference between success and failure.
- Application monitoring during deployment is mandatory!

REFERENCES

1. J. Desjardins, "How many millions of lines of code does it take?" (February, 2017) https://www.visualcapitalist.com/millions-lines-of-code/.
2. V. Antinyan, "Revealing the complexity of automotive software," *Proceedings of the 28th ACM Joint Meeting on European Software Engineering Conference and Symposium on the Foundations of Software Engineering*, pp. 1525–1528 (2020).
3. *INCOSE Systems Engineering Handbook-A Guide for System Life cycle Processes and Activities*, 4th Edn. Wiley (2015).
4. C. Shamieh, "Systems engineering for dummies," IBM Limited Edition, John Wiley & Sons, Hoboken (2012).
5. F. P. Brooks, *The Mythical Man-Month: Essays on Software Engineering*, Anniversary Edn. Addison-Wesley Longman, Reading (1995).
6. S. McConnell, *Code Complete: A Practical Handbook of Software Construction*, 2nd Edn. Microsoft Press, Redmond (2004).
7. N. Cerpa and J. M. Verner, "Why did your project fail?" *Communications of the ACM*, 52(12), pp. 130–134 (2009).
8. L. Northrop, et al., *Ultra-Large-Scale Systems: The Software Challenge of the Future*, Software Engineering Institute (2006). https://resources.sei.cmu.edu/asset_files/Book/2006_014_001_635801.pdf.
9. D. Scully, et al., "Machine learning: the high-interest credit card of technical debt," *Software Engineering for Machine Learning Workshop*, NIPS (2014).
10. R. Akkiraju, et al., "Characterizing machine learning process: A maturity framework," https://arxiv.org/abs/1811.04871 (2018).
11. A. Arpteg, et al., "Software engineering challenges of deep learning," *Proceedings of the 44th Euromicro Conference on Software Engineering and Advanced Applications (SE-AA)*, pp. 50–59 (2018).

12. S. Amershi, et al., “Software engineering for machine learning: a case study,” *ICSE-SEIP’10 Proceedings of the 41st International Conference on Software Engineering: Software Engineering in Practice*, pp. 291–300 (2019).
13. A. Horneman, A. Mellinger, and I. Ozkaya, *AI Engineering: 11 Foundational Practices*, CMU Software Engineering Institute (2019).
14. P. Santhanam, “Quality management of machine learning systems,” *In International Workshop on Engineering Dependable and Secure Machine Learning Systems, pp. 1-13. Springer, Cham, 2020.*
15. I. Ozkaya, “What is really different in engineering AI-enabled systems?” *IEEE Software*, 37(4), pp. 3–6 (July-August, 2020).
16. J. Bosch, I. Crnkovic, and H. H. Olsson, “Engineering AI systems: a research agenda,” arXiv:2001.07522 (2020).
17. W. C. Benton, “Machine learning systems and intelligent applications,” *IEEE Software*, 37(4), pp. 43–49 (July-August, 2020).
18. P. Aroonvatanaporn et al., “Reducing estimation uncertainty with continuous assessment: tracking the ‘cone of uncertainty’,” *ASE’10: Proceedings of the IEEE/ACM International Conference on Automated Software Engineering*, pp. 337–340 (2010).
19. A. Ng, “Machine learning yearning,” https://www.deeplearning.ai/programs/.
20. M. Zinkevich, “Rules of machine learning: best practices for ML engineering,” Google Blog: https://developers.google.com/machine-learning/rules-of-ml/.
21. A. Chakarov, et al., “Debugging machine learning tasks,” arXiv:1603.07292.
22. R. Lourenço, J. Freire and D. Shasha, “Debugging machine learning pipelines,” *DEEM’19: Proceedings of the 3rd International Workshop on Data Management for End-to-End Machine Learning*, pp. 1–10 (2019).
23. F. Hohman, et al., “Visual analytics in deep learning: an interrogative survey for the next frontiers,” *IEEE Transactions on Visualization and Computer Graphics*, 25(8), pp. 2674–2693 (2019).
24. E. Breck, et al., “The ML test score: a rubric for ml production readiness and technical debt reduction,” *IEEE International Conference on Big Data (Big Data)*, 1, pp. 1123–1132 (2017).
25. M. Hind, “Explaining Explainable AI,” *XRDS: Crossroads, The ACM Magazine for Students*, 25(3), pp. 16–19 (2019).
26. R. Guidotti, et al., “A survey of methods for explaining black box models,” *ACM Computing Surveys*, 51, pp. 1–42 Article no. 93 (2018).
27. V. Arya et al., “AI explainability 360: an extensible toolkit for understanding data and machine learning models,” *Journal of Machine Learning Research*, 21, pp. 1–6 (2020).
28. S. Verma and J Rubin, “Fairness definitions explained,” *IEEE/ACM International Workshop on Software Fairness (FairWare)* (2018).
29. R. K. E. Bellamy, et al., “AI fairness 360: an extensible toolkit for detecting, understanding, and mitigating unwanted algorithmic bias,” *IBM Journal of Research and Development*, 63(4/5) (2019).
30. H. Xu, et al., “Adversarial attacks and defenses in images, graphs and text: a review,” arXiv:1909.08072 (2019).
31. IBM Research Blog “The adversarial robustness toolbox: securing AI against adversarial threats,” https://www.ibm.com/blogs/research/2018/04/aiadversarial-robustness-toolbox/.
32. M. Arnold, et al., “FactSheets: increasing trust in AI services through supplier’s declarations of conformity,” *IBM Journal of Research and Development*, 63(4/5) (2019).
33. MLOps: Continuous delivery and automation pipelines in machine learning https://cloud.google.com/architecture/mlops-continuous-delivery-and-automation-pipelines-in-machine-learning.
34. D. Sato, A. Wider and C. Windheuser, “Continuous delivery for machine learning,” https://martinfowler.com/articles/cd4ml.html.

35. M. Chechik, "Uncertain requirements, assurance and machine-learning," *IEEE 27th International Requirements Engineering Conference (RE)* (2019).
36. L. Bass, I. Weber and L. Zhu, *DevOps: A Software Architect's Perspective*, SEI Series in Software Engineering, Addison Wesley (2015).
37. G. Kim, P. Debois, J. Willis and J. Humble, *The DevOps Handbook: How to Create World-Class Agility, Reliability, and Security in Technology Organizations*, IT Revolution Press (2016).
38. T. Hastie, R. Tibshirani and J. Friedman, *The Elements of Statistical Learning*, Springer (2017).
39. Cryptanalysis of the Enigma https://en.wikipedia.org/wiki/Cryptanalysis_of_the_Enigma.
40. C. Samiullah, "Monitoring machine learning models in production-a comprehensive guide," (March 14, 2020) https://christophergs.com/machine%20learning/2020/03/14/how-to-monitor-machine-learning-models/.
41. J. Verre, "Monitoring machine learning models," (May 28, 2020) https://towardsdatascience.com/monitoring-machine-learning-models-62d5833c7ecc.
42. E. Breck, et al., "Data validation for machine learning," *Second SysML Conference* (2019).
43. S. Schelter, et al., "Automating large scale data quality verification," *Proceedings of the VLDB Endowment*, 11(12) (2018).
44. S. Shrivastava, et al., "DQA: scalable, automated and interactive data quality advisor," *IEEE International Conference on Big Data (Big Data)*, pp. 2913–2922 (2019).
45. S. Qiu, et al. "Review of artificial intelligence adversarial attack and defense technologies," *Applied Sciences*, 9, p. 909 (2019).

CHAPTER 11

The Future

Prediction is very difficult, especially if it's about the future.

Niels Bohr

How do you write a single chapter on the future of AI? Futurologists, science fiction writers, brilliant academics and commentators in all different shapes and sizes have written volumes on the subject. Some of the greatest thinkers in science and technology from Kurzweil to Hawking and Musk have made bold predictions about the future of AI including bold claims about the advent of the singularity. For those not familiar with the concept of the Singularity, the theory is that there will be a point in time when AI becomes intelligent enough to create better AI. At that point, AI will just keep creating more and more intelligent versions of itself achieving levels of intelligence beyond human comprehension. The question, of course, is what happens to us when AI takes over?

Given so many fascinating, amazing and quite terrifying predictions, what can we possibly add that others haven't already covered? Perhaps, a sense of perspective and a dose of reality of AI in the Business Enterprise as opposed to the Starship Enterprise.

For a start, AI as we know it, is way off being "intelligent". Sure, we can teach it complex things, and those actions can look intelligent, but the more you understand it, the more you realise how far away from real intelligence it actually is. Society seems to have no problem in accepting an airplane's autopilot as unintelligent, even though it can take off, fly and land a 747 jet better than any human who is not a pilot. AI is what we teach it, and nothing more or less. AI is an incredibly useful asset for business, we haven't yet scratched the surface of all the complex processes we could be automating with it, but it is not about to achieve consciousness and enslave humanity.

Throughout this book, we have attempted to focus on the practical application of AI to real-world problems. Continuing in that spirit, we will finish this book by reflecting on a few significant aspects of the future of AI in the Enterprise.

DOI: 10.1201/9781003108498-12

HOW CLOSE ARE WE TO THE SINGULARITY?

The prospect of the Singularity is something that terrifies and excites me both at the same time. There have been many articles, books and YouTube videos made on the subject of the Singularity and the danger to humanity from intelligent machines.

Assuming you haven't just dropped this book and run away to live a technology-free life in the mountains, perhaps it's worth considering when we think the Singularity will be achieved. Alan Turing predicted in 1950 [1] that machines would answer questions in a way that was indistinguishable from a human expert within 50 years (i.e. by the year 2000). In 1967 Marvin Minsky said [2], "Within a generation... the problem of creating artificial intelligence will be substantially solved". I think we would all agree that both predictions have been shown to be extremely optimistic.

There have been many other predictions [3] which place the Singularity around 2045. At the time of writing, that is 24 years from now... my co-authors and I have been working in AI for over 25 years, so we have a reasonable grasp on how much progress is achievable in the predicted timeframe. Since the early 1990s, it is clear that we have come a very long way in the core sensory tasks and classification. Sadly though, machines still struggle with tasks that require higher levels of reasoning. The type of tasks that are incredibly simple for humans but are extremely challenging for machines. A really simple example is pronoun resolution in documents. We resolve pronouns because we know obscure facts or can inference about biographical information. We understand words being spoken in a noisy environment by understanding the context of the conversation, the facial expressions of the speaker and many other hidden clues.

As a simple exercise, try reading children's story and think about how you understand even the most basic of sentences. You will find that your brain is leveraging a vast knowledge base... the fact that a tortoise walks slowly or that cats like milk... and picking up clues from the accompanying pictures... such as the surprised look on the face of a butterfly. Your brain is able to fuse the text in the sentence with the clues in the pictures and use your vast knowledge base to understand the story. There isn't yet an AI system that can come close to that capability. Whilst we are making incredible progress in AI, we are still a long way from the higher-level reasoning skills required to enable the Singularity.

It's important to remember though that we can't predict the future through simple extrapolation. Technological breakthroughs tend to result in step changes rather than gradual evolutions.

For the Singularity to happen, AI needs to deliver capabilities that can fuse input from multiple sources including a vast and deep knowledge base. For me, one of the first indicators that we are making progress in this area will be when machines are able to leverage knowledge and common sense to perform basic resolution tasks in areas such as text analytics and speech recognition.

Putting aside our earlier caveat about predicting the future, 2045 does appear to be very optimistic!

James Luke

REFERENCES

1. A.M. Turing, "Computing machinery and intelligence," Mind, New Series, 59(236), pp. 433–460 (October 1950).
2. M. L. Minsky, "Computation: Finite and Infinite Machines", Prentice Hall (1967)
3. P. Rejcek, "Can Futurists Predict the Year of the Singularity?", Mar 31, 2017. https://singularityhub.com/2017/03/31/can-futurists-predict-the-year-of-the-singularity/

IT'S ALL ABOUT THE DATA – TRENDS IN THE ENTERPRISE

In Chapter 7, we talked about data and its importance in delivering real enterprise AI applications. Given the importance of data, we should expect a great deal of future AI research and development to focus on this problem.

Big Data Meets AI

Before the buzz over AI in the last decade, big data analytics was a popular topic in the enterprise. The four key 'V-attributes' of big data (i.e. Volume, Velocity, Variety, Veracity) introduced considerable challenges to the enterprise, since getting timely business value was difficult simply based on human efforts. In this context, introduction of AI to automate business use of the data can be strategic and valuable. It so happens that the creation of AI needs large amounts of data. So, this could be an ideal marriage of the technologies to produce value to the enterprise. There are some key advantages and challenges to this marriage [1,2]. On the positive side, it will facilitate new capabilities such as pattern learning at multiple granularities and causal inference based on chains of sequences. However, on the negative side, it brings complexities such as large data dimensionality, demands for model scalability, support for distributed computing and streaming data. For example, popular machine learning (ML) approaches are designed for smaller datasets, with the assumption that the entire dataset can fit in the memory. So, the full realisation of this enterprise dream will require processes and tools to enable seamless integration of the big data and ML environments.

Getting Data

The value of data has clearly been recognised by the large web companies who aim to capture as much data about their users as possible. The nature of their business is such that capturing data at a massive scale is challenging but achievable. The web companies focus huge resources on very specific problems such as enterprise search. They have tens of millions of Users who return on a daily basis and use technology that enables tracking of their behaviour. Much of the data captured and required by the web companies is 'new data'; it is generated by and then used by the web ecosystem in which the web company's applications exist.

For the business enterprise, the first challenge is to obtain data. It is not uncommon for an enterprise AI application to require data that is not at present available to the enterprise. A simple example of this is weather data. A retail organisation may wish to develop an ice cream sales prediction algorithm that requires up-to-date weather data. This data will need to be either collected from scratch or more likely purchased, and it should be no surprise that there is a growing market for data and data services [3]. We should expect a massive growth in this business as companies increasingly buy and sell data from one another.

Managing Enterprise Data

In the enterprise, the business problems are typically more complex, or at least more consequential, than their web company equivalents, and simultaneously the cleansing and normalisation of the data are more challenging. Enterprise data is often highly complex, specialised and stored in disparate systems. Enterprise data is often very old and has been captured over many decades. It should therefore be no surprise that the growing discipline of data science is delivering new tools, methods and algorithms specifically for cleansing and managing enterprise data [4].

Whilst data scientists have made a great start in developing new tooling, there is still a long way to go. We are only just starting to address the challenges of managing enterprise data. Two particular areas of concern will be privacy and scale, scale in terms of computation (see the next section to see how compute-hungry AI is becoming) and in terms of numbers of AI projects. Small AI programs, each making small autonomous decisions, will proliferate in the enterprise, this will take some looking after. As the number of AI applications grows within the enterprise, the sophistication of the tooling to manage the supporting data sets will need to increase.

From a privacy perspective, increasing public awareness and regulatory pressures will mean that enterprises need to be able to respond to regulatory requests to explain decisions or redact data. As already explained, redacting data will impact training data sets and the ability to explain historical decisions. Tooling is going to be required to manage these complex issues.

Synthetic Data

Ultimately, the best way to address the privacy issues is to use data that does not include any personal information. This can only be achieved by either anonymising real-world data or creating completely new synthetic data. In Chapter 7, we discussed the pros and cons of using synthetic data. One thing that we can predict with confidence is that arguments about the use of synthetic data will continue long into the future. The fundamental argument will continue to be that the generation of representative synthetic data is only possible if you have sufficient knowledge about the underlying system; if the knowledge exists to generate the representative data, then there is no need to use AI. Unfortunately, this argument assumes that AI is only used to model systems rather than to make decisions about those systems. If we look at other forms of engineering, the use of simulators to test and evaluate systems is both widespread and essential. We do believe that AI engineering will increasingly follow a similar strategy and there will be an increase in the use of synthetic data.

EFFICIENT COMPUTING FOR AI WORKLOADS – NEW PARADIGMS

The phenomenal growth of computing and the Information Technology industry over the past many decades has been primarily due to three intersecting trends in hardware technology:

- Moore's law of exponential reductions in the cost per transistor [5]
- Dennard scaling law [6] that allowed the possibility of smaller transistors that are also faster & lower in power
- Creative exploitations of the classic von Neumann computing architecture [7]

These resulted in the impressive performances of the microprocessors (CPUs) and Graphical Processing Units (GPUs). GPUs allow a high degree of parallelism for AI workloads and therefore significantly enhance the throughput.

In recent years, however, device size scaling has slowed down due to power and voltage considerations, while it has become simply too costly to guarantee perfect operations by the billions of transistors on a chip. At the same time, each new AI algorithm has grown more hungry for this compute power, and even with the use of ad hoc cloud computing, it may be uneconomic for all but the biggest businesses to engage in the latest AI developments. Many of the emergent methods and techniques detailed in this chapter have been chosen because they hold the promise of improving the relationship between compute power and complexity (a fancy way of saying they may allow smaller businesses to develop sophisticated AI without spending fortunes on data storage or CPU costs).

Figure 11.1 shows the computation needed for some popular AI applications over the past many decades. In contrast to the prior decades which followed Moore's law (i.e. computing operations doubling every 2 years), the computing operations for the AI applications in the last decade are doubling every 3.4 months [8]! While the earlier AI algorithms could be run on a laptop or a single server, newer algorithms require exceptionally large

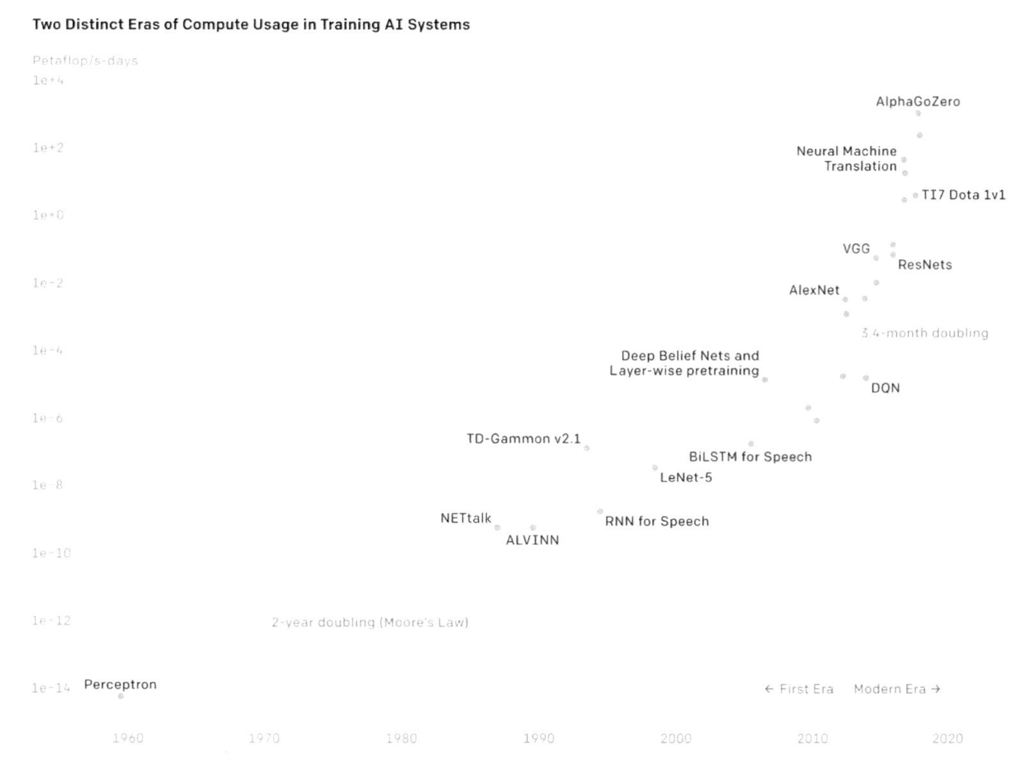

FIGURE 11.1 A plot of logarithm of PetaFlops/s-days versus calendar year for training popular AI applications. Prior to 2012, the AI computation was following Moore's law of doubling every 2 years. Since 2012, the computation in the largest AI training runs has been doubling every 3.4 months, thus increasing by more than 300,000x in that period. 1 PetaFlop/s-day = 10^{15} neural net operations per second for a day. (From reference [8] reproduced with permission.)

computing resources in data centres (i.e. special hardware, extensive cloud computing, etc.) with associated substantial energy consumption [9,10].

To meet this challenge, recent hardware developments for deep learning (DL) show a migration from a general-purpose design to more specialised hardware to improve compute efficiency using accelerators and new architectures. Here are four examples of recent advances in hardware technology that can contribute to more computing efficiency for AI workloads and/or lower power consumption.

- Reducing the numerical precision of data and computation is extremely effective for accelerating DL training workloads, while saving significant computing time and power. Sun et al. [11] proposed a hybrid 8-bit Floating Point (HFP8) format and end-to-end deep neural network (DNN) distributed training procedure. Using HFP8, they demonstrated successful training of DL models (e.g. Image Classification, Object Detection, etc.) without any loss of accuracy. This paves the way for a new generation of 8-bit hardware to support robust neural network models.
- Esser et al. [12] used a brain-inspired neuromorphic computing architecture based on spiking neurons, low-precision synapses and a scalable communication to achieve unprecedented energy efficiency. They implemented deep convolution neural networks that approach the state-of-the-art classification accuracy across eight standard datasets in vision and speech, while preserving underlying energy efficiency and high throughput. This approach makes it possible to combine powerful DL algorithms with the energy efficiency of neuromorphic processors.
- Neural network training built on traditional von Neumann computing architecture can be slow and energy intensive, due to the need for moving large volumes of data between memory and processor (the 'von Neumann bottleneck'). While the use of analogue non-volatile memory to perform the calculations *in situ* can accelerate the training by avoiding the data movement, the resulting model accuracies are generally less than those of software-based training. Ambrogio et al. [13] demonstrated mixed hardware–software neural network implementations that achieve accuracies equivalent to those of software-based training on various commonly used datasets. The estimated computational energy efficiency and throughput for their implementation exceed those of today's GPUs by two orders of magnitude, leading to a future of hardware accelerators that are both fast and energy efficient.
- When the feature space becomes large, traditional ML classification algorithms (e.g. Support Vector Machines) become computationally expensive. Quantum algorithms exploit an exponentially large quantum state space through controllable entanglement and interference. Havlíček et al. [14] demonstrated two novel methods on the superconducting quantum processor to speed up the processing, thus heralding a new class of tools that use noisy intermediate scale quantum computers to ML.

These are just examples of work going on in IBM Research to explore alternatives to the current computing paradigms to support ML workloads. It is quite possible that these approaches can provide a faster and more energy-efficient alternative to the GPU-intensive calculations of today in the near future.

ADVANCES IN ALGORITHMS – TARGETING DATA CHALLENGES AND NEURO-SYMBOLIC AI

In Chapter 3, we discussed various algorithms for AI applications. Considering that there are hundreds of research papers published every day with new algorithms for various purposes and much of it is available as open source, it is not hard to imagine that there will be plenty of new algorithms for use in business applications in the near future. Since our focus is on building real business applications for the enterprise, we want to highlight a few areas worth your attention.

Algorithmic Solutions to the Data Challenges

Much of the recent success in ML is in supervised ML where there is labelled data available of sufficient quality and quantity. In the future, we cannot assume that this will always be the case for four practical reasons:

- It may be expensive to get large amount of labelled data due to the business cost. This is particularly true where there is no existing business process to capture the data naturally. You are left with the option to have skilled business experts invest their time to do the labelling when the business value of the AI is not so clear.
- Emergent behaviour in the business can create new artefacts over time that may not be known at the training time. Think of an AI application that is designed to identify a dozen different financial fraud patterns in traditional banking, only to find out that there are new types of frauds using digital bitcoins that they did not anticipate.
- There may be rare instances that do not allow capture of adequate data. With the proliferation of internet connected devices (we mean, refrigerators, dish washers, laundry washing machines, etc., really), it is attractive to come up with an AI application to do preventive maintenance on these devices that predicts which machines are going to fail in the next month and what parts are needed to fix them. If some of these machines turn out to be very reliable (like the old days), there may not be adequate failure data to train the AI models. Another practical example of this is the challenge in creating Natural Language Processing (NLP) technology for more than 7000 languages in the world [15], when most business applications can support only a very small fraction of them in practice.
- Privacy and regulatory constraints may prevent access to the required data. A very powerful example of this exists in healthcare where massive volumes of patient data already exist; however, it is not possible for privacy reasons to consolidate this data into a single store and make it easily accessible to AI engineers.

These situations require different approaches to ML algorithms, sensitive to the availability and quantity of training data. In the sections below, we discuss examples of different algorithmic approaches that address this challenge and may help to alleviate data-related problems in AI projects. To make the discussion more concrete, we use examples of applications.

Learning with Pretrained Language Models

NLP has gone through an amazing revolution in the past 5 years [16,17]. In the good old days, NLP methods used to rely on discrete hand-crafted features for specific applications. Recent neural network-based approaches can learn universal language representations (called embeddings or neural language models) from large generic data corpuses. Such pretrained models can then be used by downstream applications such as question answering, sentiment analysis, named entity resolution and machine translation. While the progress in generic language (i.e. Wikipedia, news articles, etc.) analysis has been impressive, in the enterprise context, domain-specific language analysis is vital for business applications. Here is an example of analysis of domain-specific language content in the IT industry.

The cost of software system outages in production can be very high. Hence, predicting anomalous system behaviour ahead of failures is a very valuable goal. Good news is that IT operational environment generates many different data types (e.g. performance metrics, alerts, events, logs, tickets, etc.) to help with various preventive, debugging and support activities. However, the text data generated are very different from typical natural language texts because the vocabulary of the IT Operations domain is specialised. Liu et al. [18] used 3 million sampled logs from 64 different applications and created a neural language model to represent the log data. They showed that the ML models pretrained with log data as features outperformed those pretrained with generic natural language corpus in predicting anomalous system behaviours.

While the larger companies with dedicated data science teams, necessary data and computing infrastructure can afford to build their own domain-specific embedded language models, smaller companies may need help to make progress. We hope that there will be new technology and tool support to make this possible.

Zero-Shot Learning (ZSL)

Zero-shot learning (ZSL) is about creating ML algorithms that need *no labelled training data* for the task. Consider language translation. The idea is that you give the textual input in the source language (say, English), and the output from the application is in the target language (say, Spanish) and vice versa. Languages are complicated concepts with a lot of underlying structure, ambiguities, nuances, etc. Scope of your application can be something basic, such as assistance with elementary phrases for Londoners visiting Madrid for a few days, or a lot more demanding (i.e. translating newspaper articles, automatic translation of English language books, etc.). Typically to train the system, depending on the scope, you need to give examples of sentences in English and the corresponding proper translations in Spanish. This is what is called “labelled data” or “the ground truth”. Bigger the scope, more data you need. Clearly, the same labelled data can be used to build a model in both directions, i.e. English to Spanish and Spanish to English.

Now, let us say our Londoner also wants to visit Paris and needs help with English-French translation also. This means we need to give examples of training data and build a model for the English-French translation. Let us say we did that. Now, Londoner has a Parisian friend who wants to visit Madrid. Do we need to train a new model with the labelled data for French-Spanish translation? As the number of languages increase, the number of models for pair-wise translations increase combinatorically, six models for four languages, ten models for five languages, etc.

The most straightforward approach to translate between languages where there is no direct training data available is to use a 'bridge' language. If we want to translate between languages A <–> B, use language C as the bridge. First translate between A <–> C and then between C <–> B. The bridge language is often English since the data is more readily available. Two potential concerns about this approach are total translation time doubles due to the two translations needed, and there is a potential loss of quality introduced by the bridge language.

The algorithms in Google's Multilingual Neural Machine Translation System [19] make zero-shot translation possible between two languages for which direct labelled training data is not available. The key idea is the creation of a 'universal' language representation that helps to mediate between two languages. Obviously, the underlying model can do zero-shot translation only between languages it has seen individually before as source and target languages during training at some point but not for entirely new languages. We can think of this approach as an efficient form of transfer learning. Google Translate [20] does the translation based on this technology for more than 100 languages.

This is an example of the use of ZSL with real business impact. The technology obviously required substantial investment by Google to develop and deploy and is available as web services. How such algorithms can be created in other application areas is a topic of active research [21].

Few-Shot Learning (FSL)

Few-Shot Learning (FSL) is based on the simple idea that humans can learn from just a few examples, and we should be able to find ways to teach machines to do the same. Given the expected sparsity of data, any technical approach for FSL must implement or leverage prior knowledge in the data, model or the algorithm. We refer to [22] for a recent review of different technical approaches. We illustrate the technique with an image classification example in clinical pathology.

Use of DL techniques has revolutionised the field of biomedical imaging analysis demonstrating diagnostic accuracy at expert level [23] for many different tasks, such as diabetic retinopathy screening and skin lesion classification. To achieve such unprecedented performance, they normally require several thousands of well-labelled images for training. Typical clinical settings usually produce small datasets. In addition, obtaining large datasets might not be available for early clinical studies, for rare diseases or for new imaging modalities.

Medela et al. [24] demonstrated the use of FSL to classify tumours from the images of tissues in colon, breast and lung. They used two datasets:

i. Source domain dataset consisting of digital images of tissue samples obtained from primary tumours of colorectal cancer, divided into eight different types/classes of textures, with 625 image samples for each, with a total of 5000 samples

ii. Target domain dataset consisting of a total of 1755 digital images of healthy and tumoural samples of tissues from colon, breast and lung

The goal of the FSL was to extend the base model of eight classes to five new classes: one new type of colon tumour, two types of breast tumour and two types of lung tumour. They varied the number of target image samples for each of the new classes to see their impact on classification accuracy. Their results showed that their few-shot approach was able to obtain an accuracy of 90% with just 60 training images, even for the lung and breast tissues that were not present on the source training dataset.

FSL can also be thought of as another transfer learning technique, leveraging the information contained in the source domain to complement the information contained in the few training samples in the target domain. When the source and target domains are very different, FSL becomes less useful.

Federated Machine Learning (FML) in the Enterprise

Here are two enterprise examples [25] where having a centralised data repository of all the data needed in one location to create AI models is simply not possible:

- A multinational corporation (e.g. bank) with operations that are loosely coordinated across national boundaries subjected to local governance (e.g. General Data Protection Regulation in the European Union) on the data collection and usage
- A large hospital system with patient data in multiple sites subjected to regulations (e.g. HIPAA) that restrict access and movement of data

The clue is in the word 'Federated'. With Federated Machine Learning (FML), instead of bringing all the data to a single location to build/train your AI model, you can leave the data where they are and bring the algorithm to it. Once the local models are trained with local data, you can exchange models which are much smaller in size and do not have the same sensitivities as the raw data. Preservation of privacy in building models is a key requirement for its success. This will be huge for business in many ways, not just in network and data storage costs, but security as well.

For domains such as healthcare, this will be a game changer. It will mean that researchers can build analytics on patient data, potentially held in multiple hospitals, without ever needing to see the raw data. The patients' privacy is completely retained as their data never leaves the hospital's firewall, only the anonymous model code. If that raw patient data also happens to be large (e.g. CT scans which can be over a terabyte each), the cost savings in not having to move thousands of CT scans across a network or the internet each time you need to test a ML model will also be significant.

This is not just a pie-in-the-sky vision. Roth et al. [26] investigated the use of FML to build a model for breast density classification based on the Breast Imaging, Reporting & Data System (BIRADS) in a real-world collaborative setting with seven clinical institutions across the world. Despite substantial differences in the datasets (mammography system, class distribution, and size of data), they demonstrated that FML performs 6.3% on average better than their counterparts trained on an institute's local data alone. In addition, there was 45.8% relative improvement in the models' generalisability when evaluated on the other participating sites' testing data.

Here is an example of a tool kit that supports FML [27]. Who knows what new advancements will be made when the use of FML becomes an every-day practice across the institutions and application domains?

Edge AI

'Edge' is the brave new world of Internet of Things (IoT) devices such as mobile phones, wearable devices and automobiles which create massive amounts of data every day. Edge computing refers to the computation done locally, say on a mobile phone, rather than sending the raw data across a network for a remote server to perform the calculations. This is possible primarily due to the availability of significant processing power, memory and storage in the IoT devices. The goal of Edge AI is to leverage low latency (i.e. short response time), high privacy, improved robustness and efficient use of network bandwidth & connectivity. It can also be thought of as an extension of FML applied to the edge, with the additional challenges in communication, heterogeneity in edge system implementations and disparate data collections [28].

If you have a smart speaker at home, you can see this in action when you ask it a common question like "What's the weather like today?" It will interpret your speech on the device itself and send the interpreted command to the server on the cloud, which replies with a response "It is 75° (*in Fahrenheit of course*) and sunny". The key point here is that the raw question as speech did not have to go through the network to the cloud for processing.

Here is a business example of applying Edge AI to provide accurate dietary assessment to the users to improve the effectiveness of weight loss interventions. Liu et al. [29] implemented a "Deep Food on the Edge" system which does preprocessing and segmentation of the food images on a mobile device and uses a DNN on the cloud to do the food recognition. Their results showed that their system outperformed existing work in terms of food recognition accuracy and reduced response time to a minimum, while lowering the computing energy consumption.

It is very exciting to watch the evolution of edge AI in different business domains. We have no doubt that this will shepherd in a new generation of cheaper "smarter" decision-making consumer products.

Neuro-Symbolic AI

Moving beyond the training data challenge, there is a need to bring together two different disciplines within the field of AI.

'Symbolic' AI systems are based on knowledge representations familiar to humans, such as natural language, rules, lists and graphs, and reasoning using logical constructs. As a result, these systems are human friendly. They naturally incorporate abstractions, compositionality, cognitive models of the domain, etc. Unfortunately, teaching these systems (see examples of expert systems in Chapter 2) heavily relied on human-intensive efforts and are thus not scalable and maintainable in practice.

In contrast, the advances in artificial neural networks in the last decade have demonstrated their impressive ability to learn complex patterns directly from data. The knowledge in these 'neural' systems is captured in internal representations of the statistical models and is not easily understandable by humans. These systems are very good at learning, but poor at human-level concepts and reasoning. They lack explainability of their model outputs.

The goal of neuro-symbolic AI is to achieve the best-of-both worlds vision by combining the complementary strengths of neural and symbolic approaches in a beneficial way. As the Turing Award winner Leslie Valiant [30] points out, two most fundamental aspects of intelligent cognitive behaviour are "the ability to learn from experience, and the ability to reason from what has been learned". Thus, neuro-symbolic AI is trying to bring machines and humans closer in their intelligent cognitive behaviours. Here are some potential benefits of neuro-symbolic AI [31]:

- **Knowledge beyond Training Data**: Ability to utilise symbolic representations (e.g. background knowledge) to handle scenarios that are significantly beyond what is contained in the training data. This will also help to learn with less training data.
- **New Insights**: Ability to use logic and reasoning to create new insights not contained explicitly in the training data.
- **Explainability/Interpretability**: Neuro-symbolic systems should enhance the interpretability of system behaviour and make it easier and more transparent to a human user.
- **Error Recovery and Verification**: Proper design of neuro-symbolic systems should enable techniques to recover more easily from erroneous decisions and allow a more formal verification process.

Neuro-symbolic algorithms are generally quite complex since they need to find a knowledge representation that is suitable for both learning and reasoning. To illustrate its application, we use the example of a Neuro-Symbolic Cognitive Agent (NSCA) [32] that evaluates student drivers as they perform specific driving tasks on a training simulator employed in real-world scenarios. On the symbolic side, there is a set of rules that capture the criteria for evaluation expressed in terms of temporal logic with uncertainties. An example of such a rule is:

> If the trainee's car approaching an intersection, arrives at the intersection and stops when traffic is coming from the right, then the trainee gets a good evaluation.

Notably, in addition to the sequence of time events implied here, there is no specific mention of distance of the car from the intersection. In the study, five students participated in the driving test consisting of five test scenarios each. For each scenario, the relevant data from the simulator (i.e. relative positions and orientations of all traffic, speed, etc.) was used in the 'neural' training part. Assessment scores on several driving skills (i.e. vehicle control, traffic flow, safe driving, etc.) were provided by three human driving instructors in real time during the simulation attempts, which were simultaneously observed by the NSCA. The results showed that the NSCA was able to learn from these observations and assess the student drivers similar to the driving instructors and that the underlying knowledge can be extracted in the form of temporal logic rules.

As should be obvious from the discussion above, the neuro-symbolic algorithms of today are very specific to the use case under consideration. It will take some time for the technology to mature when the patterns of use cases can be identified, and a common framework can be established for business use.

AI ENGINEERING – EMERGENCE OF A NEW DISCIPLINE

Throughout this book, we have highlighted how AI applications are different from conventional software applications. We discussed the details of the engineering aspects of ML systems in Chapter 10. It should be clear that delivering real AI applications in the enterprise is going to take a lot more than just understanding algorithms. Hopefully, this book has contributed to that process. We summarise below the key differences once again.

- Training data defines the functionality of the ML applications. Quality and quantity of training data matter.
- An AI system can perform in an unpredictable manner when presented with previously unseen situations. Traditional testing approaches based on upfront functional specifications and expected deterministic behaviour do not work.
- Building trustworthy AI requires careful attention paid to a full range of stakeholders (both inside the development organisation and society as a whole).
- AI lifecycle includes model lifecycle for the duration of the applications deployment, plus the traditional software application lifecycle, these all add to project cost, complexity and risk.
- Detecting drifts in model performance during deployment and updating models are compulsory activities.
- Supporting IT infrastructure needs persistent versions of data, models and code for the life of the application (may be even after) for auditing purposes.

Given these differences, traditional systems engineering processes and tools must change to meet these needs. We call this new discipline, 'AI Engineering'. Luckily, there have been numerous authors discussing these very issues [33–40] recently. To highlight its importance, United States Office of the Director of National Intelligence sponsored the Software Engineering Institute at Carnegie Mellon University to lead the AI Engineering Initiative for the US government [41]. With so much interest and excitement in delivering real value through AI across the world and awareness of the need for a new discipline, we should expect AI Engineering to mature significantly in the next 5 years.

Here are at least two fundamental objectives for the new AI Engineering discipline to address:

- How do we design and build Trustworthy AI systems for a broad set of stakeholders?
- How do we specify and manage a system for mission/business-critical uses where some decision-making components can be wrong for some fraction of the time, say 20%?

As a practical matter, at the current state and maturity of the AI technology, we need to develop various techniques to improve the odds of success of an AI project in the enterprise. We hope our Doability Method contributes to this cause. Some of the more specific needs for managing the development and deployment of AI applications are:

- **Managing Data**: Tools and processes will be required to perform a whole range of functions. Obvious requirements exist in areas such as data cleansing and labelling. However, the tooling needs to go much further and enable functions for managing data provenance, redaction and anonymisation. It is important to understand exactly what data was used to develop and configure a model along with understanding the impact on the model of redacting some of that data. *The critical aspect here is sustained management of data artefacts for the life of an AI application and beyond.*
- **Governance of AI Models**: In the context of successful use or reuse of AI components within an organisation or from an external source, the context behind the creation of the AI Models is critical. A good example of emerging capabilities in this area is the concept of AI Fact Sheets [42].
- **Managing Complexity**: If an AI application needs the use of multiple AI components that interact to provide the final output (e.g. processing various sensors in a self-driving car to make a decision to stop or not), this will require sophisticated tools to enable end-to-end tracing of decisions.
- **Monitoring & Managing Change**: Continuous monitoring is critical to improve the performance of a deployed application. The aircraft industry provides a strong role model in this respect. Over a 100-year history, the aircraft industry has continuously improved safety through the rigorous investigation of every accident and the implementation of any resulting safety recommendations. Similarly, in AI, we need processes to investigate unacceptable decisions and continually improve the decision-making of the application.

In order to meet the demands to build AI business applications today, we believe AI Engineering will need to be delivered through AI Factories that integrate the expertise, tools, data and infrastructure at scale.

HUMAN–MACHINE TEAMING

In this section, we want to outline the evolution of the interaction between humans and machines since the birth of general-purpose computing and explore where this can go as the use of AI becomes more prevalent in the enterprise.

Evolution of Human–Computer Interaction

In the early days of commercial computing (e.g. IBM 360), you had to go to a computer centre and punch holes in cardboard cards to create your program that the computer could run. You gave the cards to the computer centre and came back a few hours later to pick up the cards and the output printed on a roll of paper. The idea was that the machines did special tasks that needed a lot of number crunching. If you had an error in the program and had to fix it, the process repeats. With the introduction of 'minicomputers' with interactive terminals (e.g. Digital Equipment Corporation's PDP-11), command line interfaces and the keyboard came to being. This made computing a lot more responsive to the programmer. With the advent of personal computers and the introduction of Windows-Icons-Menus-Pointers (WIMP) interface, computing got more popular because more people could use them for their daily needs such as spreadsheets for simple calculations and documents for business or personal use. You did not need to be technically trained to interact with the computer. The dawn of the internet brought a lot more data and content for the individual from around the world. The next step in the evolution of the user interaction was the introduction of smart phones. We should think of them as powerful compact computers. This resulted in the evolution of new gestures (e.g. swipes) and added multimodal aspects (i.e. text, speech and images) for performing tasks.

Two major cultural shifts happened concurrently with these trends over the decades:

i. Number of people regularly interacting with a computer increased from a few thousands to billions. The number of daily activities on the social media attests to this fact.

ii. We have moved from a transactional era to a relationship era with computing.

In other words, in the years past, we used computers when we had to do something specific. Now we live with them in our pockets. The last thing we see before we go to sleep is the phone and the first thing we see in the morning is the phone. Even if we spend a few minutes without access to the phone (e.g. when the battery needs charging), we feel helpless. Use of Alexa and Siri for doing chores is a common occurrence. This simply means our relationship with computers is moving to be more symbiotic to enrich our lives and make us more productive. While all these advances have happened in the personal lives of people, the human–machine interaction in the enterprise setting has not changed very much in the last two decades. Now enter, AI.

Augmented Intelligence

Throughout the history of AI, there has been considerable discussion on the prospect of machines replacing humans. In science fiction, it is not uncommon to be presented with an apocalyptic vision of the future where humans are battling AI for survival.

What these many different narratives seem to assume is that AI and humans will evolve separately and that problems will either be solved by humans or solved by machines. This is a huge over-simplification and almost certainly untrue. Instead, we believe that augmented human–machine systems will emerge where AI and human decision makers work in partnership to deliver more effective results.

As early as in 1960, Licklider [43] articulated the foundations of a human–machine partnership as follows:

> "Man-computer symbiosis is an expected development in cooperative interaction between men and electronic computers. It will involve very close coupling between the human and the electronic members of the partnership. The main aims are (1) to let computers facilitate formulative thinking as they now facilitate the solution of formulated problems, and (2) to enable men and computers to cooperate in making decisions and controlling complex situations without inflexible dependence on predetermined programs. In the anticipated symbiotic partnership, men will set the goals, formulate the hypotheses, determine the criteria, and perform the evaluations. Computing machines will do the routinizable work that must be done to prepare the way for insights and decisions in technical and scientific thinking."

This is an amazing prognostication 60 years ago on how we see AI in the enterprise today! The term used to describe this vision of the future was symbiotic computing. Already we are seeing existing AI applications where humans and machines work together. In such applications, the AI may perform the straightforward classification tasks whilst referring more complex and confusing cases to a human operator. However, it is not simply the case that the AI performs the simple and the human the complex. When referring the complex cases to the human, the AI may also provide an initial assessment together with supporting evidence and possibly even counter examples. The human may use other AI tools to inform his or her decision, and when the decision is made, it is fed back into the ML algorithm to improve future performance. The AI and the human work together in an augmented and symbiotic manner.

To give a further example of a physical implementation of such a partnership between humans and machines in business, we refer to the description of a Cognitive Environments Lab (CEL) by Farrell et al. [44]. CEL was equipped with motion sensors, microphones, cameras, speakers and displays. Humans interacted with the environment via speech, gestures, wands, etc. Their architecture made it possible to run cloud-based services, manipulate data, run analytics and generate spoken and visual outputs as one contiguous experience. They demonstrated an illustrative application for strategic decision-making with a business use case of mergers and acquisitions [45]. A more recent incarnation of this technology as an embodied cognitive agent [46] helps scientists to visualise and analyze exo-planet data. This includes the ability to program itself using AI planning algorithms to derive

WHAT'S IN A NAME?

If I could just invent an Arnold Swartzenegger look-alike robot to go back to 1956, I'd find John McCarthy (the person who coined the phrase "Artificial Intelligence") and give him a bit of a telling off. What was he thinking? Why couldn't he just have called it "Probabilistic Learning", or "Coding by Example", anything would be better than invoking the Dr Frankenstein spectre of "Artificial Intelligence". I doubt he had any idea how many spurious media apocalyptic scare stories or hyperbole ridden product launches those two words have caused. But there again, would we as computer scientists have striven so hard and with such passion to try and live up to the challenge, laid down in those two words?

David Porter

certain physical quantities and provide human-friendly explanations of the steps used in the derivation. It is quite possible to imagine such environments being used in places that need accumulation and assimilation of large volumes of data for decision support, such as corporate board rooms and military command centres.

This interaction between human and AI needs to be considered as part of an AI Systems Engineering approach [47] to delivering continually improving AI.

Trust and Risk in AI Decisions

As we discussed in Chapter 5, building Trustworthy AI applications is a key goal for the enterprise. If the decisions made by the AI application have serious consequences, it is only natural for humans to look for some transparency in the decision process and relevant necessary explanations before accepting the AI output. Any unexpected bias in the training data or algorithms is indeed a serious risk. Potential manipulation of the model's training data by adversaries is another concern. The additional complexity also comes from the uncertainty introduced by the underlying statistical models in ML algorithms and the need for humans to understand their implications in their decisions. A good example of this is getting stock price predictions from an AI application without a measure of its volatility, which adds to the potential investor risk. While there are tools targeting the specific areas we have mentioned here, there is no real standard of a practice yet, that helps an enterprise to manage the risk of deploying AI applications holistically. This is a particular challenge for institutions such as governmental agencies that carry the explicit public mandate for a responsible and trustworthy AI. We can only hope that with enough practical experience in the industry, the vision of a trustworthy AI will become a reality in the near future.

AI and Human Disabilities

According to a 2011 report [48], there are more than a billion people (or about 15% of the world's population) who have some form of disability that affects their daily lives. *How is this population affected by the introduction of AI in society?*

On the positive side, there are numerous advances in assistive technologies ranging from robotic arms, AI-enabled prosthetic limbs, AI route planning for the visually impaired, etc., which enrich the experiences of the disabled. For at least three decades, laws have existed [49,50] that provide requirements for information technology to support the needs of people with disabilities. Due to competing business priorities, it is fair to say actual implementations of IT systems have lagged behind in providing seamless accessibility to the disabled, usually an afterthought. With the increasing availability of multimodal user interfaces enabled by AI (e.g. text to speech, speech to text, gesture recognition, etc.), it is possible to imagine a future where an AI system will pick the appropriate mode to interact with a disabled person based on his/her profile either explicitly given or learnt through interactions. We can think of these as parameters in an Application Program Interface of Human with the AI system. This is another practical aspect of realising augmented intelligence.

Now, let us discuss the negative side. Even before the adoption of AI, despite all the good intentions, there is clear evidence for lack of consideration of the concerns and welfare of the disabled in the general psyche of the society [51]. Introducing AI to automate decisions in these contexts runs the risk of simply codifying and replicating this bias. Trewin et al. [52] described potential opportunities and underlying risks across four emerging AI application areas: employment, education, public safety and healthcare. Reference [53] provides six steps to design AI applications to treat people with disabilities fairly.

In this section, we have described various aspects of the impact of AI technology on the interactions between humans and machines. Human–Machine Teaming is no longer an esoteric topic for a few experts to discuss in annual conferences but rather a critical consideration for mainstream business applications. We expect major advances to happen here in the near future.

IN SUMMARY – SOME FINAL THOUGHTS

And so, the end is near ... we've travelled many of the highways of AI and talked at length about the practical application of AI in the enterprise. What should you do in planning for the future of AI? What are the key takeaways from this final chapter?

- In keeping with the messages in the rest of the book, data is everything. It's critical to the success of any enterprise AI project and it's critical to the future of your business. Think very hard about your data strategy and how you are going to feed the AI machine.

- Algorithms are going to keep developing, and there is a lot of fascinating work underway to address the data challenge. Neuro-symbolic approaches should bring humans and ML algorithms a little bit closer.

- To support AI workloads more efficiently and at lower power consumption, new computing paradigms are being developed.

- The new discipline of AI Engineering must mature soon to deliver the real value of AI. A key element of the discipline will be to support the creation of trustworthy systems that will facilitate the collaboration between humans and machines to solve complex problems.

For those of you who are worried about the prospect of sentient AI enslaving humanity … please rest assured that the Singularity is some way off … you can relax and sleep well.

Even if we are wrong in that prediction, you still don't need to worry because Steve Wozniak's prediction [54] for the future of AI is that intelligent robots will keep us as pets and, as all readers with pets know, that's a wonderful way to live!

REFERENCES

1. L. Zhou, et al., "Machine learning on big data: opportunities and challenges," *Neurocomputing*, 237, pp. 350–361 (2017).
2. A. L'Heureux, et al., "Machine learning with big data: challenges and approaches," *IEEE Access*, 5, pp. 7776–7797 (2017).
3. IDC Global DataSphere, Forecast: 2021-2025 The World Keeps Creating More Data - Now, What Do We Do With It All? (IDC #US46410201, March 2021).
4. N. Fishman and C. Stryker, *Smarter Data Science: Succeeding with Enterprise-Grade Data and AI Projects*, Wiley (2020).
5. G. E. Moore, "Cramming more components onto integrated circuits," *Electronics*, 38(8), pp. 114–117 (April 19, 1965).
6. M. Bohr, "A 30 year retrospective on Dennard's MOSFET scaling paper," *IEEE SSCS Newsletter*, pp. 11–13, Winter 2007.
7. Von Neumann Architecture, https://en.wikipedia.org/wiki/Von_Neumann_architecture.
8. D. Amodei and D. Hernandez, "AI and compute," (May 16, 2018) https://openai.com/blog/ai-and-compute/.
9. E. Strubell et al. "Energy and policy considerations for deep learning in NLP," *57th Annual Meeting of the Association for Computational Linguistics (ACL)*, arXiv:1906.02243 (2019).
10. W. Knight, "AI can do great things—if it doesn't burn the planet," *Wired Magazine*, (January 21, 2020) https://www.wired.com/story/ai-great-things-burn-planet/.
11. X. Sun et al., "Hybrid 8-bit floating point (HFP8) training and inference for deep neural networks," (NeurIPS 2019) *33rd Conference on Neural Information Processing Systems*.
12. S. K. Esser, et al., "Convolutional networks for fast, energy-efficient neuromorphic computing," *Proceedings of the National Academy of Sciences*, 113(41), pp. 11441–11446 (2016).
13. S. Ambrogio, et al., "Equivalent-accuracy accelerated neural-network training using analogue memory," *Nature*, 558, pp. 60–67 (2018).
14. V. Havlíček, et al. "Supervised learning with quantum-enhanced feature spaces," *Nature*, 567, pp. 209–212 (2019).
15. Ethnologue, "Languages of the world," https://www.ethnologue.com/.
16. T. Young, et al., "Recent trends in deep learning based natural language processing," *IEEE Computational Intelligence Magazine*, 13(3), pp. 55–75 (August, 2018).
17. D. W. Otter, "A survey of the usages of deep learning for natural language processing," *IEEE Transactions on Neural Networks and Learning Systems*, 32(2), pp. 604–624 (2021).
18. X. Liu et al., "Using language models to pre-train features for optimizing information technology operations management tasks," H. Hacid et al. (eds.) *ICSOC 2020 Workshops*, Springer LNCS 12632, pp. 150–161 (2021), https://www.youtube.com/watch?v=niD0UwIi-YY.

19. M. Johnson, et al., "Google's multilingual neural machine translation system: enabling zero-shot translation," *Transactions of the Association for Computational Linguistics*, 5, pp. 339–351, (2017).
20. Google Translate: https://translate.google.com/.
21. W. Wang, et al., "A survey of zero-shot learning: settings, methods, and applications," *ACM Transactions on Intelligent Systems and Technology*, 10(2), Article 13 (2019).
22. Y. Wang, et al., "Generalizing from a few examples: a survey on few-shot learning," *ACM Computing Surveys*, 53(3), Article No: 63, pp. 1–34 (2020).
23. G. Litjens, et al., "A survey on deep learning in medical image analysis," *Medical Image Analysis*, 42, pp. 60–88 (2017).
24. A. Medela, et al., "Few shot learning in histopathological images: reducing the need of labelled data on biological datasets," *IEEE 16th International Symposium on Biomedical Imaging*, pp. 1860–1864 (2019).
25. D. Verma, G. White and G. de Mel, "Federated AI for the enterprise: a web services based implementation," *IEEE International Conference on Web Services (ICWS)*, pp. 20–27 (2019).
26. H. R. Roth, et al., "Federated learning for breast density classification: a real-world implementation," In: Albarqouni S. et al. (eds.) *Domain Adaptation and Representation Transfer, and Distributed and Collaborative Learning. Lecture Notes in Computer Science*, 12444, Springer, Cham (2020).
27. IBM Federated Learning: https://ibmfl.mybluemix.net/.
28. T. Li, et al., "Federated learning: challenges, methods, and future directions," *IEEE Signal Processing Magazine*, 37(3), pp. 50–60, (May 2020).
29. C. Liu et al., "A new deep learning-based food recognition system for dietary assessment on an edge computing service infrastructure," *IEEE Transactions on Services Computing*, 11(2), pp. 249–261 (2018).
30. L. G. Valiant, "Three problems in computer science," *Journal of the ACM*, 50(1), pp. 96–99 (2003).
31. Md K. Saker, et al., "Neuro-symbolic artificial intelligence-current trends," arXiv:2105.05330 (2021).
32. H. L. H. de Penning, et al., "A neural-symbolic cognitive agent for online learning and reasoning," *Twenty-Second International Joint Conference on Artificial Intelligence* (2011); the rule cited is slightly modified from the version in the paper for clarity.
33. D. Scully, et al., "Machine learning: the high-interest credit card of technical debt," *Software Engineering for Machine Learning Workshop*, NIPS (2014).
34. R. Akkiraju, et al., "Characterizing machine learning process: a maturity framework," https://arxiv.org/abs/1811.04871 (2018).
35. A. Arpteg, et al., "Software engineering challenges of deep learning," *Proceedings of the 44th Euromicro Conference on Software Engineering and Advanced Applications (SE-AA)*, pp. 50–59 (2018).
36. S. Amershi, et al., "Software engineering for machine learning: a case study," *ICSE-SEIP'10 Proceedings of the 41st International Conference on Software Engineering: Software Engineering in Practice*, pp. 291–300 (2019).
37. A. Horneman, A. Mellinger, and I. Ozkaya, *AI Engineering: 11 Foundational Practices*, CMU Software Engineering Institute (2019).
38. P. Santhanam, "Quality management of machine learning systems," In: O. Shehory, E. Farchi and G. Barash. (eds.) *Engineering Dependable and Secure Machine Learning Systems. EDSMLS 2020. Communications in Computer and Information Science*, 1272, pp.1-13, Springer, Cham (2020).
39. I. Ozkaya, "What is really different in engineering AI-enabled systems?" *IEEE Software*,v.37, No.4, pp. 3–6 (July-August, 2020).
40. J. Bosch, I. Crnkovic, and H. H. Olsson, "Engineering AI systems: a research agenda," arXiv:2001.07522 (2020).

41. AI Engineering Initiative at Software Engineering Institute at Carnegie Mellon University https://www.sei.cmu.edu/our-work/artificial-intelligence-engineering/.
42. AI FactSheets https://www.ibm.com/blogs/research/2020/07/aifactsheets/.
43. J. C. R. Licklider, "Man-computer symbiosis," *IRE Transactions on Human Factors in Electronics*, HFE-1(1), pp. 4–11 (1960).
44. R. Farrell, et al., "Symbiotic cognitive computing," *AI Magazine*, 37(3), pp. 81–93 (2016).
45. Video of Mergers & Acquisitions scenario https://www.youtube.com/watch?v=sSMMA_yXhaM.
46. J. O. Kephart, et al., "An embodied cognitive assistant for visualizing and analyzing exoplanet data," *IEEE Internet Computing*, 23(2), pp. 31–39 (2019) https://www.youtube.com/watch?v=Fg_seQM9T0k.
47. P. McDermott, et al., "Human-machine teaming systems engineering guide," MITRE Report. (2018) https://apps.dtic.mil/sti/pdfs/AD1108020.pdf.
48. World Health Organization and World Bank, *World Report on Disability* (2011).
49. Information and Communication Technology Revised 508 Standards and 255 Guidelines: https://www.access-board.gov/ict/.
50. Americans with Disabilities Act https://www.ada.gov/.
51. P. Smith and L. Smith, "Artificial intelligence and disability: too much promise, yet too little substance?" *AI Ethics*, 1, pp. 81–86 (2021).
52. S. Trewin, et al., "Considerations for AI fairness for people with disabilities," *AI Matters*, 5(3), pp. 40–63 (2019).
53. IBM Blog: "Designing AI applications to treat people with disabilities fairly," https://www.ibm.com/blogs/age-and-ability/2020/12/03/designing-ai-applications-to-treat-people-with-disabilities-fairly/.
54. Wozniak talks: "Self-driving cars, Apple Watch, and how AI will benefit humanity," https://www.techrepublic.com/article/wozniak-talks-self-driving-cars-apple-watch-and-how-ai-will-benefit-humanity/.

Epilogue

IN A PARTICULARLY INTERESTING analysis of quantum mechanics known as the "many worlds interpretation", every fundamental event or decision in the universe splits the universe into a series of alternate realities. There are billions and billions of universes with each one representing a different reality.

In one universe you did not marry your spouse ... or you did crash your car ... or you didn't buy your dog ... and so on.

In one of those realities, your CEO read this book (actually, we're really counting on it happening in more than one reality as we've all got kids to put through college).

In those (many) realities where your CEO read this book, the scenario outlined in our prologue never happens. In fact, the conversation goes something like this ...

> The CEO is assembling a small team to consider our corporate strategy for exploiting Artificial Intelligence. There's a lot of media interest in the subject and many of our competitors are making big claims. However, the CEO wants to know what is real and what isn't. Apparently, she's read this great book that suggests we need to start with education so that we really understand the subject.
>
> She wants a team that can develop a really solid strategy. We need to start by understanding what the different types of AI are and how they really work. She's not asking us to become PhDs in the subject, but we do need to understand the core capabilities and limitations.
>
> Then we need to look at all aspects of our business and understand where AI could be applied. This can be really visionary ... there's no such thing as a bad idea ... but let's get an end-to-end view. Emerging technologies, such as AI, can add value in places we wouldn't expect and it's not always the sexiest application that should be pursued.
>
> If we can come up with a whole spectrum of ideas, then we can start to do some proper analysis to evaluate the potential value of each project and the ease with which they can be delivered. That will allow us to put together a programme that balances risk and return. It's not about blocking projects ... it's about understanding what needs to be in place for the project to be successful and then starting with the simplest actions to maximise our chances of success. To start with, we probably want to look at applications that fit within existing business processes. It's okay to come up with radical new ideas and business processes once we have a solid business case and we really know what we're doing ... but for now let's start with things we understand.

Data is going to be critical ... in fact, it's all about the data! We need to develop our Data Science skills and really look at the quality of our data. We also need to look hard at the external data we rely on and ensure it's available in the future. Whilst this will help our AI projects, it's actually a good thing to do anyway.

Whilst our competitors are making a big noise about AI, the CEO isn't sure they're really using it properly. She's seen some very high-profile demos ... the sort of things a Graduate would throw together in a few hours ... but nothing that suggests they've done this properly. We're going to do it properly with a proper infrastructure, devops and a real focus on building a trustworthy capability that our customers and the public will want to be part of.

Ultimately, this is all about skills ... AI is not engineering free and we need to put the engineering skills in place to build proper applications. Remember applications are much more than just algorithms. There's so much more to this!

So, I think the brief is quite simple ... let's get a team together ... not just techies ... we need the business and domain specialists working alongside engineers who really understand this stuff. By engineers, I mean people who build systems and understand what that means. They need to be educated in AI and systems engineering. With the right skills and the right understanding, we can start developing a proper strategy so that we can be the leaders in this field.

Index

Note: Bold page numbers refer to tables; *italic* page numbers refer to figures.